BUS

GLASS-CERAMIC TECHNOLOGY

Wolfram Höland
Ivoclar Vivadent AG

and

George Beall
Corning Incorporated

The American Ceramic Society

Published by The American Ceramic Society, 735 Ceramic Place, Westerville, OH 43081

The American Ceramic Society
735 Ceramic Place
Westerville, Ohio 43081

05 04 03 02   5 4 3 2 1

ISBN: 1-57498-107-2

Library of Congress Cataloging-in-Publication Data
A CIP record for this book is available from the Library of Congress.

For more information on ordering titles published by The American Ceramic Society or to
request a publications catalog, please call (614) 794-5890 or visit our online bookstore at
<www.ceramics.org>.

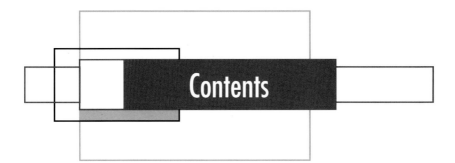

Contents

Introduction

Modern science and technology constantly require new materials with special properties to achieve breathtaking innovations. This development centers on the improvement of scientific and technological fabrication and working procedures. That means rendering them faster, economically more favorable, and better in quality. At the same time, new materials are introduced to improve our general quality of life, especially in human medicine and dentistry and daily life (e.g., housekeeping).

Among all these new materials, one group plays a very special role: glass-ceramic materials.

Glass-ceramics offer the possibility of combining the special properties of conventional sintered ceramics with the distinctive characteristics of glasses. It is, however, possible to develop modern glass-ceramic materials with features unknown thus far in either ceramics or glasses or in other materials such as metals or organic polymers. Furthermore, developing glass-ceramics demonstrates the advantage of combining various remarkable properties in one material.

A few examples may illustrate this statement. As will be shown in the book, glass-ceramic materials consist of at least one glass phase and at least one crystal phase. Processing of glass-ceramics is carried out by controlled crystallization of a base glass. The possibility of generating such a base glass bears the advantage of benefiting from the latest technologies in glass processing, such as casting, pressing, rolling, or spinning, which may also be used in the fabrication of glass-ceramics or formation of a sol–gel–derived base glass.

By precipitating crystal phases in the base glass, however, new exceptional characteristics are achieved. Among these, for example, are the machinability of glass ceramics resulting from mica crystallization and the minimum thermal expansion of chinaware, kitchen hot plates, or scientific telescopes as a result of β-quartz–β-spdumene crystallization.

Another new field consists of glass-ceramic materials used as biomaterials in restorative dentistry or in human medicine. New high-strength, metal-free glass-ceramics will be presented for dental restoration. These are examples that demonstrate the versatility of material development in the field of glass-ceramics. At the same time, however, they clearly indicate how complicated it is to develop such materials and what kind of simultaneous, controlled solid-state processes are required for material development to be beneficial.

We intend this book to make an informative contribution to all those who would like to know more about new glass-ceramic materials and their scientific–technological background or who want to use these materials and benefit from them. It is therefore a book for students, scientists, engineers, and technicians. Furthermore this monograph is intended to serve as a reference for all those interested in natural or medical science and technology, with special emphasis on glass-ceramics as new materials with new properties.

As a result of this basic idea, the first three chapters, "Principles of Designing Glass–Ceramic Formation," "Composition Systems for Glass-Ceramics," and "Microstructural Control," satisfy the requirements of a scientific–technological textbook. These three chapters supply in-depth information on the various types of glass-ceramic materials. The scientific methods of material development are clearly pointed out, and direct parallels to the applications in Chapter 4 can be drawn easily. Chapter 4 focuses on the various possibilities of glass-ceramic materials in technical, consumer, optical, medical, dental, electrical, electronic, and architectural applications, as well as uses for coating and soldering. This chapter is organized like a reference book.

Based on its contents, this book may be classified somewhere between technical monograph, textbook, and reference book. It contains elements of all three categories and thus will appeal to a broad readership. As the contents of the book are arranged along various focal points, readers may approach the book in a differentiated manner. For instance, engineers and students of materials science and technology will follow the given structure of the book, beginning at Chapter 1. By contrast, dentists or dental technicians may want to read Chapter 4 first, where they can find details on the application of dental glass-ceramics. Thus, if they want to know more details on the material (e.g., microstructure, chemical composition, crystals, etc.), they will then read Chapters 1, 2, or 3.

We carry out scientific-technological work on two continents, namely the United States and Europe. Since we are in close contact to scientists of Japan in Asia, the thought arose to analyze and illustrate the field of glass-ceramics under the aspect of glass-ceramic technology worldwide.

Moreover, we, who have worked in the field of development and application of glass-ceramic materials for several years or even decades, have the opportunity to introduce our results to the public. We can, however, also benefit from the results of our colleagues, in close cooperation with other scientists and engineers.

The authors would like to thank the following scientists who helped with this book by providing technical publications on the topic of glass-ceramic research and development:

       T. Kokubo, Y. Abe, M. Wada, and T. Kasuga, from Japan,

       J. Petzoldt, W. Pannhorst from Germany,

       I. Donald from the U.K.

       E. Zanotto from Brazil.

Special thanks go to V. Rheinberger (Liechtenstein) for supporting the book and for numerous scientific discussions; M. Schweiger (Liechtenstein) and his team for the technical and editorial advice; R. Nesper (Switzerland) for the support in presenting crystal structures; S. Fuchs (South Africa) for the translation into English; and L. Pinckney (USA) for the reading and editing of the manuscript.

History

Glass-ceramics are ceramic materials formed through the controlled nucleation and crystallization of glass. Glasses are melted, fabricated to shape, and thermally converted to a predominantly crystalline ceramic. The basis of controlled internal crystallization lies in efficient nucleation that allows the development of fine, randomly oriented grains generally without voids, microcracks, or other porosity. The glass-ceramic process, therefore, is basically a simple thermal process as illustrated in fig. H-1.

It occurred to Reamur (1739) and to many others since then that a dense ceramic made via the crystallization of glass objects would be highly desirable. It was not until about 35 years ago, however, that this idea was consummated. The invention of glass-ceramics took place in the mid-1950s by the famous glass chemist and inventor S.D. Stookey. It is useful to examine the sequence of events leading to the discovery of these materials. (Table H-1).

At the time, Stookey was not interested primarily in ceramics. He was preoccupied with precipitating silver particles in glass to achieve a permanent photographic image. He was studying lithium silicate compositions as host glasses because

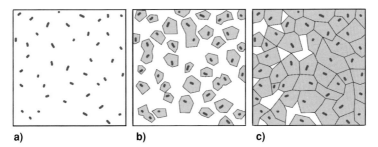

a)   b)   c)

Fig. H-1   From glass to glass-ceramic. (a) nuclei formation, (b) crystal growth on nuclei, (c) glass-ceramic microstructure.

Table H-1

Invention of Glass-Ceramics (S.D. Stookey 1950s).

- Photosensitive silver precipitation in Li_2O-SiO_2 glass; furnace overheats; $Li_2Si_2O_5$ crystallizes on Ag nuclei; first glass-ceramic.
- Sample accidentally dropped; unusual strength.
- Near-zero-thermal-expansion crystal phases described in Li_2O-Al_2O_3-SiO_2 system (Hummel, Roy).
- TiO_2 tried as nucleation agent based on its observed precipitation in dense thermometer opals.
- Aluminosilicate glass-ceramic (e.g. Corning Ware®) developed.

---

he found he could chemically precipitate silver in alkali silicate glasses, and those containing lithium had the best chemical durability. To develop the silver particles, he normally heated glasses previously exposed to ultraviolet light just above their glass transition temperature at around 450°C. One night the furnace accidentally over-heated to 850°C and on observation of the thermal recorder, he expected to find a melted pool of glass. Surprisingly, he observed a white material that had not changed shape. He immediately recognized the material as a ceramic showing no distortion from the original glass article. A second serendipitous event then occurred. He accidentally dropped the sample and it sounded more like metal than glass. He then realized that the ceramic he had produced had unusual strength.

On contemplating the significance of this unplanned experiment, Stookey recalled that lithium aluminosilicate crystals had been reported with very low thermal expansion characteristics; in particular, a phase, β-spodumene, had been described by Hummel (1951) as having a near-zero thermal expansion characteristic. He was well aware of the significance of even moderately low expansion crystals in permitting thermal shock in otherwise fragile ceramics. He realized that if he could nucleate these and other low coefficient of thermal expansion phases in the same way as he had lithium disilicate, the discovery would be far more meaningful. Unfortunately, he soon found that silver or other colloidal metals are not effective in nucleation of these aluminosilicate crystals. Here he paused and relied on his personal experience with specialty glasses. He had at one point worked on dense thermometer opals. These are the white glasses that compose the dense, opaque stripe in a common thermometer. Historically, this effect had been developed by precipitation of crystals of high refractive index such as zinc sulfide or titania. He, therefore, tried adding titania as a nucleating agent in aluminosilicate glasses and discovered it to be amazingly effective. Strong and thermal shock resistant glass-ceramics were then developed commercially within a year or two of this work as well-known products such as rocket nose cones and CORNINGWARE® cookware (Stookey 1959).

In summary, a broad materials advance had been achieved from a mixture of serendipitous events controlled by chance and good exploratory research related to a practical concept, albeit unrelated to a specific vision of any of the eventual products. Knowledge of the literature, good observation skills, and deductive reasoning were clearly evident in allowing the chance events to bear fruit.

Without the internal nucleation process as a precursor to crystallization, devitrification is initiated at lower energy surface sites. As Reaumur was painfully aware, the result is an ice-cube-like structure (Fig. H-2), where the surface-oriented crystals meet in a plane of weakness. Flow of the uncrystallized core glass in response to changes in bulk density during crystallization commonly forces the original shape to undergo grotesque distortions. On the other hand, because crystal-

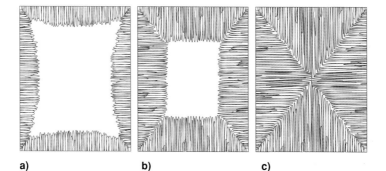

a)   b)   c)

Fig. H-2   Crystallization of glass without internal nucleation.

lization can occur uniformly and at high viscosities, internally nucleated glasses can undergo the transformation from glass to ceramic with little or no deviation from the original shape.

To consider the advantages of glass-ceramics over their parent glasses, one must consider the unique features of crystals, beginning with their ordered structure. When crystals meet, structural discontinuities or grain boundaries are produced. Unlike glasses, crystals also have discrete structural plans that may cause deflection, branching, or splintering of cracks. Thus the presence of cleavage planes and grain boundaries serves to act as an impediment for fracture propagation. This accounts for the better mechanical reliability of finely crystallized glasses. In addition, the spectrum of properties in crystals is very broad compared with that of glasses. Thus some crystals may have extremely low or even negative thermal expansion behavior. Others, like sapphire, may be harder than any glass, and crystals like mica might be extremely soft. Certain crystalline families also may have unusual luminescent, dielectric, or magnetic properties. Some are semiconducting or even, as recent advances attest, may be superconducting at liquid nitrogen temperatures. In addition, if crystals can be oriented, polar properties like piezoelectricity or optical polarization may be induced.

In recent years, another method of manufacture of glass-ceramics has proven technically and commercially viable. This involves the sintering and crystallization of powdered glass. This approach has certain advantages over body-crystallized glass-ceramics. Firstly, traditional glass-ceramic processes may be used, e.g., slip casting, pressing, and extruding. Secondly, because of the high flow rates before crystallization, glass-ceramic coatings on metals or other ceramics may be applied by using this process. Finally, and most importantly, is the ability to use surface imperfections in quenched frit as nucleation sites. This process typically involves milling a quenched glass into fine 3–15 μm particle diameter particulate. This powder is then formed by conventional ceramming called forming techniques in viscous sintering to full density just before the crystallization process is completed. Figure H-3 shows transformation of a powdered glass compact (Fig. H-3a) to a dense sintered glass with some surface nucleation sites (Fig. H-3b) and finally to a

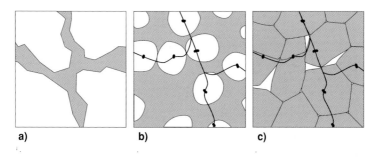

a)   **b)**   **c)**

Fig. H-3   Glass-ceramics from powdered glass. (a) powdered glass compact,
(b) densification and incipient crystallization, (c) frit-derived glass-ceramic.

highly crystalline frit-derived glass-ceramic (Fig. H-3c). Note the similarity in structure between the internally nucleated glass-ceramic in Fig. H-1c. The first commercial exploitation of frit-derived glass-ceramics was the devitrifying frit solder glasses for sealing television bulbs. Recently, the technology has been applied to cofired multilayer substrates for electronic packaging.

CHAPTER 1

Principles of Designing Glass-Ceramic Formation

1.1 ADVANTAGES OF GLASS-CERAMIC FORMATION

Glass-ceramics have been shown to feature favorable thermal, chemical, biological, and dielectric properties, generally superior to metals and organic polymers in these areas. Moreover, glass-ceramics also demonstrate considerable advantages over inorganic materials, such as glasses and ceramics. The large variety of compositions and the possibility of developing special microstructures should be noted in particular. It goes without saying that these advantageous properties assure the favorable characteristics of the glass-ceramic end products.

As the name clearly indicates, glass-ceramics are classified between inorganic glasses and ceramics. A glass-ceramic may be highly crystalline or may contain substantial residual glass. It is composed of one or more glassy and crystalline phases. The glass-ceramic is produced from a base glass by controlled crystallization. The new crystals produced in this way grow directly in the glass phase, and at the same time slowly change the composition of the remaining glass.

The synthesis of the base glass represents an important step in the development of glass-ceramic materials. Many different ways of traditional melting and forming as well as sol–gel, chemical vapor deposition, and other means of production of the base glasses are possible. Although the development of glass-ceramics is complicated and time-consuming, the wide spectrum of their chemical synthesis is useful for achieving different properties.

The most important advantage of the glass-ceramic formation, however, is the wide variety of special microstructures. Most types of microstructures that form in glass-ceramics cannot be produced in any other material. The glass phases may themselves demonstrate different structures. Furthermore, they may be arranged in the microstructure in different morphological ways. Crystal phases possess an even wider variety of characteristics. They may demonstrate special morphologies related to their particular structures as well as considerable differences in appearance depending on their mode of growth.

All these different ways of forming microstructures involve controlled nucleation and crystallization, as well as the choice of parent glass composition.

Glass-ceramics demonstrating particularly favorable properties were developed on the basis of these two key advantages, that is, the variation of the chemical composition and of the microstructure. These properties are listed in Tables 1-1 and 1-2, and are briefly outlined below:

Table 1-1

Particularly Favorable Properties of Glass-Ceramics

Processing properties

Rolling, casting, pressing, spin casting, press-and-blow method, drawing are possible

Limited and controllable shrinkage

No porosity in monolithic glass-ceramics

Thermal properties

Expansion can be controlled as desired, depending on the temperature, with zero or even negative expansion being coefficients of thermal expansion possible

High-temperature stability

Optical properties

Translucency or opacity

Photo-induction is possible

Pigmentation

Opalescence, fluorescence

Chemical properties

Resorbability or high chemical durability

Biological properties

Biocompatibility

Bioactivity

Mechanical properties

Machinability

High strength and toughness

Electrical and magnetic properties

Isolation capabilities (low dielectric constant and loss, high resistivity and breakdown voltage)

Ion conductivity and superconductivity

Ferromagnetism

1.1.1 Processing Properties

The research on the discovery of suitable base glasses revealed that the technology used in the primary shaping of glass could also be applied to glass-ceramics. Therefore, bulk glasses are produced by rolling, pressing, casting, spin casting, or by press-blowing a glass melt or by drawing a glass rod or ring from the melt. The thin-layer method is also used to produce thin glass sheets, for example. In addition, glass powder or grains are transformed into glass-ceramics.

1.1.2 Thermal Properties

A particular advantage in the production of glass-ceramics is that products demonstrating almost zero shrinkage can be produced. These specific materials are produced on a large scale for industrial, technological, and domestic applications (e.g., kitchenware).

1.1.3 Optical Properties

Since glass-ceramics are nonporous and usually contain a glass-phase, they demonstrate a high level of translucency and in some cases even high transparency. Furthermore, it is also possible to produce very opaque glass-ceramics, depending on the type of crystal and the microstructure of the material. Glass-ceramics can be produced in virtually every color. In addition, photo-induced processes may be used to produce glass-ceramics and to shape high-precision and patterned end products.

Fluorescence, both visible and infrared, and opalescence in glass-ceramics are also important optical characteristics.

1.1.4 Chemical Properties

Chemical properties, ranging from resorbability to chemical stability, can be controlled according to the nature of the crystal, the glass phase or the

Table 1-2

Particularly Favorable Combinations of Properties of Glass-Ceramics (Selection)
• Mechanical property (machinability) + thermal properties (temperature resistance)
• Thermal property (zero expansion + temperature resistance) + chemical durability
• Mechanical property (strength) + optical property (translucency) + favorable processing properties
• Strength + Translucency + biological properties + favorable processing properties

nature of the interface between the crystal and the glass phase. As a result, resorbable or chemically stable glass-ceramics can be produced. The microstructure in particular also permits the combination of resorbability of one phase and chemical stability of the other phase.

1.1.5 Biological Properties

Biocompatible glass-ceramics have been developed for human medicine and for dentistry in particular. Furthermore, bioactive materials are used in implantology.

1.1.6 Mechanical Properties

Although the highest flexural strength values measured for metal alloys have not yet been achieved in glass-ceramics, it has been possible to achieve flexural strengths of up to 500 MPa. The toughness of glass-ceramics has also been considerably increased over the years. As a result, K_{IC} values of more than 3 MPa·m$^{0.5}$ have been reached. No other material demonstrates these properties together with translucency and allows itself to be pressed or cast, without shrinking or pores developing, as in the case of monolithic glass-ceramics.

The fact that glass-ceramics can be produced as machinable materials represents an additional advantage. In other words, by first processing the glass melt, a primary shape is given to the material. Next, the glass-ceramic is provided with a relatively simple final shape by drilling, milling, grinding, or sawing. Furthermore, the surface characteristics of glass-ceramics, for example, roughness, polishability, luster, or abrasion behavior can also be controlled.

1.1.7 Electrical and Magnetic Properties

Glass-ceramics with special electrical or magnetic properties can also be produced. The electrical properties are particularly important if the material is used for isolators in the electronics or micro-electronics industries. It must also be noted that useful composites can be formed by combining glass-ceramics with other materials, for example, metal. In addition, glass-ceramics demonstrating high ion conductivity and even superconductivity have been developed. Furthermore, magnetic properties in glass-ceramics were produced similarly to those in sintered ceramics. These materials are processed according to methods involving primary shaping of the base glasses followed by thermal treatment for crystallization.

1.2 FACTORS OF DESIGN

In the design of glass-ceramics, the two most important factors are composition and microstructure (Table 1-3). The bulk chemical composition controls the ability to form a glass and determines its degree of workability. It also determines whether internal or surface nucleation can be achieved. If internal nucleation is desired, as is the case when hot glass forming of articles, appropriate nucleating agents are melted into the glass as part of the bulk composition. The bulk composition also directly determines the potential crystalline assemblage and this in turn determines the general physical and chemical characteristics; for example, hardness, density, thermal expansion coefficient, acid resistance, etc.

Microstructure is of equal importance to composition. This feature is the key to most mechanical and optical properties, and it can promote or diminish the characteristics of key crystals in glass-ceramics. It is clear that microstructure is not an independent variable. It obviously depends on the bulk composition and crystalline phase assemblage, and it also can be modified, often dramatically, by varying the thermal treatment.

1.3 CRYSTAL STRUCTURES AND MINERAL PROPERTIES

Since the most important glass-forming systems are based on silicate compositions, the key crystalline components of glass-ceramics are therefore silicates. Certain oxide minerals, however, are important, both in controlling nucleation as well as forming accessory phases in the final product.

Table 1-3

Glass-Ceramic Design

Composition
- Bulk chemical
    glass formation and workability
    internal or surface nucleation
- Phase assemblage
    general physical and chemical characteristics

Microstructure
- Key to mechanical and optical properties
- Can promote or diminish characteristics of key phase

1.3.1 Crystalline Silicates

Crystalline silicates of interest in glass-ceramic materials can be divided into six groups according to the degree of polymerization of the basic tetrahedral building blocks. These are generally classified as follows (Tables 1-4, 1-5):

- Nesosilicates (independent SiO_4 tetrahedra)
- Sorosilicates (based on Si_2O_7 dimers)
- Cyclosilicates (containing six-membered $(Si_6O_{18})^{-12}$ or $(AlSi_5O_{18})^{-13}$ rings)
- Inosilicates (containing chains based on SiO_3^- single, $Si_4O_{11}^-$ double, or multiple)
- Phyllosilicates (sheet structures based on hexagonal layers of $(Si_4O_{10})^{-4}$, $(AlSi_3O_{10})^{-5}$, or $(Al_2Si_2O_{10})^{-6}$
- Tectosilicates (frameworks of corner shared tetrahedra with formula SiO_2, $(AlSi_3O_8)^{-1}$, or $(Al_2Si_2O_8)^{-2}$)

1.3.1.1 Nesosilicates

This is the least important mineral group in glass-ceramic technology because the low polymerization of silica in these minerals does not allow glass formation at these stoichiometries (Si:O ratio = 1:4). Nevertheless, such phases as forsterite (Mg_2SiO_4) and willemite (Zn_2SiO_4) can occur as minor phases. Willemite, in particular, when doped with Mn^{2+} can create a strong green fluorescence even when present in small volume percents. Humite minerals such as chondrodite ($Mg_2SiO_4 \cdot 2MgF_2$) and norbergite ($Mg_2SiO_4 \cdot MgF_2$) are precursor phases in some fluoromica glass-ceramics.

Table 1-4

Structural Classification of Silicates Found in Glass-Ceramics

Nesosilicates

Isolated tetrahedra		
1:4 ratio of Si:O	0% sharing	
Forsterite $Mg_2(SiO_4)$		

Sorosilicates

Tetrahedral pairs		
2:7 ratio of Si:O	25% sharing	
Thorveitite $Sc_2(Si_2O_7)$		

Cyclosilicates

Ring silicates		
1:3 ratio of Si:O	50% sharing	
Beryl $Be_3Al_2(Si_6O_{18})$		

Table 1-5

Structural Classification of Silicates Found in Glass-Ceramics

Inosilicates

Single chain silicate (pyroxenes)

$\quad$ 1:3 ratio of Si:O (infinite ring)$\qquad$50% sharing
$\quad$ Enstatie $MgSiO_3$

Double chain silicates (amphiboles)

$\quad$ 4:11 ratio of Si:O (infinite ring)$\qquad$62.5% sharing
$\quad$ Tremolite $Ca_2Mg_5(Si_4O_{11})(OH)_2$

Phyllosilicates

Layer silicates (micas and clays)

$\quad$ 2:5 ratio of Si:O$\qquad$75% sharing
$\quad$ Kaolinite (china clay) $Al_2(Si_2O_5)(OH)_4$
$\quad$ Muscovite (mica) $KAl_2(AlSi_3O_{10})(OH)_2$*

Tectosilicates

Network silicates (silica and feldspars)

$\quad$ 1:2 ratio of Si:O$\qquad$100% sharing
$\quad$ Quartz SiO_2
$\quad$ Orthoclase $K(AlSi_3O_8)$*
$\quad$ Anorthite $Ca(Al_2Si_2O_8)$*

*Note that Al^{3+} sometimes substitutes for Si^{4+} in tetrahedral sites but never more than 50%. Silicates tend to cleave between the silicate groups, leaving the strong Si–O bonds intact. Amphiboles cleave in fibers; micas into sheets.

1.3.1.2 Sorosilicates

As is the case of the nesosilicates, sorosilicates are not glass-forming minerals because of their low Si:O ratio, namely 2:7. Again they are sometimes present as minor phases in slag-based glass-ceramics, as in the case of the melilite crystal akermanite $Ca_2MgSi_2O_7$, and its solid solution end member gehlenite $Ca_2Al_2SiO_7$. The latter contains a tetrahedrally coordinated Al^{3+} ion replacing one Si^{4+} ion.

1.3.1.3 Cyclosilicates

This group, often called ring silicates, is characterized by six-membered rings of SiO_4 and AlO_4 tetrahedral units which are strongly cross-linked. They are best represented in glass-ceramic technology by the important phase cordierite: $Mg_2Al_4Si_5O_{18}$, which forms a glass, albeit a somewhat unstable or quite fragile one. Because the cyclosilicates are morphologically similar to the tectosilicates and show important similarities in physical properties, they will both be included in a later section (1.3.1.6(F)).

1.3.1.4 Inosilicates

Inosilicates, or chain silicates as they are commonly referred to, are marginal glass-forming compositions with a Si:O ratio of 1:3 in the case of single chains and 4:11 in the case of double chains. They are major crystalline phases in some glass-ceramics known for high strength and fracture toughness. This is because the unidirectional backbone of tetrahedral silica linkage (see Table 1-5) often manifests itself in acicular or rodlike crystals which provide reinforcement to the glass-ceramic. Also, strong cleavage or twinning provides an energy-absorbing mechanism for advancing fractures.

Among the single-chain silicates of importance in glass-ceramics are enstatite ($MgSiO_3$), diopside ($CaMgSi_2O_6$) and wollastonite ($CaSiO_3$). These structures are depicted in Appendix Figs. 7–9. All three phases are normally monoclinic (2/m) as found in glass-ceramics, although enstatite can occur in the quenched orthorhombic form (protoenstatite) and wollastonite may be triclinic. Lamellar twinning and associated cleavage on the (100) plane are key to the toughness of enstatite, while elongated crystals aid in the increase of glass-ceramic strength where wollastonite is a major phase (see Chapter 2).

Amphiboles are a class of double-chain silicates common as rock-forming minerals. Certain fluoroamphilboles, particularly potassium fluororichterite of stoichiometry ($KNaCaMg_5Si_8O_{22}F_2$), can be crystallized from glasses of composition slightly modified with excess Al_2O_3 and SiO_2. The resulting strong glass-ceramics display an acicular microstructure dominated by rods of potassium fluororichterite of aspect ratio greater than 10. The monoclinic (2/m) structure of this crystal is shown in Appendix 10. Note the double chain $(Si_4O_{11})^{-6}$ backbone parallel to the c-axis.

Certain multiple chain silicates are good glass formers, because of even higher states of polymerization, with Si:O ratios of 2:5. These include fluorocanasite ($K_2Na_4Ca_5Si_{12}O_{30}F_4$) and agrellite ($NaCa_2Si_4O_{10}F$). Both are nucleated directly by precipitation of the CaF_2 inherent in their composition. Both yield strong and tough glass-ceramics with intersecting bladed crystals. Canasite, in particular, produces glass-ceramics of exceptional mechanical resistance, largely because of the splintering effect of well developed cleavage. Canasite has a fourfold box or tubelike backbone. Canasite is believed monoclinic (m), while agrellite is triclinic.

1.3.1.5 Phyllosilicates

Sheet silicates, or phyllosilicates, are layered phases with infinite two-dimensional hexagonal arrays of silica and alumina tetrahedra $(Si_2O_5)^{-2}$, $(AlSi_3O_{10})^{-5}$, or $(Al_2Si_2O_{10})^{-6}$. The simplest glass-ceramic crystals of this type are lithium and barium disilicate ($Li_2Si_2O_5$, $BaSi_2O_5$), both of which form glasses (Si:O = 2:5) and are easily converted to glass-ceramics. The structure of orthorhombic $Li_2Si_2O_5$ involves corrugated sheets of $(Si_2O_5)^{-2}$ on the (010) plane (Appendix 12). Lithium silicate glass-ceramics are easily melted and crystallized, and because of an interlocking tabular or lathlike form related to the layered structure, show good mechanical properties.

Chemically more complex but structurally composed of simpler flat layers are the fluoromicas, the key crystals allowing machinability in glass-ceramics. The most common phase is fluorophlogopite ($KMg_3AlSi_3O_{10}F_2$), which like most micas shows excellent cleavage on the basal plane (001). This crystal is monoclinic ($2/m$), although pseudohexagonal in appearance. It features thin laminae formed by the basal cleavage which are flexible, elastic, and tough. Because of the high MgO and F content, this mica does not itself form a glass, but a stable glass can easily be made with B_2O_3, Al_2O_3, and SiO_2 additions. Other fluoromica stoichiometries of glass-ceramic interest include $KMg_{2.5}Si_4O_{10}F_2$, $NaMg_3AlSi_3O_{10}F_2$, $Ba_{0.5}Mg_2AlSi_3O_{10}F_2$, and the more brittle mica $BaMg_3Al_2Si_2O_{10}F_2$.

The structure of fluorophlogopite is shown in Appendix 13. The individual layers are composed of three components, two $(AlSi_3O_{10})^{-5}$ tetrahedral sheets with hexagonal arrays of tetrahedra pointing inward toward an edge-sharing octahedral sheet composed of $(MgO_4F_2)^{-8}$ units. This T-O-T complex sheet is separated from the neighboring similar sheet by 12–coordinated potassium ions. This weak K–O bonding is responsible for the excellent cleavage on the (001) plane.

1.3.1.6 Tectosilicates

Framework silicates, also referred to as tectosilicates, are characterized by a tetrahedral ion-to-oxygen ratio of 1:2. The typical tetrahedral ions are silicon and aluminum, but, in some cases, germanium, titanium, boron, gallium, beryllium, magnesium, and zinc may substitute in these tetrahedral sites. All tetrahedral ions are typically bonded through oxygen to another tetrahedral ion. Silicon normally composes from 50% to 100% of the tetrahedral ions.

Framework silicates are the major mineral building blocks of glass-ceramics. Because these crystals are high in SiO_2 and Al_2O_3, key glass-forming oxides, they are almost always good glass formers, thus satisfying the first requirement for glass-ceramic production. In addition, important properties like low coefficient of thermal expansion, good chemical durability, and refractoriness are often associated with this family of crystals. Finally, certain oxide nucleating agents like TiO_2 and ZrO_2 are only partially soluble in viscous melts corresponding to these highly polymerized silicates, and their solubility is a strong function of temperature. These factors allow exceptional nucleation efficiency to be achieved with these oxides in framework silicate glasses.

A) Silica Polymorphs

The low-pressure silica polymorphs include quartz, tridymite, and cristobalite. The stable phase at room temperature is α-quartz or low quartz. This transforms to β-quartz or high quartz at approximately 573°C at 1 bar. The transition from β-quartz to tridymite occurs at 867°C and tridymite inverts to β-cristobalite at 1470°C. β-Cristobalite melts to silica liquid at 1727°C. All three of these stable silica polymorphs experience displacive transformations that involve structural contraction with decreased temperature and all can be cooled stabily or metastabily to room temperature in glass-ceramics compositions.

Quartz   The topological confirmation of the silica framework for α- and β-quartz is well-known and is shown in Fig. 1-1. The structure of α-quartz is easily envisioned as a distortion of the high-temperature beta modification. In high quartz, paired helical chains of silica tetrahedra spiral in the same sense around hexagonal screw axes parallel to the *c*-axis (Fig. 1-1a). The intertwined chains produce open channels parallel to the *c*-axis that appear hexagonal in projection. The β-quartz framework contains six- and eight-membered rings with irregular shapes and the space group is $P6_422$ or $P6_222$ depending on the chirality or handedness. When β-quartz is cooled below 573°C, the expanded framework collapses to the denser α-quartz configuration (Fig. 1-1(b) and (c)). The structural data for α- and β-quartz is shown in Table 1-6. The thermal expansion of α-quartz from 0°–300°C is approximately 150×10^{-7} K^{-1}. In its region of thermal stability, the thermal expansion coefficient of β-quartz is about -5×10^{-7} K^{-1}. Unfortunately, the β-quartz structure cannot be quenched. Therefore, pure quartz in glass-ceramics undergoes rapid shrinkage on cooling below its transformation temperature. Since α-quartz is the densest polymorph of silica stable at room pressure, $\rho = 2.65$, it tends to impart high hardness to a glass-ceramic material.

Tridymite   In his classical effort to determine phase equilibria relationships among the silica polymorphs, Fenner (1913) observed that tridymite could be synthesized only with the aid of a "mineralizing agent" or flux such as Na_2WO_4. If pure quartz is heated, it bypasses tridymite and transforms

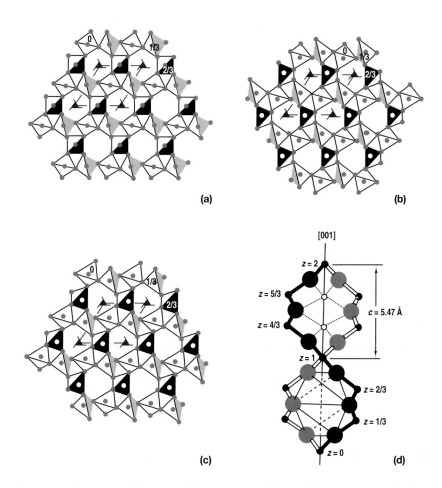

Figure 1-1 Projections of β-quartz (a) and α-quartz (b) and (c) along the c-axis. Both obverse (b) and reverse (c) settings are shown. The double helix structure of β-quartz is shown in (d).

directly to cristobalite at approximately 1050°C. A large variability in powder X-ray diffraction and differential thermal analyses of natural and synthetic tridymite led to the suggestion that tridymite may not be a pure silica polymorph. Hill and Roy (1958), however, successfully synthesized tridymite from transistor-grade silicon and high-purity silica gel using only H_2O as a flux, thus confirming the legitimacy of tridymite as a stable silica polymorph.

Tridymite in its region of stability between 867°C and 1470°C is hexagonal with space group $P6_3/mmc$. The structural data for ideal high-temperature tridymite is based upon a fundamental stacking module in which sheets of silica tetrahedra are arranged in hexagonal rings (Table 1-7 and Fig. 1-2).

Table 1-6

Unit Cell		β-Quartz		α-Quartz	
a (Å)		4.9977		4.91239(4)	
c (Å)		5.4601		5.40385(7)	
V (Å³)		118.11		112.933	
ρ (g/cm³)		2.5334		2.6495	
Space Group	$P6_422$	$P6_222$	$P3_121$	$P3_221$	
Atom Positions					
x(Si)	1/2	1/2	0.4701	0.5299	
y(Si)	0	0	0	0	
z(Si)	0	0	1/3	2/3	
x(O)	0.2072	0.2072	0.4139	0.5861	
y(O)	0.4144	0.4144	0.2674	0.7326	
z(O)	1/2	1/2	0.2144	0.7856	

Note: Data for β-Quartz at 590°C from Wright and Lehman (1981) and α-Quartz at 25°C from Will et al. (1988)

When standard tridymite is cooled below 380°C, several phase inversions occur with various changes in symmetry. These tend to produce a large shrinkage and therefore a high thermal coefficient of expansion between 0°–200°C, almost $400 \times 10^{-7}\,\mathrm{K}^{-1}$.

Cristobalite　The stable form of silica above 1470°C is cristobalite. This phase is easily formed metastably in many glass-ceramic materials and can be cooled to room temperature in the same way as tridymite and quartz. Structurally, cristobalite is also formed from the fundamental stacking module of sheets of

Table 1-7

Structural Data for HP-tridymite

Space Group	$P6_3/mmc$		
Unit Cell			
a (Å)	5.052(9)		
c (Å)	8.27(2)		
V (Å³)	182.8(3)		
ρ (g/cm³)	2.183		

Atom	x	y	z
Si(1)	1/3	2/3	0.0620(4)
O(1)	1/3	2/3	1/4
O(2)	1/2	0	0

Note: Data for 460°C from Kihara (1978)

silica with hexagonal rings, but the orientation of paired tetrahedra are in the transorientation as opposed to the cisorientation of tridymite (Fig. 1-2). This leads to a cubic instead of a hexagonal morphology. In fact, the ideal β-cristobalite is a cubic analog of diamond such that silicon occupies the same positions as carbon, and oxygen lies midway between any two silicon atoms. The space group for this structure is $Fd3m$ and the structural data for both cubic β-cristobalite and the low-temperature tetragonal alpha form are shown in Table 1-8.

The phase transition temperature between low and high modifications of cristobalite does not appear to be constant, but a typical temperature is around 215°C. The transition is accompanied by large changes in thermal expansion. The a- and c-axis of α-cristobalite increase rapidly at rates of 9.3×10^{-5} and 3.5×10^{-4} Å K^{-1}, respectively; whereas in β-cristobalite, a expands at only 2.1×10^{-5} Å K^{-1}. This behavior translates into very large, spontaneous strains of -1% along a-axis and -2.2% along c-axis during inversion.

B) Stuffed Derivatives of Silica

Buerger (1954) first recognized that certain aluminosilicate crystals composed of three-dimensional networks of SiO_4 and AlO_4 tetrahedra are similar in structure to one or another of the silicon crystalline

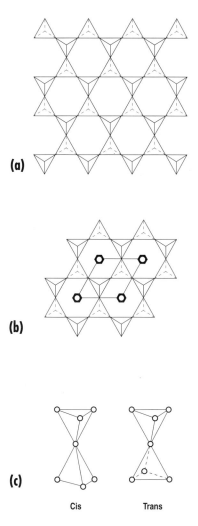

(a)

(b)

(c)

Cis   Trans

Figure 1-2
a Diagram of the tetrahedral sheet that serves as the fundamental stacking module in tridymite and cristobalite. In tridymite, the layers are stacked in a double AB sequence parallel to c, and in cristobalite the sheets create a triped ABC repeat along [111].

b Projection of the structure of ideal HP-tridymite along c. Adjacent tetrahedral layers are related by mirror symmetry, and the six-membered rings super-impose exactly.

c The cis and trans orientations of paired tetrahedra. HP-tridymite tetrahedra adopt the less stable cis orientation, which maximizes repulsion among basal oxygen ions. In β-cristobalite, the tetrahedra occur in the trans orientation (after Heaney, 1994).

Table 1-8

Structural Data for Cristobalite

		β-Cristobalite	α-Cristobalite
Space Group		$Fd3m$	$P4_12_12$
Unit Cell			
a (Å)		7.12637	4.96937
c (Å)		– – – –	6.92563
V (Å³)		361.914	171.026
ρ (g/cm³)		2.205	2.333
Atom Positions			
x(Si)	0	0.3006	
y(Si)	0	0.3006	
z(Si)	0	0	
x(O)	1/8	0.2392	
y(O)	1/8	0.1049	
z(O)	1/8	0.1789	

Note: Data for ideal β-cristobalite at 300°C and α-cristobalite at 30°C from Schmahl et al. (1992)

forms. These aluminosilicates were termed "stuffed derivatives" because they may be considered silica structures with network replacement of Si^{4+} by Al^{3+} accompanied by a filling of interstitial vacancies by larger cations to preserve electrical neutrality. As would be expected, considerable solid solution generally occurs between these derivatives and pure silica. The stable silica polymorphs cristobalite, tridymite, and quartz all have associated derivatives, as does the metastable phase keatite. Examples include the polymorphs carnegieite and nepheline ($NaAlSiO_4$), which are derivatives of cristobalite and tridymite, respectively; β-spodumene ($LiAlSi_2O_6$), a stuffed derivative of keatite; and β-eucryptite ($LiAlSiO_4$) a stuffed derivative of β-quartz.

There has been both confusion and misunderstanding concerning the nomenclature of stuffed derivatives of silica in both the lithium and magnesium aluminosilicate systems. Roy (1959) was the first to recognize a complete solid-solution series between β-eucryptite ($LiAlSiO_4$) and silica with the structure of β-quartz. Most of this series previously about $Li_2O:Al_2O_3:3SiO_2$ in silica was found metastable except very near pure silica. Roy coined the term *silica O* to describe this β-quartz solid solution. This term has been discredited largely because these phases are not of pure silica composition and, in fact, may be as low as 50 mol% silica as in the case of β-eucryptite. Moreover, the pure silica end member is β-quartz itself.

The term *virgilite* was more recently proposed (French et al., 1978) for naturally occurring representatives of lithium-stuffed β-quartz solid solutions falling between the spodumene stoichiometry $LiAlSi_2O_6$ and silica. Virgilite was further defined as including only those compositions with more than 50 mol% $LiAlSi_2O_6$. The problem with this definition is that it arbitrarily reserves a specific part of the solid-solution range for no apparent reason. Moreover, the term virgilite was coined long after these materials had been widely referred to as β-quartz solid solution in the ceramic literature.

The term *silica K* was similarly initially coined by Roy (1959) to describe another series of solid solutions along the join SiO_2–$LiAlO_2$, which are stable over a wide range of temperatures. The compositions range from below 1:1:4 to about 1:1:10 in $Li_2O:Al_2O_3:SiO_2$ proportions (Fig. 1-3) (Levin et al., 1964). Although it was initially recognized that this tetragonal series had a similar structure to the metastable form of SiO_2, namely, keatite, originally synthesized by Keat (1954) at the General Electric Company, phase equilibria studies

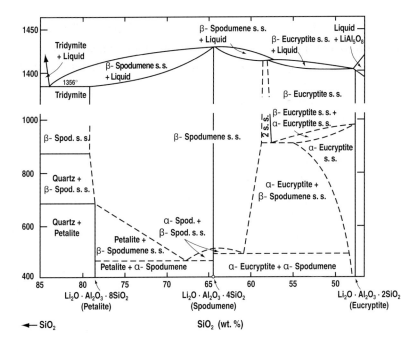

Figure 1-3 The phase diagram of the SiO_2–Li_2O·Al_2O_3·$2SiO_2$ system (after Levin et al., 1964; and Strnad 1968).

showed a large miscibility gap between pure SiO_2 keatite and the most siliceous end member of this series (Fig. 1-3). Since the term β-spodumene $LiAlSi_2O_6$ (1:1:4) was widely in use, it seemed reasonable to refer to this more limited solid-solution series as β-spodumene ss. The term stuffed keatite has also been used to describe this solid solution, but since there is no continuous composition series to silica, the mineral name β-spodumene, which identifies the general composition area, is preferred. This is consistent with standard usage as in the case of nepheline or carnegieite ($NaAlSiO_4$). These terms are preferred to stuffed tridymite or stuffed cristobalite because in these structures there is also no complete solid solution with SiO_2.

For all these reasons, the solid solution along the SiO_2–$LiAlO_2$ join are herein referred to as β-quartz for the hexagonal solid solution series and β-spodumene for the tetragonal solid solution series. The term *high-quartz solid solution* has been proposed (Ray and Muchow, 1968), instead of β-*quartz solid solution,* but the Greek letter designations are generally preferred, not only for brevity, but because more than two structural modifications are possible, as in the case of tridymite.

Another form of nomenclature was introduced by Li (1968) to differentiate between the three polymorphs of $LiAlSi_2O_6$ or spodumene. The stable phase at ambient conditions is the mineral α-spodumene, or $LiAlSi_2O_6$–I, a clinopyroxine. The first phase to form on annealing glasses of the spodumene composition is a stuffed β-quartz phase referred to as $LiAlSi_2O_6$–III, or β-quartz solid solution. Li (1968) preferred the use of the formula $LiAl_2O_6$ with a suffix denoting the structure type (I = clinopyroxine; II = keatite; III = β-quartz). This system, though it avoided the confusion between similar phases related by displacive phase transformations, e.g. α- to β-quartz, is somewhat cumbersome using formula names and is also inappropriate for a range of compositions with the same structure.

C) Structures Derived from β-Quartz (β-Quartz Solid Solutions)

Compositions and Stability   A wide range of stuffed β-quartz compositions can be crystallized from simple aluminosilicate glasses with modifying cations capable of fitting the cavities of the β-quartz structure. These include Li^+, Mg^{2+}, Zn^{2+}, and to a lesser degree Fe^{2+}, Mn^{2+}, and Co^{2+}; a range of ionic sizes from 0.6–0.8 Å. The solid solutions which have been most studied are described by the general formula $Li_{2-2(x+y)}Mg_xZn_yO·Al_2O_3·zSiO_2$ (Strnad, 1986), where $x + y \leq 1$ and z is ≥ 2. The region of proven existence of quartz solid solutions in the pseudoquaternary system SiO_2–$LiAlO_2$–$MgAl_2O_4$–$ZnAl_2O_4$ as crystallized from glass is illustrated in Fig. 1-4 (Petzoldt 1967).

Accessory solid solution components such as Li_2BeO_2 and $Al(AlO_2)_3$ have also been recognized (Beall et al., 1967). In other words, beryllium can substitute to a certain degree for silicon in the tetrahedral position and some aluminum can enter the stuffing or interstitial position in β-quartz, providing lithium is already present as the predominant stuffing ion.

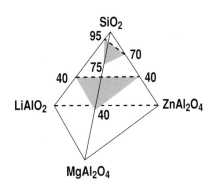

Figure 1-4 The region of proven existence of metastable solid solutions of β-quartz in the pseudoquaternary system SiO_2–$LiAlO_2$–$MgAl_2O_4$–$ZnAl_2O_4$ (wt%) (after Petzoldt 1967).

All of these β-quartz solid solution compositions are believed metastable with the exception of β-eucryptite solid solution, whose stability region is shown in Figs. 1-3 and 1-5. The latter depicts an isothermal slice of the pseudoternary system SiO_2–$LiAlO_2$–$MgAl_2O_4$ at 1230°C, about 60°C below the lowest melting eutectic in this system (see Fig. 1-6). Although all compositions in

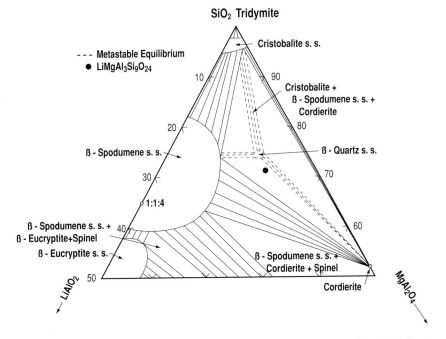

Figure 1-5 Isothermal section from the siliceous half of the system SiO_2–$LiAlO_2$–$MgAl_2O_4$ at 1230°C (wt%) (after Beall et al., 1967).

the siliceous half of this system can initially be crystallized from glass to β-quartz solid solution, only those in the lower left corner near $LiAlSiO_4$ (eucryptite) have been shown to have any range of thermodynamic stability. On the other hand, there are some very persistent metastable β-quartz solid solutions in this pseudoternary system. These can remain even when heated at 1200°C for 100 h. The most persistent composition approaches the stoichiometry $LiMgAl_3Si_9O_{24}$. There may be some structural significance to this stoichiometry, but a single crystal study would be necessary to determine if any favorable distribution of Li^+ and Mg^{2+} ions and SiO_4 and AlO_4 tetrahedra is present.

Structure and Properties   The structure of a metastable quartz solid solution of composition $LiAlSi_2O_6$ or 1:1:4 has been determined by Li (1968) using a single crystal grown from glass. The structure was confirmed as a stuffed derivative of β-quartz with Si and Al distribution in the tetrahedra completely random. Lithium ions were found to be four-coordinated and stuffed into interstitial positions, one lithium atom per unit cell. These were found randomly distributed among three equivalent sites. The lithium tetrahedra were found to be irregular and to share two edges with two Si,Al tetrahedra. The Si,Al–Li distance (2.609Å) is exceptionally short, thereby producing

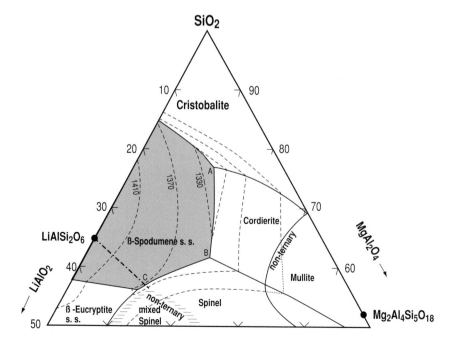

Figure 1-6 Liquidus relations in the siliceous half of the system SiO_2–$LiAlO_2$–$MgAl_2O_4$ (wt%)

strong cation repulsion. This is believed to play an important role in controlling the anomalous low thermal expansion behavior of this solid solution (Li 1968). In this structure, the *a*- and *b*-axes are functions of the Si,Al–Li distance alone, and upon heating, these axes tend to expand. On the other hand, the *c*-axis, a function of the Li–O distance, would be expected to contract, because increasing the Si,Al–Li distance decreases the shared edges and, hence, the Li–O bond length.

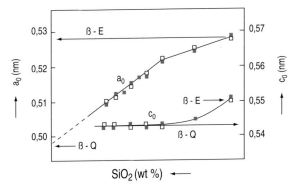

Figure 1-7 Unit cell dimensions a_0 and c_0 of β-quartz solid solutions between SiO_2 and β-eucryptite (LiAlSiO$_4$) (empty squares). Full squares correspond to peraluminous β-quartz solid solutions between $LiAl_{1.17}SiO_{4.25}$ and SiO_2 (after Nakagawa and Izumitani, 1972).

Figure 1-7 illustrates the lattice parameters a_0 and c_0 of the solid solutions of β-quartz between silica and eucryptite (LiAlSiO$_4$) (Nakagawa and Izumitani, 1972). Figure 1-8 shows the corresponding coefficients of thermal expansion of these solid solutions (Petzoldt 1967). Note that this coefficient is heavily negative near β-eucryptite, plateaus slightly negative from 50 to 80 wt% SiO_2, approaches zero above 80 wt%, and then becomes strongly positive. It is evident that the β-quartz solid solutions are incapable of persisting to room temper-

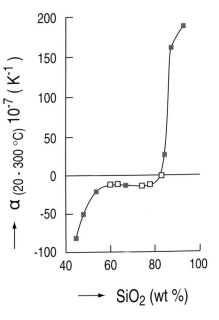

Figure 1-8 Coefficient of thermal expansion of solid solutions of β-quartz crystallized from glasses in the SiO_2–$LiAlO_2$ system (after Petzoldt 1967).

ature when their composition is as siliceous as 82 wt%. This is illustrated in Fig. 1-9, where compositions of 15, 10, and 5 mol% LiAlO$_2$ exhibit increasing inversion temperatures associated with the alpha-to-beta transformation

(Petzoldt 1967). Evidently, there is insufficient lithium to prop up the structure much above 85 mol% SiO_2. The range of ultra-low expansivity important from practical melting considerations ranges from 52 to 75 wt% SiO_2.

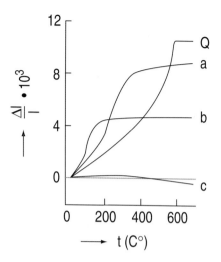

The coefficients of thermal expansion of more complex solid solutions of β-quartz in the pseudoquaternary system SiO_2–$LiAlO_2$–$MgAl_2O_4$–$ZnAl_2O_4$ show a general decrease with increasing Li^+ and Zn^{2+}. On the other hand, Mg^{2+} increases the thermal expansion behavior (Strnad 1986). For example, the thermal expansion of stuffed β-quartz at the cordierite stoichiometry $(Mg_2Al_4Si_5O_{18})$ from 25°–870°C averages 47.2×10^{-7} K^{-1}. The three cations Li^+, Mg^{2+}, and Zn^{2+} can therefore be used to tailor a thermal expansion coefficient very close to zero or, indeed, any value between -50×10^{-7} K^{-1} and $+50 \times 10^{-7}$ K^{-1} over a wide range of silica contents.

Figure 1-9 Dilatometric curves of quartz solid solutions crystallized from glasses in the SiO_2–$LiAlO_2$ system: mol% Li_2O (a) 5%, (b) 10%, (c) 15%, (Q) 0% = pure quartz (after Petzoldt 1967).

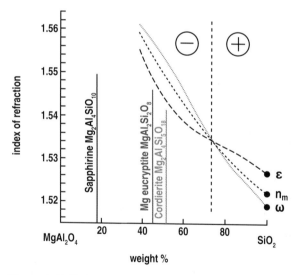

Figure 1-10 Refractive indices of metastable β-quartz solid solutions in the SiO_2–$MgAl_2O_4$ system (after Schreyer and Schairer 1961).

Similarly, the refractive index and birefringence of β-quartz solid solutions can be varied over a considerable range. Figure 1-10 shows the indices of refraction of quartz solid solutions along the SiO_2–$MgAl_2O_4$ join (Schreyer and Schairer, 1961). It can be seen that optically positive quartz becomes isotropic at 72 wt% and negative at lower levels of silica. The existence of optically isotropic quartz suggests the possibility of transparency in polycrystalline β-quartz ceramics.

Structure of β-Eucryptite   Considerable structural understanding has developed concerning the most highly stuffed derivative of β-quartz, namely β-eucryptite (LiAlSiO$_4$). The structure of β-eucryptite was first shown to be based on that of β-quartz by Winkler (1948). A complete ordering of Al and Si over the tetrahedral sites produces a doubling of the *c*-axis as revealed by weak "superlattice" reflections of the type *hkl.* *l* = odd. Loewenstein's aluminum avoidance rule is, therefore, satisfied because each Si is surrounded by four Al tetrahedra and vice versa. Winkler also determined that the Li atoms reside in the same planes as the Al atoms and that they are tetrahedrally coordinated by oxygen.

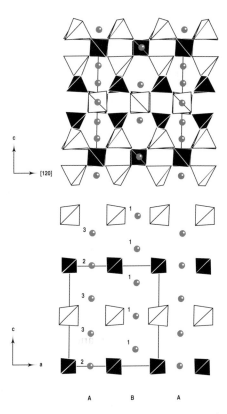

Figure 1-11 Projections of the structures of β-eucryptite. Si-containing tetrahedra are plotted in black; Al tetrahedra are white. The view directions are: [001] (upper diagrams) and [100] (lower). At low temperatures the Li ions reside on three distinct sites, Li^1–$_3$ (numbered). Above T_c ~ 755 K, the Li ions become distorted over the two sites Li$_1$ and Li$_2$. Structural data from Guth and Heger (1979).

The LiO$_4$ tetrahedra share edges with adjacent AlO$_4$ tetrahedra (Fig. 1-11). At low temperatures, the Li ions reside on three distinct sites, Li 1–3 (numbered) (Fig. 1-11). Above a critical temperature of 755 K, the lithium ions become distorted over the two sites Li1 and Li2.

The temperature dependence of lattice constants is well known. On increasing temperature, there is an expansion in the (001) plane, but a corresponding contraction along *c* so that the net overall volume of the cell decreases. This peculiar behavior can be explained by the edge sharing of Li- and Al,Si-containing tetrahedra. At room temperature, the Li–(Al,Si) distance is very small (2.64Å). At this stage, the four atoms (Li,Al,Si and two O atoms) that are associated with the edge sharing are all coplanar. The repulsive force between Li and Al,Si is reduced by thermal expansion in the xy plane, and to maintain the Li–O and Al,Si–O bond distances, the shared O–O edge length must decrease. The resulting decrease in the c cell edge is illustrated in Fig. 1-12.

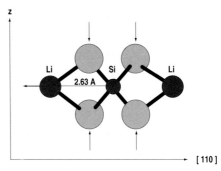

Figure 1-12 Structural rationalization of the thermal expansion behavior for β-eucryptite. The Li and Al,Si atoms are coplanar normal in the (001) plane, and the LiO_4 and $(Al,Si)O_4$ tetrahedra share edges. Thermal expansion in the (001) plane reduces the repulsive force between Li and Al,Si. However, maintaining the metal-oxygen bond distances requires a contraction along [001]. Thus, a and b increase with increasing temperature, while c decreases (after Palmer 1994).

Aside from the solid solution between β-eucryptite and SiO_2 described earlier, there is considerable solid solution with $AlPO_4$. Up to 48 mol% $AlPO_4$ has been observed by Perrotta and Savage (1967) to enter the β-eucryptite structure. $AlPO_4$-rich compositions show some split diffraction peaks, suggesting a symmetry-reducing phase transition. The ionic conductivity, already high in β-eucryptite, increases with the phosphate content. Tindwa et al. (1982) suggested that this increase in conductivity was related to the reduction of strength of the bonding between Li and the surrounding channel O.

D) Structures Derived from Keatite (ß-Spodumene Solid Solution)

Keatite, sometimes referred to as *silica K*, is a high-pressure form of SiO_2, which has neither been recognized in nature nor appears to have any field of thermodynamic stability without the additions of alkalis or water. The phase may be synthesized at 0.1 GPa and 800 K from silica gel.

Composition and Stability   Although keatite is a metastable form of SiO_2, its stuffed derivatives along the SiO_2–$LiAlO_2$ join are stable over a wide range of temperatures, at least at relatively low pressure. The composition range of β-spodumene solid solutions also coincides with practical glass-forming areas. Figure 1-3 shows the stable solid solution region from below 60 wt% to almost 80 wt% silica. The region of thermal stability is believed to be between 500°C and the solidus temperature which, for the congruent melting $LiAlSi_2O_6$ composition, is maximized near 1425°C. A surprisingly large amount of MgO can be substituted for Li_2O in this solid solution, as is shown in Fig. 1-5, but a decrease in thermal stability results due to a lower solidus temperature (see Fig. 1-6). Unlike the metastable stuffed β-quartz solid solutions, however, little ZnO is permitted into the keatite (β-spodumene) structure: more than one weight percent results in the presence of gahnite ($ZnAl_2O_4$).

Structure and Properties   The β-spodumene framework has been confirmed to be isotypic with keatite (Li and Peacor, 1968). The Li ions are coordinated to four O atoms. The silicon and aluminum distributions in the tetrahedra are random. The structure consists of interlocking five-membered rings of Si,Al tetrahedra. These rings run parallel to either the (010) or (100) planes, thus helping to create channels that can be filled with lithium. As the temperature is increased, the c-axis expands while a- and b-axes are seen to contract. A structural explanation involving release of strain in the five-membered rings resulting in a change of tetrahedral orientation was offered by Li and Peacor (1968). The thermal expansion behavior of both tetragonal axes, as well as the average volume expansion of the unit cell, are given in Figs. 1-13 and 1-14 (Ostertag et al., 1968). Note that the more siliceous compositions have the minimum average expansion. These compositions are preferred in ceramic applications where very low thermal expansion must be combined with thermal stability for both long-term high temperature use and thermal shock resistance.

Skinner and Evans (1960) initially investigated the stability of $LiAlSi_2O_6$ β-spodumene. A full crystal structure refinement by several authors indicated a space group $P4_32_12$. β-Spodumene has an obvious channel structure, and ionic conductivity reflects this structural feature. The conduction is thermally activated with an activation energy

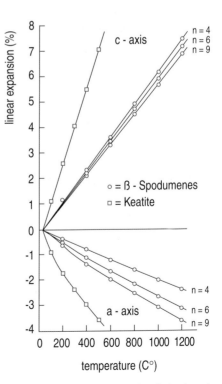

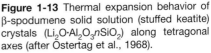

Figure 1-13 Thermal expansion behavior of β-spodumene solid solution (stuffed keatite) crystals ($Li_2O·Al_2O_3nSiO_2$) along tetragonal axes (after Ostertag et al., 1968).

of 0.81 eV, similar to that for β-eucryptite. A space-filling model of the 114 β-spodumene is shown in Fig. 1-15.

E) Structures Derived from Tridymite

The most important stuffed derivatives of tridymite fall in the binary system $NaAlSiO_4$–$KAlSiO_4$. The sodium end-member is nepheline ($NaAlSiO_4$), and the potassium end-member, kalsilite. Natural nepheline occurs at the $Na_3KAl_4Si_4O_{16}$ point, 25% of the way from soda-nepheline to kalsilite. Solid solution has been observed between natural nepheline and $Na_3(AlSiO_4)_4$. Experimental solid solutions in the nepheline area are shown in Fig. 1-16, which also illustrates the framework silicate feldspar and leucite solid solutions. The effect of temperature on nepheline and kalsilite solid solutions is shown in Fig. 1-17,

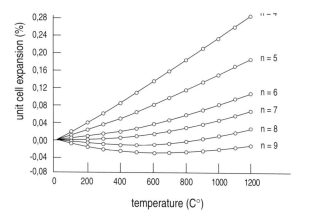

Figure 1-14 Volume thermal expansion in crystals of $Li_2O \cdot Al_2O_3 \cdot nSiO_2$ β-spodumene solid solution (after Ostertag et al., 1968).

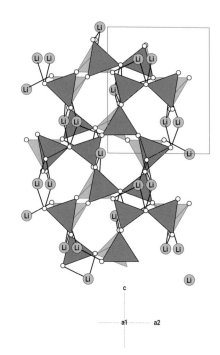

Figure 1-15 β-spodumene structure viewed along [100]. Tetrahedra are centered by Al,Si atoms. Li sites occur in pairs (50% occupancy in each), at the edges of fivefold rings. The Li-O coordination is a distorted tetrahedron. See Appendix Fig. 6 for [001] projection showing tetrahedra arranged along tetragonal screw axes (after Li and Peacor, 1969).

which is an isobaric section at 0.2 GPa.

The crystal structure of nepheline was first determined by Buerger et al. (1954), and later refined by Hahn and Buerger (1955). The refinements confirmed the earlier speculation that nepheline is indeed a stuffed derivative of tridymite. The alkali cations are ordered in two different sites: the K ions reside in open hexagonal rings in ninefold coordination with oxygen, while the Na sites are distorted, oval rings with a coordination of eight oxygens around each Na ion. The ratio of oval to hexagonal rings is thus 3:1 (Fig. 1-18).

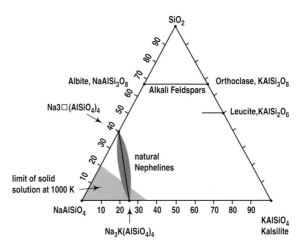

Figure 1-16 Ternary diagram showing compositions of phases in the nepheline–kalsilite– SiO_2 system.

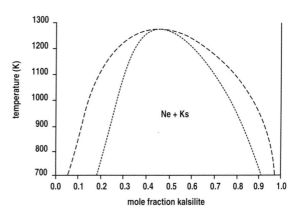

Figure 1-17 Solvus (dashed curve) and spinodal (dotted curve) for the nepheline–kalsilite solid solution at 0.2 GPa. (After Ferry and Blencoe, 1978).

For synthetic soda nephelines containing more than six Na ions per unit cell, the excess Na must be accommodated on the large alkali site. Since ions smaller than K do not achieve full contact with the surrounding channel oxygens, Na substitution must involve either the collapse of the hexagonal channels or an off-centering of Na toward the channel wall. In any case, unusual microwave absorption has been observed with such high soda nephelines. The thermal expansion coefficient for sodium nepheline, the end-member, is approximately 90×10^{-7} K^{-1}, while the natural potassic material is higher, approximately 120×10^{-7} K^{-1}.

Kalsilite, the potassium end-member $KAlSiO_4$, is a stuffed tridymite derivative with adjacent layers rotated through 180°. Kalsilite shows Al,Si ordering in adjacent sites causing a reduction in symmetry from the idealized tridymite $P6_3/mmc$ to $P6_3mc$. Potassium is surrounded by nine nearest oxygens, three apical oxygens linking the upper and lower tetrahedral sheets, plus two sets of three basal oxygens from the upper and lower ditrigonal rings (Fig. 1-19). The specific volume of kalsilite is about 8.3% greater than that of soda nepheline. The thermal expansion coefficient is also higher, approaching 150×10^{-7} K^{-1}. Both nepheline and kalsilite are common phases in aluminosilicate glass-ceramics containing Na_2O and K_2O.

F) Structure Property Relationships in Ring Silicates

A subset of the framework silicates or tectosilicates is the ring or cyclosilicates. Important examples of these minerals in glass-ceramics are cordierite, especially in its hexagonal form, sometimes referred to as indialite ($P6/mcc$) and the double-ring silicate osumilite, of similar space group.

Cordierite The structure of cordierite is shown in Fig. 1-20, projected on the basal plane (001). The hexagonal rings are formed of

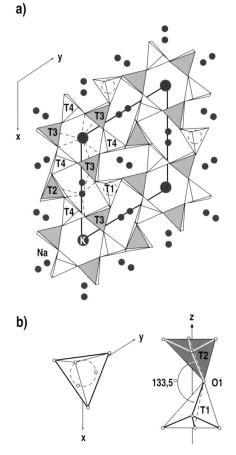

a)

b)

Figure 1-18 The crystal structure of nepheline plotted using the coordinates of Hahn and Buerger (1955). (a) Projection along [001]. Large black spheres represent K; small black spheres represent Na. The bonds to nearest-neighbor O atoms are indicated by the dashed lines. SiO_4 tetrahedra are shown in black, pointing downward; AlO_4 tetrahedra are white and point upward. A triad axis passes vertically through the T1 and T2 sites; oxygen is off-center from this axis (represented by the dashed circles) and disordered over three sites. (b) Details showing the effects of apical oxygen (O1) off-centering. The T1 (Al)–O–T2 (Si) bond angle is reduced from 180° to 133.5° (after Palmer 1994).

six tetrahedra, of which five are silicon and one is aluminum. The two Mg ions in the unit formula form octahedral units with O tying the rings together. In addition, the other three tetrahedral aluminum groups are interspaced between the magnesia octahedra separating the predominantly silicate rings. The rings are distributed in reflection planes parallel to the base at heights 0, c/2, and c. The juxtaposition of layering of the aluminosilicate rings produces large cavities along the channels enclosed by the sixfold rings. Large molecular or ionic species such as H_2O, Cs^+, K^+, and Ba^{2+} are known to occupy these cavities. Cordierite is orthorhombic pseudohexagonal with space group *Cccm*. On heating near the melting temperature, cordierite is transformed into the hexagonal form sometimes referred to as indialite. This transformation is continuous and of the order–disorder kind involving both the Si_5Al (ring) and the $AlMg_2$ (inter-ring).

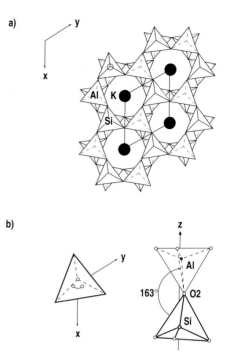

Figure 1-19 The crystal structure of kalsilite. (a) Projection of the average, $P6_3$ structure along [001], plotted using the coordinates of Perrotta and Smith (1965). (b) The effect of apical oxygen (O_2) displacement from the triad axis. The off-centered O_2 atom can occupy one of three possible sites, indicated by open circles, at a distance of 0.25 Å from the [001] triad. The off-centering reduces the Al–O_2–Si bond angle from 180° to a more energetically favorable 163° (after Palmer 1994).

The most simple solid solutions involving low-pressure cordierites consist of divalent substitutions for octahedral magnesium in the structure. Typical examples are Fe^{2+} for Mg^{2+} and Mn^{2+} for Mg^{2+}. Both these substitutions have the adverse effect of reducing the refractoriness or the temperature of initiation of melting in the crystalline material.

A more interesting substitution involves replacing two tetrahedral Al^{3+} ions by one Be^{2+} and one Si^{4+} ion. This results in the stoichiometric end-member $Mg_2BeAl_2Si_6O_{18}$ from normal cordierite $Mg_2Al_4Si_5O_{18}$ and produces changes in lattice parameters as well as in thermal expansion behavior (see Table 1-9). From room temperature to 800°C, the average linear thermal expansion coefficient increases from 11.7×10^{-7} K^{-1} in stoichiometric

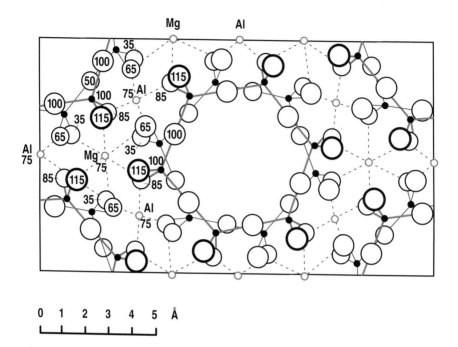

Figure 1-20 A comparison of the structures of beryl, $Be_3Al_2Si_6O_{15}$, and cordierite, $Al_3Mg_2Si_5AlO_{18}$, viewed on the plane of the sixfold rings. In cordierite the b-axis is vertical and the a-axis horizontal (after Byström 1941).

Table 1-9

Composition, Lattice Parameters, and Thermal Expansion of Cordierites

Composition	Lattice Parameters (25°C)			CTE (22°C–800°C) $\times 10^{-7}\ K^{-1}$	Thermal Expansion Anisotropy (22°C–800°C) $\left(\dfrac{\Delta a}{a} - \dfrac{\Delta c}{c}\right)$
	a (Å)	c (Å)	V (Å3)		
$Mg_2Al_4Si_5O_{18}$ (1000°C – 2 h)	9.784	9.349	775.05	11.7	2600
$Mg_2BeAl_2Si_6O_{18}$ (1000°C – 2 h)	9.609	9.276	741.67	19.2	2560
$KMg_2Al_5Si_4O_{18}$ (950°C – 2 h)	9.811	9.470	789.42	20.6	4600
$CsMg_2Al_5Si_4O_{18}$ (950°C – 2 h)	9.804	9.457	787.21	22.0	900

Note: All glasses melted at 1650°C, 16 h ground, and crystallized at indicated schedules (Data after Evans et al., 1979)

cordierite to $19.2 \times 10^{-7} \, K^{-1}$ in the substituted beryllium version. Further beryllium substitution produces the hexagonal mineral beryl, $Be_3Al_2Ai_6O_{18}$, which is isostructural with cordierite. Beryl, however, is not known to have been internally nucleated and crystallized in glass-ceramics. Crysoberyl, $BeAl_2O_4$, with the olivine structure, is the phase that is nucleated upon heat treatment of "beryl" glass.

A broader form of solid solution in cordierite relies on stuffing the large hexagonal ring cavities with alkali or alkaline earth ions. This stuffing must be accompanied by a substitution of tetrahedral aluminum for silicon in the ring position to preserve charge balance. Thus, stoichiometric cordierite $Mg_2Al_4Si_5O_{18}$ may have its hexagonal cavities stuffed with appropriate large alkali cations like potassium and cesium to give $KMg_2Al_5Si_4O_{18}$ or $CsMg_2Al_5Si_4O_{18}$. Similar stuffing with large alkaline earth cations such as Ba^{2+} will result in the end-member $BaMg_4Al_{10}Si_8O_{36}$. A key effect of these large stuffing ions on the cordierite composition is increased glass stability allowing sintering prior to crystallization. Table 1-9 lists the lattice parameters and the modifications in thermal expansion of the various cordierite end-members as crystallized from glass. It can be seen that the linear expansion coefficients cover a range from $12 \times 10^{-7} \, K^{-1}$ to $22 \times 10^{-7} \, K^{-1}$ over the range of $0°\text{--}800°C$. The cesium-stuffed cordierite gives the closest match to silicon, a point of some significance in electronic packaging.

Another concern in glass-ceramic materials formed by surface crystalliza-tion of frits (particulate glass) is the stress that can develop between grains where anisotropic thermal expansion in different lattice directions is high. Clearly, this stress is a function of grain size as well as anisotropy in thermal expansion coefficient. For a given grain size, however, low anisotropy, such as observed with cesium-stuffed cordierite (Table 1-9), will produce the mini-mum intergranular stress. Since the anisotropy in cesium-stuffed cordierite is only one-third that of stoichiometric cordierite, the grain size may be larger without exceeding the intergranular stresses present in standard cordierite.

Osumilite   Osumilite is a member of the milarite mineral group characterized by double hexagonal rings of composition $(Si,Al)_{12}O_{30}$. Two sixfold rings are joined by the apices of their tetrahedra to form a double ring. Since each tetra-hedron has three corners in common with other tetrahedra and one free corner, Si, Al and O are in the ratio of 2:5 corresponding to the formula $(Al,Si)_{12}O_{30}$. The most important member of the osumilite group crystallizing in glass-ceramics corresponds to the formula $MgAl_2Si_4O_{12}$ $[Mg_5Al_6(Al_4Si_{20})O_{60}]$, which corresponds to one Mg^{2+} ion per double unit formula in a channel site, filling only one-quarter of the available channel sites. This metastable phase is readily crystallizable from simple ternary glasses (Schreyer and Schairer, 1962) and its X-ray pattern corresponds to that of natural osumilite: $(K,Na,Ca)Mg_2Al_3(Si,Al)_{12}O_{30}$, where half the channel sites are filled. Other osumilite-type phases found in glass-ceramics include roedderite $K_2Mg_5Si_{12}O_{30}$ and its sodium analog, where alkali completely fill the channel sites.

The coefficient of thermal expansion of osumilite is quite low, about $20 \times 10^{-7} K^{-1}$, and, being more siliceous in composition than cordierite, it readily forms a stable glass. This glass can be powdered and surface crystallized as a devitrifying frit, or it can be internally nucleated and crystallized in bulk by certain nucleating agents like silicon. As in the case of cordierite, various stuffed varieties of osumilite can be formed using calcium, barium, or alkali ions in solid-solution schemes like $Ba^{2+} + 2Al^{3+} \leftrightarrow 2Si^{4+}$. In fact, one of these stuffed osumilites, barium osumilite ($BaMg_2Al(Al_3Si_9O_{30})$), has been used as a matrix for silicon oxycarbide (Nicalon®) reinforced glass-ceramics. It is refractory, can be sintered from glass powder, and closely matches SiC and related oxygen-containing compounds in coefficient of thermal expansion (Beall et al., 1984).

G) Other Tectosilicates

In addition to stuffed derivatives of the silica structures and ring silicates, a number of other framework minerals can be crystallized in glass-ceramics. Among these are the feldspars, in particular celsian and anorthite, and the complex feldspathoids, leucite, and pollucite. Despite their wide natural occurrence and excellent glass-forming capacity, the feldspars have not been proven to be particularly useful glass-ceramic phases. The alkali feldspars such as albite, $NaAlSi_3O_8$, and sanidine or orthoclase, $KAlSi_3O_8$, form exceptionally stable glasses that will not crystallize in practical periods of time. Anorthite, $CaAl_2Si_2O_8$, the triclinic ($P\bar{1}$) feldspar forms a good glass and crystallizes easily, but it is not readily internally nucleated. It does, however, form good powder processed (fritted) glass-ceramics possessing an intermediate coefficient of thermal expansion near $50 \times 10^{-7} K^{-1}$. The barium equivalent of anorthite, celsian, $BaAl_2Si_2O_8$, monoclinic with space group $I2/c$, possesses a somewhat lower coefficient of thermal expansion near $40 \times 10^{-7} K^{-1}$ and is readily internally nucleated with titania. This crystal has an extremely high melting point, 1685°C, compared with 1550°C for anorthite, and is therefore impractical in a monophase glass-ceramic. It has found applications as a secondary phase, however.

There are four feldspathoid minerals that are key constituents in certain glass-ceramics. Nepheline solid solutions and kalsilite have already been discussed in Section 1.3.1.6 (B). The leucite family has more complex structures that include leucite itself ($KAlSi_2O_6$) and pollucite ($CsAlSi_2O_6$). In their simplest form, these minerals are cubic with a large unit cell containing 96 oxygen atoms. The silica tetrahedra are linked in a system of four- and six-member rings to form a three-dimensional framework in such a way that four-membered rings are linked onto each side of a six-membered ring (see Fig. 1-21). The threefold symmetry axes pass through the six-membered rings and are nonintersectional so there is only one such axis in each cube of side $a/2$.

The channels along these axes house the largest species, Cs^+ and K^+, in the case of pollucite and leucite, respectively. Unlike pollucite, leucite is only cubic at high temperatures and inverts to a tetragonal form on cooling toward room temperature. The inversion is believed to be due to a collapse of the framework caused by the relatively small potassium ions occupying the cesium sites along the triad axis channels. Above the inversion temperature T_i, the minerals become cubic but there is still some rotation or unwinding of tetrahedra until a discontinuity temperature T_d is reached, at which point the structure reaches a state of maximum expansion, above which the thermal expansion is very low. As seen from Fig. 1-22, the inversion temperatures for synthetic leucites and pollucite made from a gel are 620°C, 320°C, and below room temperature for leucite, Rb-leucite, and pollucite, respectively.

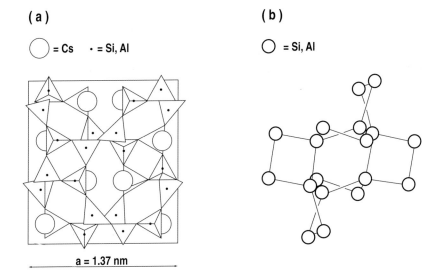

(a)

◯ = Cs   • = Si, Al

a = 1.37 nm

(b)

◯ = Si, Al

Figure 1-21 Structure of pollucite ($CsAlSi_2O_6$). (a) Upper half of unit cell projected on (001). (b) Linkage showing four rings on each side of the six-membered ring.

Because of the propping effect of the large Cs ion, pollucite shows a very low thermal expansion coefficient averaging below 15×10^{-7} K^{-1} from 0–1000°C and almost zero above the 200°C discontinuity temperature (Taylor and Henderson, 1968). This result conflicts with somewhat higher coefficients measured on sintered natural

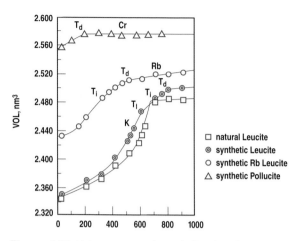

Figure 1-22 Volume expansion of "leucites," including Cs-leucite (pollucite) (after Taylor and Henderson, 1968).

pollucite (24×10^{-7} K^{-1}) and pollucite made through Cs ion exchange of lithium and sodium aluminosilicates (30×10^{-7} K^{-1}) (Richerson and Hummel, 1972). Obviously, pollucite has a low thermal expansion behavior which is, however, highly dependent on its mode of synthesis.

Leucite is cubic above 325°C, with space group $Ia3b$ and unit cell edge 1.34 nm. At room temperature, however, it is tetragonal with space group $I4_1/a$, with a = 12.95 Å and c = 13.65 Å. Because of the uniquely high thermal expansion of leucite, this mineral is used to raise the thermal expansion coefficient of glass-ceramics including dental materials (see Sections 2.2.9 and 4.4.2), and other glazable porcelains. Both leucite and pollucite are refractory minerals, the former with a melting point of 1685°C and the latter with a melting point above 1900°C.

1.3.2 Phosphates

1.3.2.1 Apatite

In their publication, LeGeros and LeGeros (1993) outline the different apatites, ranging from natural apatite (minerals) to biological (human dentin, enamel, and bone) and synthetic (chemically synthesized) apatite. This publication clearly shows that apatite is a group of crystalline compounds. The most important of these compounds is calcium hydroxyapatite. All the related crystal structures, such as fluoroapatite, chloroapatite, and carbonate apatite are derived from it.

The crystal structure of hydroxyapatite, $Ca_{10}(PO_4)_6(OH)_2$, has been studied by Beevers and McIntyre (1956), Kay et al. (1992), Elliott (1994), and Young et al. (1966). Hydroxyapatite has the hexagonal $P6_3/m$ space group. This designation refers to the sixfold c-axis perpendicular to three equivalent a-axes (a_1, a_2, a_3) at angles of 120° to each other. Ten Ca^{2+} ions are located at two different sites in the unit cell. These ions are referred to as Ca1 and Ca2 in this section. The four Ca1 ions of the unit cell are hexagonally coordinated. They are located at point 0 and 0.5 on the c-axis ($c = 0$, $c = 0.5$). They are coordinated in an a-b axial direction at a considerable distance to the Ca2 ions. The six Ca2 ions of the unit cell are at point 0.25 or 0.75 on the c-axis ($c = 0.25$, $c = 0.75$). They are coordinated in a triangle at the same height of the c-axis. The OH groups occupy the corners of the unit cell. The PO_4 tetrahedra demonstrate the stable network of a helix structure of $c = 0.25$ to $c = 0.75$. According to LeGeros and LeGeros (1993), the lattice parameters of mineralogical hydroxyapatite (found in Holly Springs, MS, USA) are $a = 9.422$ Å, $c = 6.880$ Å + 0.003 Å.

Fluoroapatite (F-apatite), chloroapatite (Cl-apatite) and carbonate apatite (CO_3-apatite) are derived from hydroxyapatite. In F-apatite and Cl-apatite, the F^- and Cl^- ions assume the position of the OH^- ions. When F^- and Cl^- ions are inserted in the Ca triangle, their position in relation to the OH^- in hydroxyapatite changes (Fig. 1-23). As a result, the lattice parameters change compared with those of hydroxyapatite (Young and Elliot, 1966). When F^- ions are inserted in place of OH^- ions, the a-axis is reduced and the c-axis remains constant (F = apatite: $a = 9.382$ Å, $c = 6.880$ Å). The insertion of Cl^- ions enlarges the unit cell (Cl⁻ apatite: $a = 9.515$ Å, $c = 6.858$ Å). The crystal structure of fluoroapatite is shown in Appendix 19.

When $(CO_3)^{2-}$ ions are inserted in apatite, two types of structures result. In type A, the larger planar CO_3 groups replace the smaller OH groups, causing the a-axis to expand and the c-axis to contract. In type B, the smaller CO_3 groups replace the larger PO_4 tetrahedra, causing the a-axis to contract and the c-axis to expand. Both crystal types were identified by LeGeros and LeGeros (1993). In addition to the substitution of structural groups in carbonate apatite, the partial substitution of Ca^{2+} ions by Na^+ ions has also been observed. This substitution balances the charges when $(CO_3)^{2-}$ is inserted.

The following ions are generally known to be suitable for insertion into the apatite structure:

- for Ca^{2+}: Mg^{2+}, Ba^{2+}, Pb^{2+} (among others)
- for $(PO_4)^{3-}$: $(BO_3)^{3-}$, vanadate, manganate, $(CO_3)^{2-}$, $(SiO_4)^{4-}$
- for $(OH)^-$: F^-, Cl^-, $(CO_3)^{2-}$

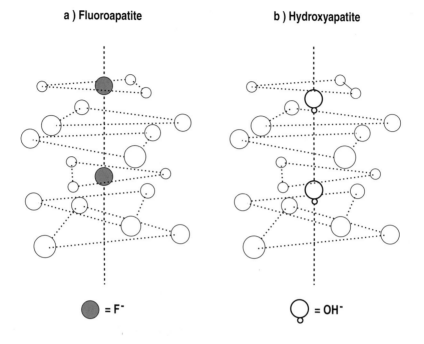

Figure 1-23 Positions of F⁻ and OH⁻ in Ca₂ triangles of apatites (*c*-axis channel):
a) Fluorapatite
b) Hydroxyapatite
according to Elliott (1994).

The ion substitution in hydroxyapatite primarily influences the morphology of the crystals and the different properties of the various types of apatite. LeGeros and LeGeros (1993), for example, observed a change in the morphology of CO_3-apatite from acicular crystals to rods and to equiaxed crystals when increasing amounts of $(CO_3)^{2-}$ were inserted. Another noteworthy change in the properties concerns the chemical solubility, which decreases according to the following sequence of apatites:

- CO_3-apatite
- Sr- or Mg-apatite
- OH-apatite
- F-apatite

Scientifically speaking, biological apatites, such as apatite crystals of human dentin, enamel, and bone must be classified as carbonate apatites. These substances also demonstrate different levels of solubility because of their various degrees of crystallinity. Therefore, enamel carbonate apatite is chemically more stable than the dentin carbonate apatite because of a higher degree of crystallinity. All these apatites, however, demonstrate higher chemical solubility than do hydroxyapatite and fluoroapatite.

The following minor elements and substances are contained in biological apatites: Na^+, K^+, Mg^{2+}, $(HPO_4)^{2-}$, Cl^-, and F^-. Trace elements such as Sr^{2+}, Pb^{2+}, Ba^{2+} are also contained.

While the molar ratio of Ca/P in synthetic hydroxyapatite is 1.667 (or Ca/P wt ratio = 2.151), that of various commercial hydroxyapatite ceramic products is between 1.57 and 1.70. As a result, the following crystalline phosphates have also been identified in synthetic hydroxyapatite: β-tricalcium phosphate, $Ca_3(PO_4)_2$; tetracalcium phosphate, $Ca_4P_2O_9$; anhydrous dicalcium phosphate, $CaHPO_4$; or calcium pyrophosphate, $Ca_2P_2O_7$.

1.3.2.2 Orthophosphates and Diphosphates

A) SiO$_2$-Type Orthophosphates and Diphosphates

These crystals are known as aluminium orthophosphate crystals. There are three polymorphous main modifications of these crystals (Kleiber 1969). One of these is berlinite, which is an isotype to quartz as a trigonal-trapezoidal SiO_2-type. Another is $AlPO_4$ as a cristobalite structure of cubic to pseudocubic SiO_2. The third main crystal phase of $AlPO_4$ is of the tridymite-type of hexagonal to pseudohexagonal structure. In addition to these main phases of $AlPO_4$, high-temperature modifications of the SiO_2-type (e.g., β-quartz, β-cristobalite, etc.) are known.

It is interesting to note that $AlPO_4$ crystals of the tridymite-type generally show defects in the lattice structure. $AlPO_4$, however, crystallizes more easily into the ideal two-layer structure of tridymite than does SiO_2. The reason is the higher polarization of the phosphate tetrahedron compared with that of the $(SiO_4)^{4-}$-tetrahedron.

Berlinite is the α-quartz modification of $AlPO_4$. Its tetragonal structure is characterized by a = 4.9429 Å, c = 10.9476 Å, V = 231.6 Å and space group $P3_121$ (Schwarzenbach 1966).

Despite the highly isotypical relationship between phosphates and SiO_2, the crystallochemical relationship is not close enough to enable mixing of Si(IV) and P(V) in the crystal lattice of $AlPO_4$. This principle of diadochy, i.e., in which P(V) is partially replaced by Si(IV), is possible in apatite. This reaction, however, does not occur in $AlPO_4$.

Because of this isotypical relationship of the berlinite modification of $AlPO_4$ to quartz, the question now arises: Have stuffed derivatives been observed in $AlPO_4$-based crystals as they have been in silica quartz? At this point, however, it must be noted that these types of crystals are not yet known in crystallography.

B) Other Orthophosphates

Apart from $AlPO_4$ crystals, other phosphates with structural similarities to silicates have also been determined. For example, $Mg_2(P_2O_7)$ is isotypical with thortveitite, $Sc_2(Si_2O_7)$. A similar isotypical relationship is also valid for Na_6 (P_6O_{18}) and β-wollastonite, $Ca_6(Si_6O_{18})$.

In addition to apatite and the $AlPO_4$–modifications, phosphates of the nasicon-type structure are also important in silica-free phosphate glass-ceramics. Winand et al. (1990), have shown that $NaTi(PO_4)_3$ belongs to the nasicon-type structure of $Na_{1+x}Zr_2P_{3-x}Si_xO_{12}$. The space group is $R3c$. Divalent ions may also be inserted into these structures without changing the nasicon structure. According to Schulz et al. (1994), the complex structural formula can be written as follows:

$$M^I_x M^{II}_{1-x} M^{IV}_{1+x} M^{III}_{1-x} (PO_4)_3$$

If Na(I), Ca(II), Zr(IV), and Al(III) are inserted into this structure, crystals with the following structural formula are created:

$$Na_x Ca_{1-x} [(Zr,Ti)_{1+x} Al_{1-x}] \cdot (PO_4)_3$$

Orthophosphates that do not occur as main crystal phases but rather as secondary crystal phases in glass ceramics may also have very complex structures. For example, $Na_5Ca_2Al(PO_4)_4$ exists monoclinically (Alkemper et al., 1994). It has lattice parameters of a = 11.071(3) Å, b = 13.951(4) Å, c = 10.511(3) Å, β = 119.34(1)°, V = 1415.2 Å^3.

It was possible to precipitate lithium orthophosphate as a main crystal phase in glass-ceramics (Section 2.6.5). The structures of low- and high-temperature forms of Li_3PO_4 crystals are similar. Keffer et al. (1967), determined that both Li and P are in tetrahedral coordination and are tightly held together by cornersharing of the tetrahedra. Each oxygen is tetrahedrally coordinated with three lithium and one phosphorus atom. The high form of Li_3PO_4 shows an orthorhombic structure and is characterized by a = 6.115 Å, b = 10.475 Å, c = 4.923 Å and space group $Pmn2_1$. The high-temperature form is demonstrated in Fig. 1-24. Keffer et al. (1967), could show that the high-temperature form was precipitated by crystallization from melts above the transition range, while precipitation from an aqueous solution at or near room temperature results in the low-temperature form. When the low-temperature form is heated, it transforms irreversibly at 502°C to the high-temperature form.

C) Diphosphates (Pyrophosphates)

Diphosphates of the $Na_{27}Ca_3Al_5(P_2O_7)_{12}$-type demonstrate an equally complex structure. The crystals are trigonal (a = 25.438 Å, c = 9.271 Å, V = 5195.6 Å^3). They can be classified in the space group $R3$ (No. 148) (Alkemper et al., 1995).

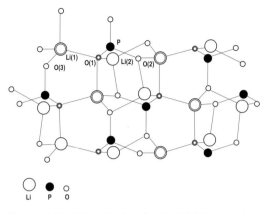

Figure **1-24** 0kl projection for the high-temperature forms of lithium phosphate, according to Keefer et al. (1967).

1.3.2.3 Metaphosphates

Calcium metaphosphate is the most significant of the metaphosphate glass-ceramics with regard to its applications. The structure of calcium metaphosphate, which has been studied in detail by Rothammel et al. (1989), is shown in Fig. 1-25. It is monoclinic and consists of the $P2_1/a$

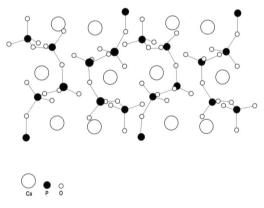

Figure **1-25** Structure of calcium metaphosphate, according to Rothammel et al. (1989).

space group. The lattice parameters of the structure are a = 16.960 Å, b = 7.7144 Å, c = 6.9963 Å, β = 90.394° Å when V = 915.40 Å^3. The crystal structure is composed of (PO_4) chains that are arranged along the [001]-axis in meandering form. The chains are connected by Ca^{2+} ions. The distance between P and O measures an average 1.584 Å in the case of bridge atoms and 1.487 Å^3 for terminal atoms. The calcium ions are located at two points. The Ca1 ions possess the coordination number 8 when the bond length of Ca to O measures 2.381–2.696 Å. This polyhedron is similar to that of a tetragonal antiprism. The Ca2 ions possess the coordination number 7 when the bond length measures 2.339–2.638 Å. This polyhedron is similar to that of a trigonal prism.

1.4. NUCLEATION

As described earlier, controlled crystallization of glasses is an important prerequisite in the development of glass-ceramics. Without controlled crystallization, glass-ceramics with very special properties could not be produced. Nucleation is a decisive factor for controlled crystallization, as the development of crystals in a base glass generally takes place in two stages: formation of submicroscopic nuclei, and their growth into macroscopic crystals. These two stages are called nucleation and crystal growth. The temperature dependence of nucleation and crystallization of glasses has been described by Tammann (1933). Stookey (1959) used this function to develop glass-ceramics and, therefore, expanded the theory of nucleation and crystal growth. The nucleation rate (I) and crystal growth rate (V) as function of reduced temperature (T/T_l) are shown in Fig. 1-26.

In most cases, controlled crystal growth produces various types of crystals, rather than just one crystal phase, giving the glass-ceramic product special properties. The crystals must also demonstrate a specific arrangement in the microstructure of the glass matrix. Furthermore, certain glass-ceramics contain glass phases in addition to several crystal phases.

The processes in question are thus of a highly complex nature. A comprehensive mathematical theory describing all the different processes does not yet exist. As a result, with today's standard of knowledge concerning the development of glass-ceramics, nucleation is assumed to be influenced by two general factors:

1) Appropriate selection of the chemical composition of the base glass, with the usual addition of a nucleating agent

2) Controlled heat treatment of the base glass, with time and temperature as process variables. Additional synthesis steps such as treatment with a salt melt may precede subsequent heat treatment.

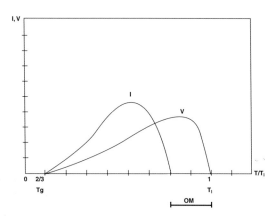

Figure 1-26 Nucleation rate and crystal growth rate as function of reduced temperature (T/T_l). (T_l) represents liquidus temperature. OM is Ostwald–Miers range (see Section 1.5). OM is related to T_l–T_s, where T_s is the solidus temperature.

The successful development of glass-ceramics is thus dependent on highly controlled synthesis procedures and thorough analysis of phase development processes with high-resolution analytical methods (e.g., electron microscopy, X-ray diffraction, thermal and chemical methods).

Findings from the theories of nucleation have also contributed to the optimization of these comprehensive experimental investigations. As a result, glass-ceramics with improved properties have been produced, specifically glass-ceramics with specific solid-solution limits like β-spodumene and stuffed β-quartz or with stoichiometric compositions like cordierite and lithium disilicate. Initial results for the successful application of the theory of nucleation to multicomponent glass-ceramics with a nonstoichiometric composition were achieved with mica and anosovite glass-ceramics.

At this stage, therefore, significant basic theories should be discussed regarding the practical applications of the theories developed for nucleation. The individual theories will not be addressed in detail, as Kingery, Bowen, Uhlmann (1975), Uhlmann (1977), Zanotto (1994), Gutzow and Schmelzer (1995), Weinberg et al. (1997), and Meyer (1968) have already done so in their publications.

According to Volmer's (1939) classical definition, a nucleus is an entity that already belongs to the new phase but is in unstable equilibrium with respect to the supersaturated parent phase. Thermodynamics and kinetics must be considered for further description of the formation of this nucleus. Several of these fundamentals will be discussed briefly.

The thermodynamic driving force of the glass-crystal transition is the chemical potential or free energy between the melt and the crystal. The reaction rate of nucleation must be examined with regard to the kinetics of nucleation in particular. Conclusions drawn from these examinations are useful in the development of materials.

A distinction must be made between homogeneous and heterogeneous nucleation in the glass-crystal transformation. In the process of homogeneous nucleation, a new phase develops in the absence of any foreign boundaries due to local fluctuations of density and kinetic energy. In heterogeneous nucleation, foreign boundaries such as substrates and grain boundaries are involved. This is also called catalyzed nucleation. Heterogeneous nucleation is the typical mechanism used in the development of glass-ceramics, as boundaries cannot be excluded and are indeed generally effective in the development of most glass-ceramics.

In the subsequent text, only a few selected basic theories for the two mechanisms of homogeneous and heterogeneous nucleation will be addressed.

Reference will be made to additional theoretical work. By describing the kinetic processes, criteria are compiled that will be of practical use in the development of glass-ceramics. Several examples of specialized glass-ceramics will be used to illustrate these findings.

1.4.1 Homogeneous Nucleation

Based on the work of Volmer (1939), the free-energy change must first be examined to establish the glass-crystal phase transition. Then the following information is valid for ΔG (free enthalpy) in the phase transition (Meyer 1968):

$$\Delta G = -\frac{4}{3}\pi r^3 \Delta G_V + 4\pi r^2 \gamma + \Delta G_E \qquad \text{(Eq.1-1)}$$

For reasons of simplicity, spherical particles are assumed with a radius of r. The interfacial enthalpy that corresponds to the required energy for the formation of the new surface of the nucleus is represented by γ. ΔG_V is the free-enthalpy change per unit volume that is produced by the formation of nuclei. ΔG_E is the fraction representing elastic distortion energy during a structural change. From a mathematical point of view, this contribution of ΔG_E can be ignored for melt–crystal and vapor–crystal transformations. The figure should, however, be taken into account in the crystallization of glasses and controlled surface crystallization in particular (Meyer 1968).

In addition to the classical theory of the equilibrium for the glass-crystal transition, nonclassical theories were developed by including irreversible thermodynamics (Cahn 1969 and Charles 1973). These theories take into account nonspherical nuclei, high degrees of supercooling, and small interfacial energies. An attempt has been made here to consider a theory of three-dimensional nucleation.

In general, the following statements of classical theories are valid concerning ΔG. When ΔG is negative, an important prerequisite has been achieved for the development of particles capable of growing. Therefore, referring to Eq. 1-1, the surface term ($4\pi r^2 \gamma$) and the elastic strain term (ΔG_E) are smaller than the volume term $-\frac{4}{3}\pi r^3 \Delta G_V$). This is the case if a particular minimum size r^* (critical nucleus size) is achieved. Kingery et al. (1975), established the theoretical derivation of nuclei with the critical radius (r^*), without considering ΔG_E, as

$$r^* = -\frac{2\gamma}{\Delta G_V} \qquad \text{(Eq. 1-2)}$$

The critical nucleus size r^* corresponds to the critical free enthalpy ΔG^* (see Fig. 1-27). The critical free energy is determined according to Eq. 1-3.

$$\Delta G^* = -\frac{16\pi\gamma^3}{3(\Delta G_V)^2} \qquad \textbf{(Eq. 1-3)}$$

Thus newly formed particles measuring r^*, are called critical nuclei. Particles of this size and larger (supercritical nuclei) are capable of growing. Particles that are smaller than r^* are called embryo or subcritical particles and are incapable of growing and eventually disintegrate. Kinetic processes are useful for investigating the balance between the formation and dissolution of nuclei. If reaction rates are characterized accordingly, then steady- and non-steady-state processes must be distinguished. In a steady-state process, the reaction rate is invariant with time. In a non-steady-state process, a time lag occurs before the steady-state reaction rate is reached.

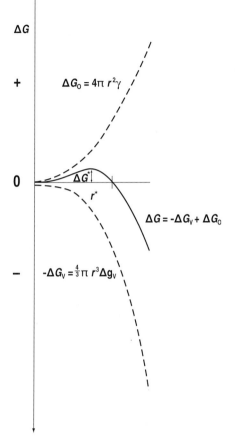

Figure 1-27 Free enthalpy (ΔG) as function of nucleus size (r).

When examined from a physical point of view, many nucleating processes should possess non-steady-state characteristics, since the formation of critical complexes containing n^* molecules with overall radius r^* is the result of formation and dissolution reactions. It is generally assumed that n^* is approximately 500 molecules or more (Gutzow 1980).

Thus a certain amount of time (non-steady-state time lag) is expected to elapse before a steady dispersion of clusters is attained in a medium that does not possess critical nuclei. This time lag is usually connected with the time required for molecules to diffuse and form clusters.

The respective mathematical equations for the nucleation rate (I) as a function of time t (in each case at constant temperature) and the nuclei number (N) as a function of t will be discussed in Section 1.4.3.

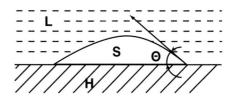

Figure 1-28 Model for heterogeneous nucleation. H represents the heterogeneous substrate/catalyst, S represents the nucleus/solid and L the mother phase, (θ) shows the contact angle.

1.4.2 Heterogeneous Nucleation

Heterogeneous nucleation involves phase boundaries, special catalysts, and foreign substrates that are distinct from the parent phase. This type of situation occurs when the driving forces involved in the formation of a new phase are stronger than those required by the parent phase for its transformation into a crystal. Figure 1-28 shows the model on which heterogeneous nucleation is based. The theory for the formation of critical free enthalpy in heterogeneous nucleation (ΔG_H^*) is derived from the contact angle (θ) relationship of Young's equation and is given by

$$\Delta G_H^* = \Delta G^* \, f(\theta) \qquad \qquad \text{(Eq. 1-4)}$$

$$\text{where } f(\theta) = \frac{(2 + \cos\theta)(1 - \cos\theta)^2}{4}$$

Based on Fig. 1-27, three special situations will be discussed.

First, if the heterogeneous substrate (H) is not wetted, the contact angle θ equals 180° and $f(\theta)$ equals 1. This phenomenon returns to a homogeneous nucleation process. Second, if the surface of H is completely wetted and the contact angle (θ) is close to 0°, then $f(\theta) \geq 0$ and ΔG_H^* are very small. Thus for $\theta < 180°$, heterogeneous rather than homogeneous nucleation will occur. Furthermore, nuclei with the critical size of r^* are preferentially formed.

The relationships between the different interface energies of the three phases, that is, of the formed crystal (S), the heterogeneous substrate (H), and the liquid/melt (L) will also provide several criteria for determining the effectiveness of the nucleating agent.

The interface energy of the heterogeneous substrate and the melt, γ_{HL} is defined according to Eq. 1-5.

$$\gamma_{HL} = \gamma_{SH} + \gamma_{SL} \cos\theta \qquad \qquad \text{(Eq. 1-5)}$$

Three examples of the desirable criteria for efficient heterogenous nucleation mentioned are presented below:

1) Small γ_{SH} indicates a low interface energy between the heterogeneous catalyst and the newly formed crystal.
2) Large γ_{HL} indicates a major mismatch of the thermal expansion coefficient compared with SH.
3) Similarity of the lattice parameters of the heterogeneous crystal and the newly formed crystal permits the determination of solid-state reactions based on epitaxy.

The phenomenon of epitaxy takes place if the lattice geometry of the different crystals is similar. In a first approximation, it is obvious that epitaxy takes place if the difference between the lattice parameters is less than 15%.

In addition to the simplified rule concerning the orientation of lattice parameters, the following parameters are also responsible for the occurrence of epitaxy:

- Similarity of lattice structures (structural adjustment of the intergrowing lattice planes)
- Bonding state in the host and guest crystals
- Real structure (influence of defects)
- Degree of coverage of the crystal surface with foreign nuclei

In various examples, it was demonstrated that epitaxy occurs when extensive differences in the lattice distances are evident. For example, it is also possible for epitaxy to initiate dislocations beyond interfaces. Thus the real distances between the host and guest crystals are larger than 15% (Meyer 1968).

1.4.3 Kinetics of Homogeneous and Heterogeneous Nucleation

Weinberg (1992a), Weinberg et al. (1997), and Zanotto (1997), reported in detail on transformation kinetics via nucleation and crystal growth. The standard theory of this type of phase transformation kinetics was developed by Johnson and Mehl and Avrami and Kolmogorov (see Weinberg et al., 1997). Therefore, this theory is called the JMAK theory. The JMAK equation (Eq. 1-6) is universal and applicable to glass-ceramics.

$$X(t) = 1 - \exp[-X_e(t)] \qquad \text{(Eq. 1-6)}$$

$X_e(t)$ is the extended volume fraction transformed and t is time. $X_e(t)$ can be expressed as (Eq. 1-7).

$$X_e t = kt^n \qquad \text{(Eq. 1-7)}$$

k is a constant and n is the Avrami exponent. For continuous, homogeneous nucleation and three-dimensional spherical growth, $k = \frac{\pi}{3}$ and $n = 4$.

For fast heterogeneous nucleation, the Avrami coefficient becomes 3. The quantity of the Avrami exponent can be determined as the slope in a plot of (Eq. 1-8).

$$\ln[\ln(1 - X(t))^{-1}] \text{ versus } \ln t \qquad\qquad \text{(Eq. 1-8)}$$

The JMAK theory has been applied to different glass-ceramic systems, such as BaO·2 SiO$_2$ glass (Zanotto and James, 1988), Na$_2$O·2 CaO·3 SiO$_2$ glass (Zanotto and Galhardi, 1988) and lithium disilicate glass (Zanotto 1997). The lithium disilicate glass exhibited a composition close to the stoichiometric molar ratio of Li$_2$O:2 SiO$_2$. The crystal phase formation was well defined by homogeneous nucleation based on the Avrami coefficient of 4 in the JMAK equation (Zanotto 1997). However, homogeneous nucleation was not observed in glasses of the Na$_2$O·2 SiO$_2$ and PbO·SiO$_2$ type (Zanotto and Weinberg, 1988).

Weinberg et al. (1997), showed that the JMAK theory is invalid for non-spherical particles, e.g., needlelike crystalline phases. Therefore, Weinberg and Birnie III (2000), proposed theoretical models describing the kinetics of the crystallization of highly anisotropic particles. These models took into account the influence of blocker particles and shaped aggressors on the precipitation of anisotropic crystals.

The results of the kinetics of nucleation are primarily functions of time (i.e., the nucleation rate $I = f(t)$ and the nucleus size $N = f(t)$ at constant temperature. The development of theories enabling comparisons of steady- and non-steady-state processes are of particular importance in this case.

The following Zeldovich equation is valid for I in the formation of a critical cluster

$$I = I_o \exp\left(-\frac{\tau}{t}\right) \qquad\qquad \text{(Eq. 1-9)}$$

where I_o represents the steady-state nucleation rate and the non-steady-state time lag.

The parameters required for determining I_o and τ are given by the following approximations.

I_o is defined by

$$I_o = A \exp\left(\frac{\Delta G^* + \Delta G_D}{kT}\right) \qquad\qquad \text{(Eq. 1-10)}$$

where ΔG^* is the Gibbs free energy change for the critical nucleus and represents the thermodynamic barrier, ΔG_D is the kinetic barrier for nucleation, k is Bolzmann's constant, and A the pre-exponential factor. The further development of Eq. 1-10 as well as the theory beyond the Zeltovich equation (Eq. 1-9) has been published by James (1982).

From mathematical derivations, I_o could then be determined according to Eq. 1-11 (Gutzow, 1980).

$$I_o = \text{const } Z \left(\frac{1}{\eta}\right) \exp\left(-\frac{\text{const } \gamma^3 \phi}{T^3 \Delta T^2}\right) \qquad \text{(Eq 1-11)}$$

The parameters η and ϕ are given in the following description of τ. Thus the nucleation rate at low temperatures is limited by high viscosity and at high temperatures by vanishing undercooling ($\Delta T \rightarrow 0$).

The non-steady-state time lag τ is given by

$$\tau = \frac{\text{const } \gamma \, \xi \, \eta}{\Delta\mu^2 Z} \qquad \text{(Eq. 1-12)}$$

In this case, γ represents the interfacial energy, ξ is a function of the activity ϕ for foreign substrates, which is different for different substrates. Z is the attachment probability (sticking coefficient) of molecules of the parent phase to the nucleus. Therefore, Z is related to epitaxy. For crystallization in glass, Z is approx. 10^{-3}. The viscosity of the parent phase is represented by η. The difference between the chemical potentials of the melt and the crystal is represented by μ. It is determined by Eq. 1-13.

$$\Delta\mu = \frac{\Delta S_1 \cdot T \cdot \Delta T}{T_1} \qquad \text{(Eq. 1-13)}$$

In this equation T_1 represents the liquidus temperature (melting temperature), ΔS_1 the molar entropy of the melt, and ΔT the degree of supercooling. On a graph, I may be represented as a function of t when I and τ are determined as described above.

The nucleation number (N) may be established by integrating I according to Eq. 1-14.

$$N = \int_0^t I(t)\, dt \qquad \text{(Eq. 1-14)}$$

The nucleation rate and the nucleus number as a function of time have been schematically presented by Gutzow (1980) for four cases:

- homogeneous steady state

- homogeneous non-steady state
- heterogeneous steady state
- heterogeneous non-steady state

These functions are shown in Fig. 1-29.

During synthesis and experimentation in the development of glass-ceramics, the determination of $N–t$ functions and comparison with the model functions in Fig. 1-29, for example, are of importance in trying to establish the mechanism involved. Determination of the non-steady-state time lags may be of particular significance.

The non-steady-state time lag may approach zero when supercritical nuclei are already present. These nuclei are athermal. In this case, subsequent crystallization is determined by the crystal growth rate of the athermal crystals.

τ may range from a few minutes to a few hours, if a correspondingly long aging process is required until primary particles are formed.

1.4.4 Examples for Applying the Nucleation Theory in the Development of Glass-Ceramics

Predominantly simplified model glasses with stoichiometric compositions were used to test the different nucleation theories. Multicomponent glasses were examined in few cases. Thus simple glass systems that are significant for the development of glass-ceramics will be discussed at this stage.

In the development of glass-ceramics, two mechanisms are generally used: volume and surface nucleation. The mechanism of volume nucleation will be examined in detail. The role of non-steady-state processes, phase separation reactions, and heterogeneous nucleating agents are critical. Surface nucleation is evaluated for controlling nucleation processes.

1.4.4.1 Volume Nucleation

The mechanism of volume crystal nucleation is predominantly applied in the development of glass-ceramics. The development of the first glass-ceramic by Stookey (1959) demonstrated how crystals of uniform size can be precipitated in the glass matrix with controlled volume nucleation.

A) Non-Steady-State Processes

The recognition of non-steady-state processes permits control over phase formation processes in glass-ceramics. If a significant time lag in the range of several hours occurs, for example, the type of processes taking place during this time must be examined. Whether the rate-controlling step can be influenced in any way must also be investigated.

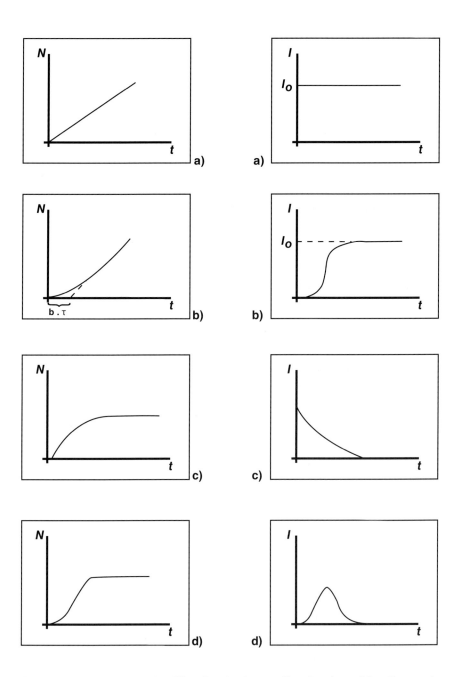

Figure 1-29 Nucleation number (*N*) and nucleation rate (*I*) as functions of time (*t*), according to Gutzow 1980. (a) homogeneous steady state, (b) homogeneous non-steady state, (c) heterogeneous steady state, (d) heterogeneous non-steady state.

The experiments of James (1982) have established that in a near-stoichiometric lithium disilicate glass that demonstrates nucleation at 440°C with subsequent crystallization, a time lag of 8 h occurs. Similarly long time lags of 5–12 h were also determined by Deubener et al. (1993) during the investigation of a lithium disilicate glass. A more detailed analysis of the phase development processes during the non-steady-state time lag using high-resolution methods also revealed a previously unknown transition phase e.g., Li_2SiO_3, prior to the formation of lithium disilicate crystals. With these results, however, the validity of earlier findings that demonstrated that the nucleation of lithium disilicate crystals involved a homogeneous process has been rendered questionable. It is quite clear that heterogeneous nucleation would be involved.

In the development of an anosovite type $(Al_2O_3, MgO)\cdot(TiO_2)$ glass-ceramic, Gutzow (1977) analyzed non-steady-state processes in a non-stoichiometric glass of the $SiO_2–Al_2O_3–CaO–MgO–TiO_2$ system. Time lags were determined at 855°C in the range of 2–4 h.

In Section 1.5.1, the crystallization of non-steady-state processes will be examined using mica glass-ceramics. The relationship between the observed non-steady-state time lag and phase separation will be demonstrated. The basic relationship between nucleation and microimmiscibility will be discussed in the following section.

B) Heterogeneous Nucleation and Microimmiscibility

Based on the development of photosensitive glasses, Stookey (1959) successfully used metals such as gold, silver, and copper as heterogeneous nucleating agents to achieve volume crystallization in base glasses. The metals were incorporated into the base glass in ionic form, e.g., Au^+, Ag^+. Metal nuclei, e.g., Au^0, Ag^0, in the presence of Ce^{3+} were reduced in the glass with ultraviolet light. In the process, Ce^{3+} was oxidized to Ce^{4+}. The crystalline nucleus, e.g., lithium metasilicate, then grew heterogeneously on the metal nucleus in the $SiO_2–Al_2O_3–Li_2O–K_2O$ base glass.

An epitaxial effect based on similar lattice parameters (McMillan 1979) presented a theoretical explanation for the favorable nucleating action of metals. Furthermore, mechanical strain was present at the substrate-glass interface, producing a high interfacial energy, as the coefficients of thermal expansion of the metal and the new nucleus were substantially different. As a result, catalyzation of nucleation could also be expected.

Since 1959, however, the principle of heterogeneous nucleation with metals has been successfully applied in the development of only a few glass-ceramics. To produce lithium disilicate glass-ceramics, McCracken et al.

(1982) and Barrett et al. (1980), for example, incorporated heterogeneous additions of metal into the base glasses. The preferred metal in the glasses developed by Barrett et al. (1980) was platinum. It was added to the raw material in the form of 0.003 wt% to about 0.01 wt% $PtCl_6$. The main crystal phase formed on Pt^0 nuclei. This type of glass-ceramic was developed for dental applications.

For most of the commercially significant glass-ceramics, heterogeneous nucleation of the base glass has been achieved with the targeted development and utilization of microimmiscibility. As a result, the glass structure has been decisively influenced. According to Beall and Duke (1983), the following heterogeneous nucleating agents in particular have permitted the formation of primary crystals in the SiO_2–Al_2O_3–Li_2O and SiO_2–Al_2O_3–MgO glass systems by special microphase separation processes: TiO_2, ZrO_2, P_2O_5, Ta_2O_5, WO_3, Fe_2O_3, and F. These nucleating agents either accumulate in a specific microphase of the phase-separated base glass or promote phase separation. Heat treatment initiates nucleation of a primary crystal phase. This phase is the basis for the heterogeneous growth of additional phases and the main crystal phase of the respective glass-ceramics in particular. Several particularly significant examples will be addressed in this book. Particular attention will be given to TiO_2 and ZrO_2 as single nucleating agents or double nucleating agents when they are combined.

With the development of the first glass-ceramics, Stookey (1959) determined the extraordinarily advantageous nucleating effect of TiO_2 in base glasses of the SiO_2–Al_2O_3–Li_2O system. Based on this finding, further base-glass systems were discovered and developed (Beall 1972, McMillan 1979). Thus 2 to 20 wt% TiO_2 was incorporated in SiO_2–Al_2O_3–Li_2O glasses in the development of glass-ceramics with minimal thermal expansion and in SiO_2–Al_2O_3–MgO base glasses in the production high-strength glass-ceramics (see Sections 2.2.2 and 2.2.5).

In the SiO_2–Al_2O_3–Li_2O system, the formation of a primary phase of the titanate type ($Al_2Ti_2O_7$) was achieved with a TiO_2 nucleating agent (approximately 4 mol%) in the microphase-separated base glass heat treated at 825°C. In the subsequent crystallization process, β-quartz solid-solution crystals, and at a later stage, β-spodumene solid-solution crystals grew heterogeneously on this primary phase (Beall and Duke, 1983).

As epitaxial processes in the heterogeneous formation of the main crystal phases on the primary titanates played an important role, the lattice parameters of the heterogeneous substrate could be tested by using TiO_2 as well as by incorporating mixed oxides of the TiO_2 and/or ZrO_2 type. Thus Petzoldt and Pannhorst (1991) demonstrated the epitaxial relationship between the nucleation phase ($ZrTiO_4$) and the β-quartz solid-solution phase. In these investigations, they were able to determine that new crystals grew on the nucleating phase and that new nucleating agents also formed during the growth process of the β-quartz crystals. The latter phenomenon took place in a phase boundary reaction.

In the SiO_2–Al_2O_3–MgO system, the effect of the TiO_2 nucleating agent was similar to that in the system containing lithium. Thus phase separation and primary crystallization of titanates played an important role. The development of high-strength glass-ceramics was achieved by amorphous phase separation of the base glass and the precipitation of the nucleating phase, $MgTi_2O_5$. The main crystal phase cordierite was precipitated as a result of very complex solid-state reactions (Beall 1992). When TiO_2 was added to SiO_2–Al_2O_3–MgO glasses, nucleation could also be influenced by specific reduction and oxidation ratios. In this way, the formation of Ti(III) building blocks as nucleating agents could be used by reducing melting conditions in the development of high-strength glass-ceramics (Höland et al., 1991a).

Headley and Loehmann (1984) discovered the epitaxial growth of lithium metasilicate and lithium disilicate on Li_3PO_4 crystals in a special glass-ceramic. In accordance with the proposed application, the glass-ceramic was heat treated at 1000°C for 20 min before crystallization for the purpose of sealing. This is not the normal way of processing glass-ceramics. Additional heat treatment at 650°C and 820°C resulted in the epitaxial growth of lithium metasilicate and lithium disilicate on Li_3PO_4 as shown in Figs. 1-30 and 1-31. At 650°C lithium metasilicate was formed on the surface of Li_3PO_4 (Fig. 1-30), and at 820°C lithium disilicate was precipitated epitaxially (Fig. 1-31). The epitaxial growth was determined by electron diffraction. The misfit of the atomic planes of these types of crystals (see Appendix 12 for structure of lithium disilicate, and Section 1.3.2 for Li_3PO_4) was in the

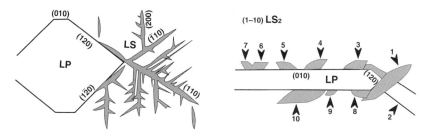

Figures 1-30 and 1-31 Epitaxial growth of Li_2SiO_3 (LS) (Fig. 1-30) and $Li_2Si_2O_5$ (LS$_2$) (Fig. 1-31) on Li_3PO_4 (LP), according to Headley and Loehmann (1984).

range of −5.3% to +3.8%. Therefore, this misfit is within the acceptable range for epitaxial growth of ±15%.

Particularly tough glass-ceramics have been developed by incorporating ZrO_2 in the subaluminous portion SiO_2–Al_2O_3–MgO base-glass system. Zirconia played an equally significant role as the nucleating agent in this system as it did in the SiO_2–Al_2O_3–Li_2O system. In different compositions of the same base glass, phase separation initially took place in the range of 800°–900°C. Subsequent nucleation was followed by rapid crystallization to tetragonal zirconia crystals (Beall 1992). Enstatite was precipitated as a main crystal phase. Partridge et al. (1989) also used ZrO_2 to develop enstatite-based glass-ceramics.

P_2O_5 is another oxide nucleating agent. It plays an important role particularly in low-alumina glasses with stoichiometric compositions as well as in nonstoichiometric multicomponent glasses in combination with other additions (nucleating agents) such as alkaline ions. As shown below, phase separation can be controlled with P_2O_5.

P_2O_5 nucleating agents of approximately 1 mol% in glasses with a stoichiometric composition will be discussed first. For example, P_2O_5 was used as a nucleating agent in lithium disilicate glasses. There are different explanations for the nucleating effect of P_2O_5 in lithium disilicate. It has been assumed that P_2O_5, like TiO_2, reduces the interfacial energy (γ) between the nucleus and the glass matrix in Eqs. 1-1 through 1-3 (James 1982). A transient phase has also been determined. James and McMillan (1971) have shown the formation of a phase containing lithium prior to the crystallization of lithium disilicate, which was suggested as Li_3PO_4. As a result, it is likely that P_2O_5 initiates microphase separation which, in turn, induces the formation of a transient phase. The crystallization of lithium disilicate was initiated by this transient phase.

The primary formation of lithium phosphates also plays a significant part in the multicomponent glass-ceramic of the SiO_2–LiO_2–ZrO_2–P_2O_5 system. Furthermore, high ZrO_2 contents of 15 wt% to approximately 30 wt% also have a major influence on nucleation. Because of the high diffusion rates of Li ions, phase separation primarily takes place in the glass, and lithium phosphate and zirconia crystals grow at high speeds. Moreover, the presence of epitaxial interactions should not be excluded from this process (Höland et al., 1995a).

Fluorine ions have achieved great practical importance in the controlling of nucleation by phase separation. The development of mica glass-ceramics in particular must be addressed at this point. The initiation of phase separation that represents the beginning of mica glass-ceramic development has been shown by incorporation of fluorine, for example, in base glasses of the $SiO_2-(B_2O_3)-Al_2O_3-MgO-Na_2O-K_2O$ system. Na_2O, K_2O, and fluorine enriched in the glass phase such that when the glass was heat treated, norbergite formed a primary crystal phase in the glass phase by homogeneous nucleation.

The final product was a machineable glass-ceramic having a typical house of cards microstructure (Beall 1971a; Vogel and Höland, 1982; Höland et al., 1991). The kinetics of the subsequent crystallization process is presented in Section 1.5.1. It was also possible to produce a new type of mica glass-ceramic by precipitating curved mica crystals. One of the keys to developing such a glass-ceramic was the reduction of the nucleation process by minimizing the phase separation in the base glass (Höland et al., 1981).

When CaO and P_2O_5 were incorporated in the $SiO_2-Al_2O_3-MgO-Na_2O-K_2O-F$ mica glass-ceramic system, nucleation of apatite could be achieved parallel to the formation of mica. Homogeneous nucleation of apatite occured in a second glass phase. By controlling multiphase separation, two types of nucleation occurring almost simultaneously have become possible (Höland et al., 1983).

The major points concerning the relationship between nucleation and microimmiscibility are summarized below. These points also take Uhlmann's (Uhlmann and Kolbeck, 1976; Uhlmann 1980) findings on this subject into account.

- By utilizing phase separation in the base glass, volume crystallization can be achieved at an earlier stage or delayed because of changing the composition of the matrix phase. Thus, surface crystallization or uncontrolled volume crystallization can be suppressed or avoided.
- Phase separation may lead to the formation of a highly mobile phase demonstrating homogeneous crystallization, while the matrix crystallizes heterogenously, either in parallel, or later in the process.
- Phase separation processes may lead to the formation of interfacial areas that may demonstrate preferred crystallization.

Important findings, based on the theory of nucleation, have clearly demonstrated why phase separation of nucleation can be controlled and why primary crystals formed with phase separation permit further controlling of nucleation processes.

With phase separation, material transport delivers the molecular building blocks for the nucleus to the nucleation site (droplet glass phase). The term ΔG_D in Eq. 1-10 appears to have been reduced to allow nucleation to progress directly and rapidly to the desired primary nucleus in the droplet phase. The controlled process of phase separation reactions thus already represents an important part of nucleation.

Nucleation thus achieved within the droplet glass phase can now result in the homogeneous nucleation of a special crystal phase. This crystal phase, in turn, has a heterogeneous effect on the surrounding glass matrix phase. Titanate or mixed phases of the TiO_2 and ZrO_2 type in the SiO_2–Al_2O_3–LiO_2 system, for example, initiate active centers in the volume of the glass. They increase the nucleation rate and reduce the interfacial energy (γ) between themselves and the new phase. According to Eq. 1-11, therefore, the nucleation rate is increased.

The growth front can be a catalytic reaction rather than the nucleation of active sites in the volume. Barry et al. (1970) studied nucleation in glasses of the SiO_2–Al_2O_3–Li_2O system. They concluded that nucleation may be continuonsly triggered at the β-quartz crystalline growth front as growth proceeds and TiO_2 is enriched at the moving boundary, rather than occuring at each titania-rich nucleous simultaneously. Cracking is often observed in this type of nucleation since shrinkage developed at the interface.

1.4.4.2 Surface Nucleation

Controlled nucleation in the volume of the base glass has opened many possibilities for producing new materials. Most glass-ceramics are manufactured according to this procedure. There are also base glasses in which controlled volume nucleation cannot be initiated. In these glasses, controlled crystallization can only be achieved with surface nucleation. These processes, however, are more difficult to control. Furthermore, the driving forces of nucleation are still being investigated.

It is quite certain that nucleation and crystallization can be accelerated and controlled by tribochemical activation of the glass surface (Meyer 1968). A general description of the phenomenon has been given in the history section. Surface activation of powdered glass is achieved by fine grinding. In the grinding process, chemical bonds are broken. In addition, reactive OH^- groups are formed in interactions with atmospheric humidity. In certain situations, however, radicals may also be produced. During heat treatment of the powdered glass, these reactive surfaces form the basis for nucleation. Also sharp high-energy "points" can be induced with accelerating nucleation. The initiation of nucleation could also be influenced by other chemical and physical reactions, which will be shown in this section.

One must first consider general problems with respect to the application of surface nucleation in a technologically applicable process. When using surface crystallization in the fabrication of monolithic samples, sintering and nucleation are partially separated for better control of the processes. Forming was thus achieved during the first part of the heat treatment of the powdered glass producing viscous sintering, and nucleation and crystallization from ghost interfaces were achieved in the second part.

On the basis of this glass-ceramic preparation, the kinetic reactions of nucleation in particular have been examined in different material systems. Particular emphasis was placed on the possibilities of controlling the nucleation process (Weinberg 1992b). Zanotto (1994) demonstrated that "surface nucleation sites" were saturated in the early stages of crystallization. Furthermore, he found that the number of nuclei (N) in Eq. 1-14 appeared to be constant. In most cases, the surface nucleation rates were too high. As a result, they were difficult to control. Therefore, in 1989 a special research program of the Technical Committee 7 (TC 7 crystallization of glasses) of the International Commission on Glass (ICG World Forum) was founded to investigate the driving forces of surface crystallization. As a result of this research program, the processes of surface nucleation were examined on a stoichiometric or near-stoichiometric base glass for cordierite glass-ceramics (Pannhorst 2000). Schmelzer et al. (1995), Zanotto (1994), Donald (1995), and Höland et al. (1995) found that seeding of the cordierite glass surface with fine-powdered glass of the same composition or heterogeneous particles, e.g., Al_2O_3, could initiate nucleation. Figure 1-32 shows the growth of cordierite crystals produced by heterogeneous nucleation with powdered glass.

Further investigations on the kinetics of cordierite nucleation revealed that the elastic strain term (ΔG_E) in Eq. 1-1, could also be of significance in surface nucleation. In a specialized experiment, Schmelzer et al. (1995) demonstrated that ΔG_E could be reduced to accelerate nucleation. In the experiment, a crack was made in a smooth

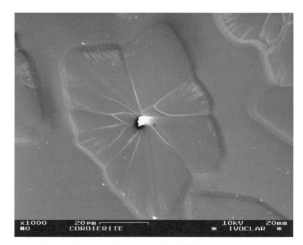

Figure 1-32 Formation of β-quartz solid solution (μ-cordierite) from a seed glass particle on the surface of cordierite glass.

glass surface. After heat treatment, nucleation was studied.

While preferred growth of nuclei was observed at the edge of the crack at the exposed surface, healing had unexpectedly commenced at the crack tip (see Fig. 1-33). This phenomenon may be explained by the fact that the elastic strain in this area is smaller than at the crack tip. Thus another means of initiating surface nucleation with controlled surface damage (e.g., activation with powder) is available, in addition to others such as the tribochemical activation of glass surfaces (Meyer 1968).

Sack (1965) pursued the objective of producing sintered glass-ceramics by nucleation from glass powders of nonstoichiometric composition. The nucleation was carried out from "ghost interfaces" of glass granules after sintering. He developed sintered glass-ceramics of the

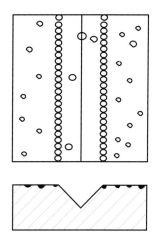

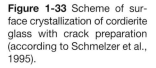

surface nucleation

Figure 1-33 Scheme of surface crystallization of cordierite glass with crack preparation (according to Schmelzer et al., 1995).

SiO_2–Al_2O_3–MgO system and succeeded in forming cordierite, anorthite, spinel, and forsterite main crystal phases.

In addition to cordierite glass-ceramics, special glass-ceramics have been produced using surface nucleation in the development of biomaterials for bioactive implants and glass-ceramics for dental applications (Kokubo 1982, Höland 1995). Apatite-wollastonite glass-ceramics and glass-ceramics containing leucite or zirconia have been produced. Nucleation and further phase development in these three types of glass-ceramics are addressed in Sections 2.2.9, 2.4.2, 2.4.6, 2.6.5, 4.4.1, and 4.4.2.

1.4.4.3 Temperature–Time Transformation Diagrams

Glass-ceramic compositions are originally produced as glasses and are subsequently crystallized in the subsolidus thermal region; that is, below the first point of equilibrium melting of the stable crystalline assemblage conforming to the particular composition. One might expect that the crystals formed would therefore correspond in composition and proportion to that defined by subsolidus phase equilibria. In many cases, however, metastable crystalline phases result, presumably because of similarity in atomic structural arrangement: linkage, coordination, etc., between the parent glass and the initial major crystalline phases. Sometimes these phases cannot be formed under any conditions from conventional ceramic powders. Perhaps the best example of this phenomenon is the precipitation of β-quartz solid solutions from a wide variety of lithium, magnesium, and zinc aluminosilicate glasses heated near 900°C for a relatively short time, like a few hours. These β-quartz glass-ceramics are not only metastable at these temperatures but will

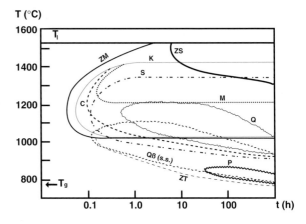

Figure 1-34 Time–temperature–transformation diagrams for an MgO–Al$_2$O$_3$–3SiO$_2$ glass-ceramics: Qβ(solid solution): β-quartz solid solution, K:cordierite, P: Mg-petalite, S: spinel, C: cristobalite, M: mullite, ZT: tetragonal ZrO$_2$, ZM: monoclinic ZrO$_2$, ZS: ZrSiO$_4$, Q: α-quartz.

convert, under prolonged heat treatment at these or higher temperatures, to stable phases: β-spodumene solid solution, cordierite, or gahnite in the zinc aluminosilicate system.

   A convenient technique for displaying the relationship between metastable and stable phases for a given glass-ceramic composition is through the use of a "TTT" diagram, which plots phase transformation boundaries as a function of temperature and time. Beall and Duke (1969) investigated a diagram for a multicomponent lithium aluminosilicate glass-ceramic nucleated with a mixture of titania and zirconia, where the major phases are capable of broad solid solution. Because of this capability, a relatively simple diagram results, although this is in part due to the difficulty of measuring separation temperatures and times of the oxide nuclei.

   By contrast, a TTT plot for an apparently simpler composition: MgO·Al$_2$O$_3$·3SiO$_2$ with a larger amount of nucleating agent (6 mol% ZrO$_2$) is shown in Fig. 1-34 (Conrad 1972). Ironically, this simple composition resulted in many more phases than the multicomponent glass (Beall and Duke, 1969) most of them stoichiometric, yielding a far more complex diagram. No fewer than eight metastable phases (glass, two quartz solid solutions, magnesium-petalite, spinel, mullite, and both tetragonal and monoclinic zirconia) were observed, as well as the stable three-phase assemblage: cordierite, silica (cristobalite over most of the thermal range), and zircon.

TTT diagrams have also provided useful information in the estimation of glass formation potential of liquids that are kinetically prone to premature crystallization upon cooling (Uhlmann 1980). The ratio of the temperature at the nose of the TTT curve for a crystal appearing on the liquidus to its liquidus temperature (T_n/T_l) is an indication of the crystal growth potential of this phase. The critical cooling rate for glass formation, $(dT/dt)c$, is roughly inversely proportional to this ratio. In other words, the closer the temperature of the nose to that of the liquidus, the more unstable is the glass.

TTT diagrams for needlelike apatite formation in a multicomponent system were demonstrated by Höland et al. (2000c).

1.5 CRYSTAL GROWTH

Once the nucleus has attained the critical size r^* (Eq. 1-2), crystal growth can commence. As shown in Fig. 1-26 (Section 1.4), the I (nucleation rate) and V (crystal growth rate) curves overlap, so that crystals with r^* grow. In the range of metastable supercooling (Ostwald–Miers range), however, nucleation is absent. As a result, only nuclei formed at low temperatures in the range of I grow into crystals. An entirely different situation results, however, if a glass is cooled from a high temperature of T_l, for example, and heat treated in the Ostwald–Miers range. Although the crystal growth rate is still high, nuclei and consequently crystals cannot develop. The direct relationship between nucleation and crystallization is thus underlined again.

The crystallization rate of glasses is determined by the extent to which material transport to the interface between the nucleus and the surrounding glass matrix is achieved. Consequently, the interface is of great importance with regard to kinetic and morphological processes of crystal growth. In theoretical models of crystallization, the entropy change at the interface is considered as the primary thermodynamic property. Information is thus given on the degree of order at the interface during crystallization. Uhlmann (1982) examined crystallization processes in different model substances, such as lithium disilicate, sodium disilicate, and anorthite glass. He demonstrated that different crystal growth rates resulted according to the degree of order at the interface.

Three basic models were developed to describe crystal growth rates: normal growth, screw dislocation growth, and surface crystallization.

In the development of glass-ceramics, normal growth with the application of the volume crystallization mechanism proved to be particularly important. Surface crystallization of glass-ceramics proceeds according to the mechanism implied by the name. The development of monolithic glass-ceramics according to the screw dislocation growth mechanism, observed during the crystallization of thin vapor deposition coatings, for example, has not yet been identified.

A possible significance of this mechanism in the development of ring-shaped mica that grow around nucleation centers (see Section 3.2) must still be investigated.

The model of normal growth considers a microscopically rough interface and gives the crystal growth rate (V) as Eq. 1-15:

$$V = \mathrm{v}a\left(1 - \exp\left(\frac{-\Delta G}{kT}\right)\right) \qquad \text{(Eq. 1-15)}$$

where v is the frequency factor of material transport to the interface, a is a distance compared with a molecular diameter, and ΔG is the free-enthalpy change.

The screw dislocation growth model expands on Eq. 1-15 by an additional factor f. This factor takes the fraction of preferred growth sites at the dislocation point into account.

The surface crystallization model, in contrast to that of normal growth, is based on a relatively smooth surface. V is determined according to Eq. 1-16:

$$V = \left(C \cdot \mathrm{v} \, \exp\left(\frac{-B}{T \cdot \Delta T}\right)\right) \qquad \text{(Eq. 1-16)}$$

where C and B are functions of the time required for the formation of the nucleus relative to that required for its propagation across the interface. Uhlmann (1982) conducted observations of parameters C and B.

Significant aspects of crystal growth in the development of glass-ceramics, that is, primary growth, anisotropic growth, surface growth, and secondary growth, are discussed in the following sections. Primary growth refers to a process that is characterized in the primary growing process to "growth to impingement." The secondary grain growth decreases the surface area of the new crystals after complete growth. In primary and secondary growth, crystals do not demonstrate oriented growth. The growth follows the theory of normal growth.

Anisotropic growth involves special orientations with higher growth rates. Therefore, anisotropic growth is a special phenomenon of the theory of normal growth.

In surface growth, the process of surface crystallization including reactions of "ghost interfaces" in glass powders is preferred over other mechanisms.

These different types of crystallization together have permitted the development of glass-ceramics with special microstructures and previously unknown properties.

1.5.1 Primary Growth

Primary grain growth was used to produce one of the most important glass-ceramics, that is, β-quartz/β-spodumene glass-ceramic. Thus the crystallization mechanism in the SiO_2–Al_2O_3–Li_2O system with nucleating agents of TiO_2 presented in Section 1.4.4.1.B will be further investigated from the point of view of the crystal growth mechanism. The formation of the heterogenous substrate, the $Al_2Ti_2O_7$ phase, was initiated with phase separation and a nucleating agent. The metastable β-quartz solid solution phase subsequently grew on $Al_2Ti_2O_7$ by heterogeneous interaction. Beall and Duke (1983) determined that the stable mixed crystal of the β-spodumene type could be produced above 950°C.

The primary growth of β-quartz solid solution of "growth to impingement" is characterized by a process similar to isotropic growth beginning in the center of the nucleus. Thus crystal phase formation takes place, and the secondary growth of β-quartz solid solution follows. This secondary growth (see also Section 1.5.5) is characterized by a special kind of "ripening" of crystals. The secondary grain growth of β-spodumene crystals was studied by Beall and Duke (1983) as a direct functional relationship between temperature (T) and time (t) of heat treatment. Figure 1-35 illustrates this type of function.

A typical microstructure of an SiO_2–Al_2O_3–Li_2O–TiO_2 glass-ceramic is shown in Fig. 1-36 (see Section 2.2). During heat treatment at 950°C, for example, a maximum crystallite size of approximately 0.1 μm was achieved. The resulting glass-ceramic demonstrated both minimal thermal expansion and a high degree of transparency. The microstructure formed at 1000°C is shown in Fig. 1-37.

Controlled crystal growth also involves a functional relationship between heat treatment and the properties of the glass-ceramic such as the linear thermal expansion coefficient and optical properties. By using fundamental theories of optical spectroscopy on β-spodumene glass-ceramics by Beall and Duke (1983), it was established that the

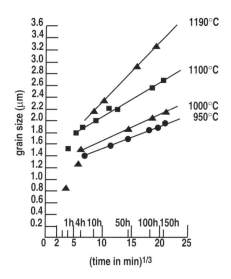

Figure 1-35 Grain size as a function of temperature T and time t for a typical β-spodumene solid solution glass-ceramic.

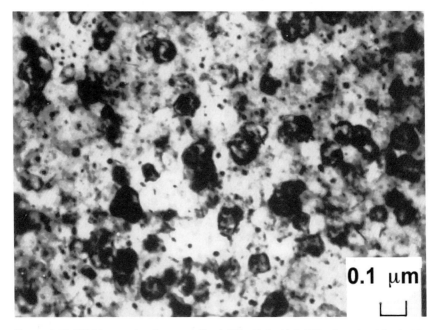

Figure 1-36 TEM image showing crystallized SiO_2–Al_2O_3–Li_2O–TiO_2 glass, heat treated to 950°C. β-quartz solid solution precipitated.

best transparency of glass-ceramics could be achieved when twice the crystallite size was smaller or equivalent to the wavelength of visible light.

The recognition and analysis of the relationship between crystal growth and material properties have formed a basis for demonstrating how a normal growth process can be used to develop specific properties.

1.5.2 Anisotropic Growth

The anisotropic growth in the formation of glass-ceramics is typical for sheet silicate crystals of mica-type, as well as for chain-silicate crystal and chain-phosphate crystal structures. Thus, this crystal formation was used to produce glass-ceramics, having special properties, e.g., high mechanical strength and/or high toughness or good machinability. Potassium fluorrichterite and fluorcanasite (Beall 1991) belong to the group of chain silicates and metaphosphates of $Ca(PO_3)_2$-type (Abe et al., 1984) belong to this group of phosphate crystals with chain structures.

In this section, the kinetics of anisotropic precipitation of mica crystals will be investigated as an example of anisotropic growth. Because of their three-layered structure, mica crystals feature pronounced preferential growth along the (001) plane. This preferred anisotropic growth of mica crystals of phlogopite

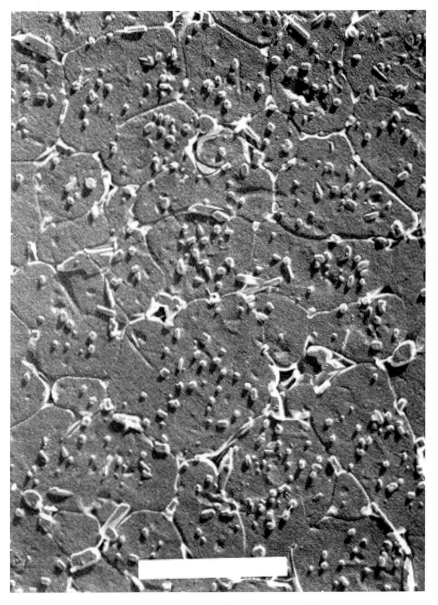

Figure 1-37 TEM image showing crystallized SiO_2–Al_2O_3–Li_2O–TiO_2 glass, heat treated to 1000°C for 45 min. β-spodumene solid solution was formed in the glass-ceramic (bar = 1 µm).

type in glasses was used by Beall et al. (1971a) to develop machineable glass-ceramics with a house of cards microstructure (see Section 3.2).

The kinetics of the anisotropic growth of flat mica crystals which form a glass-ceramic of house of cards structure was also investigated by Höland et al. (1982) in a glass of the composition: 50.6 mol% SiO_2, 16.6 mol% Al_2O_3, 18.0 mol% MgO, 3.2 mol% Na_2O, 2.0 mol% K_2O, and 9.6 mol% F.

If the glass is heat treated as a monolithic sample at 980°C for 15 min, platelet-shaped mica crystals of the phlogopite type $(Na/K)Mg_3(AlSi_3O_{10})F_2$ grow in a secondary stage, after the precipitation of norbergite $Mg_3[SiO_4F_2])$. Figure 1-38 demonstrates that the orientation of this growth is preferentially anisotropic along the [00l] axis of the crystals. In the electron micrograph of Fig. 1-38 (a TEM replica), the anisotropic growth fronts are clearly visible.

Now the question arises, as to what driving forces are responsible for this growth process and how the kinetic process should be described. Thus the following two points should be addressed.

First, what is the connection between the norbergite primary crystal phase and the phlogopite? Are non-steady-state reactions included in the nucleation process? Second, how can the mica crystal growth be kinetically determined?

To answer these questions, two types of specimens were produced: mono-lithic specimens, and thin glass films with a thickness of approximately 0.3 mm. After heat treatment of the mentioned base glass between 765° and 910°C crystal phase development was evaluated, using light-optical and elec-tron micrographic images. As shown in Fig. 1-39, at lower temperatures

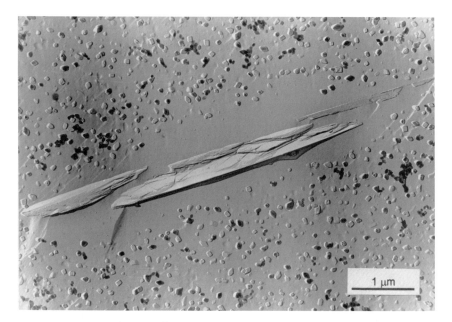

Figure 1-38 Electron replica micrograph of an SiO_2–Al_2O_3–MgO–Na_2O–K_2O–F glass after heat treating the glass at 980°C for 15 minutes. The preferred anisotropic growth proceeds along (001) planes.

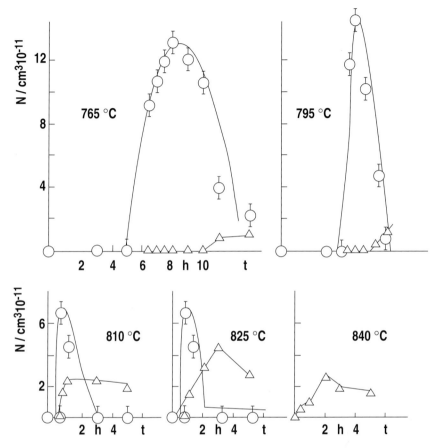

Figure 1-39 Number of norbergite crystals (circles) and phlogopite crystals (triangles) in the base glass in dependence of time and temperature of heat treatment.

(765°C and 795°C) a non-steady-state time lag of 4–5 h is observed before the first norbergite crystals become visible.

A closer examination of the partial process taking place at these temperatures (765°C and 795°C) demonstrates that no real time lag, in the sense of delayed crystal growth, occurs during this period. Instead, liquid–liquid phase separation begins and then proceeds in time. Thus the nucleation that permits the formation of norbergite is initiated.

The next important process that becomes evident in crystal growth is shown in Fig. 1-39. In each case, the formation of mica (represented by triangles) takes place because of norbergite decomposition. The ions and structure units of norbergite provide substances required by the growing mica crystals. Thus the chemical reaction equation can be formulated so that the mica in the reaction equation is formed by norbergite and components of the glass phase: $Mg_3[SiO_4F_2] + [\frac{1}{2}(Na/K)_2O + 2SiO_2 + \frac{1}{2}Al_2O_3] \rightarrow$ (Na/K) Mg_3 $(AlSi_3O_{10})F_2$.

It is surprising that the residual glass matrix consists of leucite stoichiometry. The anisotropic linear growth of mica crystals was determined in the kinetic process between 765° and 910°C. The length was established as the average value (*d* in μm) according to the heat treatment used. Thus Fig. 1-40 illustrates the particular growth curves. The rate of crystal growth of various temperatures was calculated from the rise in these curves. These values of the crystal growth rate, in turn, produce curve 1 in Fig. 1-41. This figure includes the curve representing the formation of norbergite (curve 2 in Fig. 1-41) according to Chen (1963) as well as the theoretical curve for the decomposition of phlogopite (curve 3 in Fig. 1-41).

Kinetic examinations of crystallization in glass, according to Gutzow (1980) for example, demonstrate the relationship between crystal growth rates and viscosity in more detail. In these studies, the growth rate (*V*) in the crystalline phase of a viscous glass melt is determined as a function of temperature.

The viscosity (η) of the melt as well as supercooling (Δ*T*) are included in the growth rate (*V*) equation.

$$V = C\frac{1}{\eta}\Delta T \qquad \text{(Eq. 1-17)}$$

Thus, *V* could be expressed with sufficient accuracy on the basis of Eq. 1-17 (Gutzow 1980) where C is a constant and Δ*T* is defined as

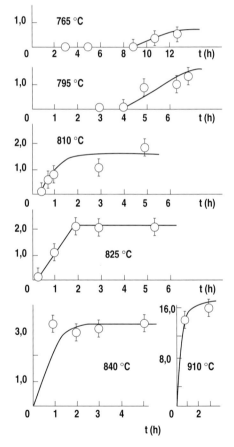

Figure 1-40 Crystal size of phlogopite in the glass ceramic after different heat treatments of the base glass. Results of 765°C, 795°C, 810°C, 825°C, 840°C are related to thin films, results for 910°C were determined for monolithic samples.

$\Delta T = T_1 - T$. T_1 represents the metastable liquidus temperature for the growth phase.

The experimentally established viscosity curve of the SiO_2–Al_2O_3–MgO–Na_2O–K_2O–F glass (Fig. 1-42) was adapted according to a regression analysis of the Vogel–Fulcher–Tamann equation that corresponds to the values $A = -2.105$; $B = 6780$ and $T_0 = 177.9°C$.

$$\log \eta = \frac{A + B}{T - T_0} \qquad \text{(Eq. 1-18)}$$

For the composition mentioned, T_1 equals 1593K (liquidus temperature). For small temperature ranges (ΔT), η can be represented by the following function with sufficient accuracy:

$$\eta = C \cdot \exp\left(\frac{U}{RT}\right) \qquad \text{(Eq. 1-19)}$$

U is the activation energy for the viscous flow process. As T must be taken as a constant in comparison with η (T), $\log \eta$ and also $-\log V$ must result in a linear function of $1/T$. This was confirmed by determining the course of the curves $(\log \eta) - (1/T)$ and $(-\log V)$ as function of $(1/T)$ in Fig. 1-43. Thus the $(-\log V) - (1/T)$ function was calculated from the values in Fig. 1-41 and the parameter for $\log \eta$ as function of $1/T$ was determined from the viscosity curve of the glass (Fig. 1-42).

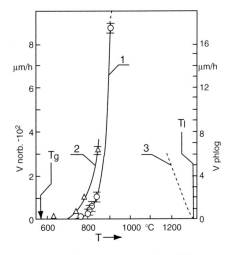

Figure 1-41 Crystal growth rates of the main crystal phases in dependence of T.
Curve 1. Crystal growth rate of phogopite according to Fig. 1-40.
Curve 2. Crystal growth rate of the intermediate phase according to Chen (1963).
Curve 3. Theoretical curve of phlogopite decomposition close to the liquidus temperature (T_l).

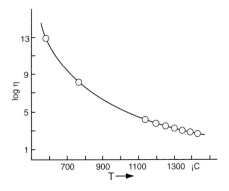

Figure 1-42 Viscosity of the SiO_2–Al_2O_3–MgO–Na_2O–K_2O–F glass as a function of the temperature. The circles represent experimental datas and the curve was calculated after the Vogel–Fulcher–Tamann equation.

The activation energy for the growth of phlogopite crystals and for the viscous flow process can be determined from Fig. 1-43. As both curves are almost parallel, the value for both processes comes to approximately 314 kJ/mol (Höland et al., 1982a).

As a result, the activation energy for the crystal growth of phlogopite is the same as for the viscous flow process. Thus it can be concluded that the viscous flow process deter-mines the rate of crystal growth.

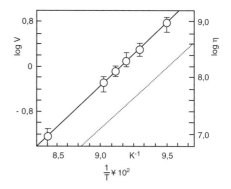

Figure 1-43 Crystal growth rate of phlogo-pite crystals and viscosity as a function of temperature.

In other words, crystal growth of mica in glass-ceramics is diffusion con-trolled. By closely examining the electron micrograph in Fig. 1-38, the kinetic process identified also becomes evident. A type of diffusion halo has clearly formed around the "rapidly" growing phlogopite crystal. Near the growing mica, norbergite crystals decompose and provide species for the mica crystals. Anisotropic crystal growth results. The preferred orientation of growth, however, is not primarily determined by the previously formed phase.

1.5.3 Surface Growth

This crystal growth mechanism is linked to surface nucleation. The success-ful application of the mechanisms of surface nucleation and crystallization in the development of cordierite (Semar and Pannhorst, 1991), apatite-wollastonite (Kokubo 1991), and leucite glass-ceramics (Höland et al., 1995a) has already been mentioned in Section 1.4.

These mechanisms are presented by using the examples of cordierite and leucite glass-ceramics. To clearly demonstrate the mechanisms, experiments were conducted with monolithic test specimens rather than the fine-powdered glasses required in the industrial manufacturing of glass-ceramics. The following descriptions, therefore, refer only to the mechanism as demonstrated in monolithic specimens.

Following the initiation of surface nucleation, the crystal growth mecha-nism demonstrated at least three extraordinary features in cordierite (Zanotto 1994; Schmelzer et al., 1995; Donald 1995) as well as leucite glass-ceramics (Höland et al., 1995 a,b).

1) Crystal growth of the main crystal phases begins at the surface and progresses into the volume of the crystal. It is interesting to note that leucite crystals in contrast to cordierite demonstrate preferred growth. The c-axis of the leucite crystal is vertically oriented with respect to the surface of the base glass.

2) The beginning of the growth process (cordierite or leucite glasses) is determined by a type of "two-dimensional" crystal formation. In other words, very flat crystallites initially grow on the surface of the base glass (Fig. 1-44). In a following reaction, this crystal phase was converted into a main crystal phase.

3) Very small seeding particles with the same composition as the base glass (cordierite or leucite glass) were used as a heterogenic nucleating agent to produce the main crystals phase. Leucite crystals grow in seeded areas as well as on unseeded, virtually flat surfaces (Fig. 1-45). The crystal growth rate was almost the same in both cases. Seeding, however, considerably increased the number of nuclei and thus the number of crystallites in both leucite and cordierite glass-ceramics.

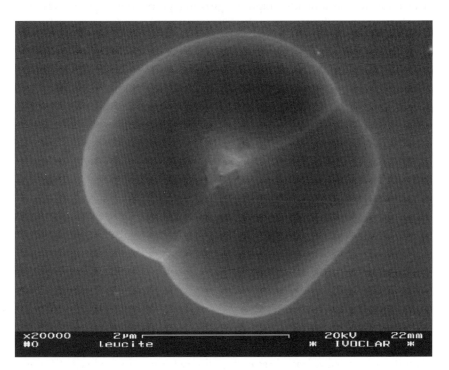

Figure 1-44 SEM image showing crystallization of leucite by seeding monolithic glass surfaces with glass dust. One glass particle initiates growth of a highly disordered primary crystal during heat treatment at 720°C for 12 h.

Figure 1-45 Growth of formed leucite crystals without seed particles with simultaneous precipitation of highly disordered primary crystals formed by seed particles. SEM.

These mechanistic analyses of growth processes could also be applied in the industrial production of glass-ceramics according to the surface crystallization mechanism. As already mentioned, fine-powdered glasses are used for the industrial manufacturing of cordierite (Semar et al., 1989) and leucite glass-ceramics. The differences in the crystallization process for the monolithic test specimens were determined, for example, with thermal analyses (Donald 1995). Most important, however, was that by finely grinding the base glass, tribochemically activated surfaces could be produced. Furthermore, the controlled production of very fine particles achieved a type of seeding.

According to the results mentioned, crystallization could be influenced in two ways by using a specially produced fine-powdered base glass. First, the reactive powdered surface had a catalytic effect on crystallization. Second, a type of autocatalysis was achieved by the smallest particles of a size fraction in which the smallest particles promoted the crystallization of larger ones by a heterogeneous reaction. Both processes influenced the crystal growth rate of the main phase.

A leucite glass with an average grain size of approximately 25–30 µm demonstrated a considerably higher crystal growth rate than a monolithic test sample (5 × 5 × 1 mm). In the first case, the crystal growth rate registered 2.2 µm/h at 920°C; in the second case 4 µm/h at 920°C.

In addition to producing cordierite, apatite-wollastonite, and leucite glass-ceramics, it was also possible to produce lithium-phosphate-zirconia glass-ceramics (Höland et al., 1996a) by taking advantage of the surface crystal growth mechanism. These particular glass-ceramics demonstrated a combination of special properties. One of these properties is very good individual shaping, using a viscous flow process. Another special property is a high mechanical strength of approximately 300 MPa, due to transformation toughening of tetragonal zirconia in glass-ceramics. Transformation toughening in glass-ceramics was also studied by Echeverría and Beall (1991) and Sarno and Tomozawa (1995).

1.5.4 Dendritic and Spherulitic Crystallization

1.5.4.1 Phenomenology

When glass-forming melts crystallize to a phase of identical composition to the glass, either well-faceted (euhedral) or non- or poorly-faceted (anhedral) crystals result, often depending upon whether the entropy of melting is large or small, respectively (Uhlmann 1971). On the other hand, if the glass is significantly undercooled or crystallizes initially to a phase significantly different in composition from the bulk glass, a fibrillar or skeletal type of crystallization is common. This is generally referred to as dendritic if the morphology is treelike or sheaflike with significant branching or as spherulitic if it is composed of sub-parallel needlelike crystallites radiating from a distinct center and forming an essentially spherical mass (see Fig. 1-46).

Where both dendrites and spherulites occur in a given glass composition, the latter are generally formed at lower temperatures and higher viscosities than the former (Carr and Subramanian, 1982). A great deal of work has been devoted to describing and explaining these crystallization habits (Keith and Padden, 1963; Vogel 1985; Uhlmann 1982; Caroli et al., 1989, Duan et al., 1998). Some of the key factors can be summarized as follows:

- Crystal growth usually proceeds more rapidly in certain lattice directions and these may become exaggerated as dendritic or spherulitic arms
- Chemical species that exist in the glass but cannot enter the crystal are rejected along the boundaries of the growing crystal, thereby preventing its widening and encouraging its growth into fresh glass at extremities

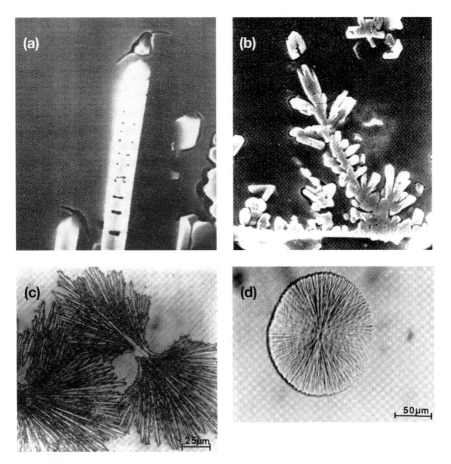

Figure 1-46 Dentrites and spherulites as crystallized from a glass: (a) anisotropic growth of crystobalite along [100] in a 0.15 K_2O – 0.85 SiO_2 glass crystallized at 763°C. (b) dendritic crystals in the same glass showing branching in [100] and [111] directions. (c) sheaf form of dendritic/spherulitic growth at 550°C in P_2O_5-nucleated lead silicate glass, and (d) spherulites in some glass grown at 450°C (after Scherer and Uhlmann, 1977; Carr and Subramanian, 1982).

- The latent heat given off by crystallization is most effective on broad faces, tending to warm these areas relative to edges and particularly corners, thus slowing crystallization on faces as compared to corners, and increasing anisotropic growth. The phenomenon of branching, essential to dendrites but either absent or of low angle in the case of spherulites, is typically a function of the crystallography involved. In the case of cristobalite dendrites growing in a binary SiO_2–K_2O glass (Scherer and Uhlmann, 1977), the crystallites growing in the [100] direction tend to branch in the [010] or [111] directions, yielding branching angles of 90° and 54.7°, respectively (Fig. 1-46a,b).

Several studies of the growth rate of dendritic layers from the surface of glass or of the kinetics of spherulitic growth have shown a linear advance with time, i.e., a growth rate independent of time (Vogel 1983; Uhlmann 1982; Caroli et al., 1989; Duan et al., 1998; Scherer and Uhlmann, 1977). Initially it was thought that this suggested an interface-controlled rather then diffusion-controlled growth. Linear growth with time is, however, consistent with a fibril always moving into fresh material of constant composition, or with the fibril growing under steady-state conditions with a constant radius of curvature at the growth front (Uhlmann 1971). Indeed, large aspect ratios are observed for the fibrils, and the independence of the growth rate on time probably reflects the crystal length being large in comparison with the relevant diffusion length. With decreasing crystallization temperature, an increasingly fine fibrillar morphology would be expected and is in fact generally observed.

1.5.4.2 Dendritic and Spherulitic Crystallization Applications

Dendritic and spherulitic growth and resulting morphologies are observed in many glass-ceramic systems, and are the basis of at least two commercial materials: chemically machinable photosensitive glass-ceramics (Fotoform® and Fotoceram® of Corning, and Foturan® of Schott/mgt mikroglas technik AG), and the architectural glass-ceramic Neoparies® of Nippon Electric Glass. In fact, fibrillar morphologies are common in glass-ceramics of the SiO_2–Li_2O, SiO_2–BaO, SiO_2–PbO–P_2O_5, SiO_2–CaO–Na_2O–K_2O–F (canasite, miserite) (Beall 1991, 1992; Pinckney and Beall, 1999), and basalt (Beall and Rittler, 1976) systems, among others.

The dendritic microstructre of the Corning chemically machinable Code 8603 glass-ceramic is the basis of its unique features. Figure 3-8 in Section 3.2.4 illustrates this microstructure with soluble lithium metasilicate crystals forming a connective dendritic network capable of rapid attack in dilute hydrofluoric acid. The hexagonal nature of the intersecting dendrites can be observed. Since the crystalline dendrites are nucleated on particles of silver that are produced through ultraviolet exposure, a complex etchable pattern can be transferred and induced into the glass.

Interlocking and loose spherulites can be useful in achieving high strength and fracture toughness in glass-ceramics. Thus, in chain-silicate glass-ceramics based upon acicular crystals, interlocking dendritic sheaves or spherulites help to enhance fracture toughness (Fig. 1-47). This has been demonstrated for several chain silicates including canasite ($Na_4K_2Ca_5Si_{12}O_{30}F_4$), miserite ($KCa_5(Si_2O_7)(Si_6O_{15})F$), and wollastonite ($CaSiO_3$).

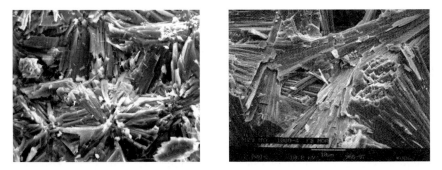

Figure 1-47 Interlocking spherulites and sheaves of chain silicates in tough glass-ceramics: (a) loose spherulites of canasite, (b) sheaves of miserite (after Beall, 1991b; Pinckney and Beall, 1999).

Commercial glass-ceramics based on the latter phase and used as architectural sheet show an attractive texture due to the polished appearance of countless spherulites large enough to be perceived by the naked eye.

1.5.5 Secondary Grain Growth

The term *secondary grain growth* is used in glass-ceramic technology to refer to increases in crystal size that occur after maximum crystallinity is achieved. In most cases of high-volume-percent crystallinity, this grain growth follows crystal impingement, and is slower than the *primary grain growth* that occurred between nucleation and impingement.

For the sake of comparison, primary grain growth as a function of time at constant temperature is characterized by the following rates:

$$d = d_0 + kt^{1/2} \qquad\qquad \text{(Eq. 1-20)}$$

in case of dissusion controlled growth, and

$$d = d_0 + kt \qquad\qquad \text{(Eq. 1-21)}$$

for interface controlled growth. The former case is more prevalent where the crystals do not have similar compositions to the parent glass, while the latter is common where the crystal composition closely matches that of the glass.

With secondary grain growth in both glass-ceramics and sintered ceramics, the grain size increased as:

$$d = d_0 + kt^{1/r} \qquad\qquad \text{(Eq. 1-22)}$$

where r typically varies between 2 and 10. In the case of β-spodumene glass-ceramics, Chyung (1969) has shown that the value of r is 3. Certainly, secondary grain growth would be expected to be more sluggish than primary grain growth which, after all, is essentially a devitrification phenomenon. Primary grain growth is driven by the Gibbs free energy change between crystal and glass, a substantial driving force below the liquidus, whereas secondary grain growth is basically a recrystallization phenomenon driven by energy differences associated with a reduction in surface area. The decrease in grain boundary energy over four orders of magnitude from 1 μm grains to 1 cm grains is typically only 0.1–0.5 calories/mg.

Factors that can decrease secondary grain growth rates even further are inclusions or minor phases along grain boundaries, and films of viscous residual glass surrounding grains. In both cases, diffusing species from the major phase must trespass around or through these phases to allow large crystals of the major phase to grow at the expense of smaller ones.

Examples of minor phases that hamper grain growth in glass-ceramics are the siliceous residual glass in internally nucleated β-spodumene glass-ceramics (Chyung 1969), and mullite along grain boundaries in surface-nucleated, powder-formed β-spodumene glass-ceramics like Cercor® heat exchanges.

The importance of controlling secondary grain growth in glass-ceramics is related to improved thermal stability. Not only does mechanical strength deteriorate with grain size increases above 5 μm (Grossman 1972), but also anisotrophy in thermal expansion coefficient in several key glass-ceramic crystals (β-spodumene, β-quartz solid solution, mica, and cordierite) may produce high intergranular strain if the crystals are allowed to grow. Such strain often results in microcracking, a generally undesirable phenomenon that always decreases mechanical strength, and may even affect the integrity of the glass-ceramic article.

CHAPTER 2

Composition Systems for Glass-Ceramics

2.1 ALKALINE AND ALKALINE-EARTH SILICATES

2.1.1 SiO_2–Li_2O (Lithium Disilicate)

As mentioned in the History section, lithium disilicate glass-ceramic was the first glass-ceramic that Stookey (1953, 1959) developed. The fundamental research conducted by Stookey provided a basis for the large-scale development of glass-ceramics in a variety of chemical systems. Furthermore, other materials systems based on lithium disilicate have been also developed according to his findings.

The initial basis for the development of a material in the lithium disilicate system were compositions that were derived from the stoichiometric composition of phyllosilicate crystals, $Li_2Si_2O_5$. The structure is shown in Appendix Fig. 12.

2.1.1.1 Stoichiometric Composition

As shown by the binary phase diagram in Fig. 2-1 according to Levin et al. (1964), the lithium disilicate crystal phase (or lithium phyllosilicate, using the nomenclature of Liebau (1961) is a compound that melts congruently at 1033°C. A polymorphous transformation of the crystal phase was observed at approximately 939°C.

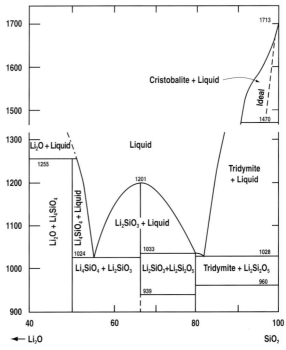

Figure 2-1 Phase diagram of SiO_2–Li_2O (Levin et al., 1964), (wt%, T in °C).

The glass formation and crystal behavior in binary SiO_2–Li_2O glasses have been studied in detail. The stoichiometric composition of 33.33 mol% Li_2O and 66.66 mol% SiO_2 (or 19.91 wt% Li_2O and 80.09 wt% SiO_2) received particular attention. The investigations revealed a certain opacity to be present in base glasses containing between 5 mol% and almost the stoichiometric amount of Li_2O in lithium disilicate. Vogel (1963) determined this phenomenon to be glass in glass phase separation (immiscibility of glasses). These results clearly showed that phase separation processes are likely to play a part in the nucleation of lithium silicate glass-ceramics. If glass-ceramics are produced from base glasses with the exact stoichiometric composition, however, phase separation does not occur.

Nucleation of base glasses with the stoichiometric composition of lithium disilicate for glass-ceramic manufacture has been examined in detail. McMillan (1979) and James (1985), in particular, have conducted comprehensive studies in this field. Nevertheless, for many years after Stookey's (1953, 1959) fundamental findings, the mechanisms of nucleation were never completely determined. Section 1.4 addresses the reaction mechanisms of nucleation of lithium disilicate, according to findings from the 1980s and 1990s. It is important to note at this stage that prior to the crystallization of $Li_2Si_2O_5$, the formation of primary crystals of different compositions plays an important part. Even the formation of "crystal embryos" above the melting point of glasses with a close-to-stoichiometric composition of lithium disilicate was considered to be a potential nucleation mechanism (Ota et al., 1997). Zanotto (1997) discovered a metastable phase in lithium disilicate glasses.

In addition to the examination of the nucleation mechanism of lithium disilicate glass-ceramics, the 1980s and 1990s have seen the analysis of the microstructure of glass-ceramics with a stoichiometric composition as well as the improvement of their chemical durability. Schmidt and Frischat (1997), for example, managed to generate images of various structures by using scanning electron microscopy in conjunction with atomic force microscopy. Furthermore, they were also able to control the development of these structures.

Barret and Hench (1980) and Wu (1985) improved the chemical durability of lithium disilicate glass-ceramics to a significant extent by incorporating additions such as Al_2O_3 and K_2O to the stoichiometric base glass. The objective of improving the chemical durability of this glass-ceramic was to render the material suitable for use as a biomaterial in human medicine and, in particular, as a restorative material in dentistry.

It must be noted that a significant improvement of the chemical durability of lithium disilicate glass-ceramics was achieved later in the development of glass-ceramics with nonstoichiometric compositions.

2.1.1.2 Nonstoichiometric Compositions

Fundamental research on certain glasses near the lithium disilicate composition were carried out by Stookey (1959). He incorporated metal ions, such as those of silver, Ag^+, in glasses as nucleating agents for controlling the crystallization of base glasses. The following composition is typical: 80 wt% SiO_2, 4 wt% Al_2O_3, 10.5 wt% Li_2O, 5.5 wt% K_2O, 0.02 wt% CeO_2, 0.04 wt% AgCl.

Exposure to UV light initiates the following reaction:

$$Ce^{3+} + Ag^+ \rightarrow Ce^{4+} + Ag^0$$

The neutral silver is formed by this reaction. During the subsequent heat treatment of the glass at approximately 600°C, colloids of metallic silver are formed (Beall 1992). This colloidal silver forms heterogeneous nuclei of approximately 80 Å for the subsequent crystallization of a lithium metasilicate primary crystal phase, Li_2SiO_3. Lithium metasilicate possesses a chain silicate structure. The crystallization of this compound proceeds dendritically. The dentritic growth of crystals in glass-ceramics is discussed in Section 3.2.4.

Lithium metasilicate crystals are characterized by the ease with which they dissolve from the glass-ceramic in dilute hydrofluoric acid. Thus, Stookey (1953) developed a high-precision patterned glass-ceramic, the structure of which results from the etching. High-precision structural parts in different shapes are produced by placing a mask on the material and exposing the open areas to UV light. The finished product, predominantly composed of a glass matrix, was marketed under the brand name Fotoform®. If these products are exposed to additional UV and thermal treatment, a lithium disilicate main crystal phase is produced. This type of glass-ceramic carries the brand name Fotoceram®. It was produced and distributed by Corning Glass Works, now Corning Incorporated.

Apart from the possibilities of shaping the glass-ceramic as desired, Stookey (1954) and McMillan (1979) discovered additional properties of these glass-ceramics, which were very promising for industrial applications of the material. These properties include high flexural strength of 100–300 MPa (Hing and McMillan, 1966; Borom et al., 1975), outstandingly high fracture toughness of 2–3 MPa·m$^{0.5}$ (Mecholsky 1982), and high electrical resistivity of 3×10^9 ohm·cm. These electrical properties combined with a low loss factor of 0.002 at 1 MHz and 25°C are impressive for a glass-ceramic with a high alkaline ion content. Therefore, these properties presented ideal prerequisites for applications in electrical engineering.

Lithium disilicate glass-ceramics demonstrate a relatively high linear coefficient of thermal expansion of approximately 105×10^{-7} K^{-1}. This property is favorable for the fabrication of special composite materials, e.g., for sealing to metal substrates in the electrical industry (Beall 1993).

While retaining and, in certain instances, optimizing the particularly advantageous mechanical, thermal, and electrical properties of lithium disilicate glass-ceramics with a stoichiometric composition, new nonstoichiometric glass-ceramics demonstrating improved chemical properties were developed. Nonstoichiometric implies that the SiO_2:Li_2O molar ratio deviates greatly from 2:1 and that the system is rendered considerably more complex with numerous additional components.

Beall (1993) and Echeverría (1992) achieved notable results in the development of a new lithium disilicate glass-ceramic. The new material is distinguished by the following three characteristics:

- First, the particular ratios of SiO_2 and Li_2O, which are responsible for the formation of the main crystal phase
- Second, the nucleating agents
- Third, the components of the glass matrix. The chemical components were selected to confer good chemical durability. Typical compositions are shown in Table 2-1. In accordance with these compositions, the residual glass content can be divided into the following composition groups:
  - Chkalovite glass with excess SiO_2 (Types A, B, and C in Table 2-1). The crystalline compound of chkalovite has the chemical formula $Na_2O \cdot ZnO \cdot 2SiO_2$. In this case, however, a crystal does not form, and the glass phase is maintained. Therefore, it must be written as $R_2O \cdot ZnO \cdot nSiO_2$ where R = Na, K, and where n can range between 2 and 10.
  - Potassium feldspar glass (Type D). A glass-ceramic with a glass matrix similar to that of feldspar, $K_2O \cdot Al_2O_3 \cdot 6SiO_2$, can contain 10–20 wt% glass phase.
  - Alkaline earth feldspar glass with excess SiO_2 (Types E and F). Thirty to fifty wt% of an aluminosilicate glass matrix of the $RO \cdot Al_2O_3 \cdot nSiO_2$ type where R = Ca, Ba, Sr and n = 2 can be incorporated for alkaline earth feldspars.

Table 2-1

Composition of Lithium Disilicate Glass-Ceramics with Excellent Chemical Durability (wt%)

Glass	A	B	C	D	E	F
SiO_2	69.3	76.3	75.3	74.2	68.6	71.8
ZrO_2	-	-	-	-	2.01	2.4
Al_2O_3	-	-	-	3.54	8.56	8.95
ZnO	5.28	2.92	1.9	-	-	-
CaO	-	-	-	-	1.01	3.83
Li_2O	15.4	13.4	17.4	15.4	11.2	10.7
K_2O	6.05	3.38	2.2	3.25	-	-
P_2O_5	3.84	3.84	2.9	3.37	4.18	4.37

The excess SiO_2 in the glass matrix above that of the mineral composition of the three types of glass-ceramics measures approximately 10 wt%. All three types of lithium disilicate glass-ceramics contain P_2O_5. Based on different glass compositions, the optimum temperature for the heat treatment of the base glasses is different. The initiation of nucleation in a glass-ceramic featuring a chkalovite glass matrix is approximately 500°C. The optimum nucleation temperature for a glass-ceramic with a feldspar glass matrix is higher at 650°C.

Apart from containing lithium disilicate crystals, the glass-ceramic containing alkaline earth aluminosilicate (Types E and F) also contains cristobalite, SiO_2; wollastonite, $CaSiO_3$; and β-spodumene, $Li_2O \cdot Al_2O_3 \cdot nSiO_2$. The highest crystal content is achieved in this glass-ceramic.

The following points regarding the mechanisms of nucleation and crystallization of these glass-ceramics containing P_2O_5 must be summarized before the individual microstructure formations of glass-ceramic end products and the resulting properties are introduced:

- The structure of the crystalline compound $Li_2Si_2O_5$ was studied by Liebau (1961), Dupree et al. (1990), and Smith et al. (1990). Its typical glass-ceramic morphology is reflected in its layered structure, i.e., it is either tabular or in bladed laths.
- The lithium disilicate layered silicate phase, $Li_2Si_2O_5$, is the primary crystal phase. In addition, the growth of lithium disilicate may be initiated by the primary crystallization of the precursor chain silicate, lithium metasilicate, Li_2SiO_3.
- Nucleation may be initiated by phase-separation processes. In a special glass-ceramic heated at 1000°C for sealing applications and additionally

heat treated at 650°C for 15 min and 820°C for 20 min, the formation of Li_3PO_4 crystals takes place as a primary crystal phase. Lithium disilicate grows on Li_3PO_4 based on epitaxial interactions (Headley and Loehman, 1984) (see Section 1.4).

The microstructure of these glass-ceramics is characterized by well-cleaved crystals and interlocking crystal aggregates. The size of the crystallites can be controlled by the selection of the heat treatment, while the composition remains the same. After four hours of heat treatment at 800°C, the crystals measure approximately 3 μm in length. When heat treated at 750°C for four hours they measure less than 1 μm.

In addition, these glass-ceramics are characterized by special properties, such as translucency, high mechanical stability of 300–400 MPa, and fracture toughness of approximately 3 MPa·m$^{0.5}$ (see Table 2-2).

The good mechanical properties result from the high fraction of crystalline phase of approximately 80 wt% and the characteristic microstructure. After mechanical fracture tests, the interlocking trabecular crystals exhibit a step-like fracture, which indicates that the interlocking crystals redirect the fracture front. As a result, the growing crack meets with a high degree of resistance, which inhibits fracture propagation. In the evaluation of the mechanical stability in relation to the chemical composition, the glass-ceramic with the feldspar glass matrix demonstrated the highest values.

The coefficient of thermal expansion can also be controlled by the glass matrix and the microstructure. For a feldsparlike glass matrix, expansion values of 80×10^{-7} K^{-1} are achieved; for a K-chkalovite-resulted glass, up to 120×10^{-7} K^{-1} values are achieved.

For the successful fabrication of Type A–F glass-ceramics (Table 2-1), Echeverría (1992) developed a process in which a "self-glazing" effect is

Table 2-2

Mechanical Properties of Type A Glass-Ceramics According to Table 2-1

Glass	Strength, Ksi, MPa	Toughness MPa·m$^{0.5}$
A	28.5, 197	2.83
B	27.6, 190	2.29
C	33.9, 234	3.0
D	31.4, 217	3.3
E	22.1, 153	-

produced during the crystallization of the glass-ceramic at temperatures between 650° and 850°C. The high quality of the surface of the glass-ceramics enables the material to be used for a variety of applications. Apart from producing a smooth, shiny surface, the procedure is also very economical.

Using a special hot-press procedure, Schweiger et al. (1998) and Frank et al. (1998) also developed a powder-processed lithium disilicate glass-ceramic. To optimize the viscous properties for the hot-press procedure at approximately 920°C, components such as La_2O_3, MgO, and pigments were added to the main components SiO_2, Li_2O, P_2O_5, K_2O, and ZnO. The fabrication process and the reactions and mechanisms involved in the development of these glass-ceramics are summarized below:

- Glasses in the composition range according to Table 2-3 are fabricated and milled to powder.
- Additives such as pigments are added to the glass powder. A sintering process is initiated during heat treatment at temperatures of approximately 500°–800°C to produce a raw glass-ceramic in the form of a cylindrical ingot.
- This ingot is transformed into a viscous state of approximately 10^5–10^6 Pa·s in a special hot-press apparatus (EP 500, Ivoclar Vivadent AG, Liechtenstein). Thereafter, it is pressed at approximately 920°C for 5–15 min to form a glass-ceramic body. This glass-ceramic of the lithium disilicate type does not require additional heat treatment. The glass-ceramic end product contains main crystal phases of $Li_2Si_2O_5$ and Li_3PO_4.

Table 2-3

Composition of a Lithium Disilicate Glass-Ceramic for the Hot-Press Procedure

Composition	wt%
SiO_2	57–80
Al_2O_3	0–5
La_2O_3	0.1–6
MgO	0–5
ZnO	0–8
K_2O	0–13
Li_2O	11–19
P_2O_5	0–11
Additives	0–6
Color substances	0–8

- Given the assumed mechanisms, nucleation must be initiated by P_2O_5. It is still unclear to what extent crystalline orthophosphate has an epitaxial effect on the formation of lithium silicates. This topic is being addressed in an ongoing research program.
- The preferential crystallization mechanism is that of volume crystallization. However, surface reactions cannot be neglected when considering crystallization and nucleation in powder compacting and subsequent sintering and crystallization. In these processes, water has a special effect on the production of lithium disilicate glass-ceramics, as demonstrated by Helis and Shelby (1983) and Davis (1997).

It must be noted that the reaction mechanisms in glass/glass-ceramic powders are usually of a very complex nature. Jacquin and Tomozawa (1995) addressed the complex sintering process and the various ways of controlling it in the fabrication of lithium silicate glass-ceramics.

- In some multicomponent glass-ceramics (containing Al_2O_3), the crystallization of the $Li_2Si_2O_5$ main crystal phase, however, clearly occurs via the precursor phase of lithium metasilicate, Li_2SiO_3, as shown by the DSC curve (Fig. 2-2) in thermal analysis and X-ray diffraction investigations. The microstructure of the resulting glass-ceramic is shown in Fig. 2-3.

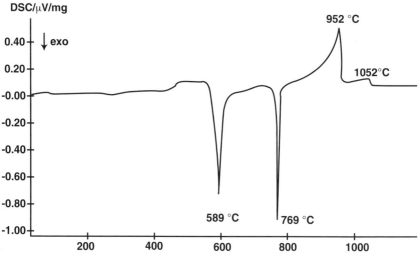

Figure 2-2 DSC curve of a base glass from the SiO_2–Li_2O system (Schweiger et al., 1999). Crystallization peaks of Li_2SiO_3 (at 589°C) and $Li_2Si_2O_5$ (at 770°C) appear.

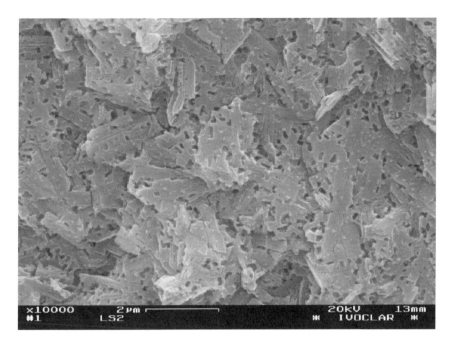

Figure 2-3 Microstructure of a lithium disilicate glass-ceramic after hot pressing, (Schweiger et al., 1999).

- On the other hand, complex reactions of parallel and secondary solid-state reactions were studied in an Al_2O_3-free lithium disilicate glass-ceramic of the composition 63.2 mol% SiO_2, 29.1 mol% Li_2O, 2.9 mol% K_2O, 3.3 mol% ZnO, 1.5 mol% P_2O_5 (Höland et al., 2000a). Li_3PO_4 crystals were formed after the crystallization of Li_2SiO_3 and $Li_2Si_2O_5$ (see Section 3.3 high temperature X-ray diffraction of this glass-ceramic), a characteristic also confirmed in glasses of similar composition (Holland et al., 1998). Therefore the nucleation mechanism may be based on steep compositional gradients. The crystals of $Li_2Si_2O_5$ were precipitated already at low temperatures (approximately 520°C) in parallel to Li_2SiO_3. The intensity of growth of lithium disilicate increased at 680°C after dissolution of Li_2SiO_3 and SiO_2/cristobalite (intermediate phase). Thus, a solid-state reaction of $Li_2SiO_3 + SiO_2 \rightarrow Li_2Si_2O_5$ may be valid.
- The properties of this glass-ceramic are characterized by high strength of approximately 350 MPa, fracture toughness of approximately 3.3 $MPa \cdot m^{0.5}$, and translucency (Schweiger et al., 2000).

In Japan, Goto et al. (1997) developed a lithium disilicate glass-ceramic for magnetic memory disk substrates. The composition of the glass-ceramic is characterized by 65–83 wt% SiO_2, 8–13 wt% Li_2O, 0–7 wt% K_2O, 0.5–5.5 wt% (sum of MgO, ZnO, PbO), with 0–5 wt% ZnO, 0–5 wt% PbO, 1–4 wt% P_2O_5, 0–7 wt% Al_2O_3 and 0–2 wt% (As_2O_3, Sb_2O_3).

2.1.2 SiO_2–BaO (Sanbornite)

2.1.2.1 Stoichiometric Barium-Disilicate

The development of glass-ceramics with a barium disilicate primary crystal phase was conducted in two different ways. The first type of glass-ceramic was produced from base glasses with a composition almost corresponding to the stoichiometric composition of barium disilicate. The second type of glass-ceramic with sanbornite crystals was developed from multicomponent glasses using controlled crystallization.

Barium-silicate glass-ceramics were developed from base glasses of the SiO_2–BaO system without adding nucleating agents. MacDowell (1965) and James (1982) investigated this type of homogeneous nucleation. The primary crystal phases of barium disilicate, $BaO \cdot 2SiO_2$, and $BaSi_2O_5$ (sanbornite), were precipitated in the base glass by heat treating them and forming a fine-grained glass-ceramic microstructure. In contrast to lithium disilicate crystals, barium disilicate crystals demonstrate a quite different glass-ceramic microstructure in their early stage. Barium disilicate grows as spherulite crystals, and lithium disilicate exhibits a rod-shaped morphology.

The formation of the spherulites was reported by Burnett and Douglas (1971) as a process of primary phase formation of the high-temperature modification of barium disilicate. Lewis and Smith (1976) determined this primary crystal formation to be crystal aggregates of many extremely fine crystals.

The chemical composition of the SiO_2–BaO glasses and the heat treatment influence the crystal phase formation processes in the glass-ceramic material. The microstructure of the base glass is the basis for this process of controlled crystal formation. Based on the fundamental research of glass microstructure, Seward et al. (1968) demonstrated that the main processes of glass microstructure formation of barium silicate glasses proceeded according to phase separation. Metastable immiscibility extended from approximately 100 mol% SiO_2 to approximately 30 mol% BaO.

James (1982) determined the influence of phase separation on the kinetics of nucleation and crystallization in base glasses. He showed that glasses of the stoichiometric composition of barium disilicate (33.3 mol% BaO, 66.7 mol% SiO_2) demonstrated the highest nucleation rate compared with other barium disilicate crystals. In glasses having chemical compositions close to the stoichiometric composition of barium disilicate, however, the morphology of the microstructure of the base glass prior to crystallization had a remarkable influence on the kinetics of nucleation and crystallization. James (1982) explained this phenomenon by investigating a glass with a composition of 26 mol% BaO, 74 mol% SiO_2. He carried out different heat treatments of the glass prior to its crystallization at 700°C. In the process, he determined that the crystallization rate at 700°C depended on the pretreatment of the glass. Clearly the base glass demonstrating the highest degree of phase separation, resulting from pretreatment, also demonstrated the highest degree of crystal formation (measured as the crystal number per volume at 700°C at a specific time). The net result of this investigation showed that controlled phase separation in $BaO–SiO_2$ glasses formed a droplet glass phase with a composition close to that of stoichiometric barium disilicate. In contrast to glasses with no phase separation (below 33 mol% BaO, according to Seward et al., 1968), this droplet phase induces homogeneous nucleation as suggested by MacDowell (1965).

As discussed, the formation of the high temperature modification of barium disilicate is the primary process of crystal formation. This type of crystal grows as spherulites. The morphology of the spherulites, however, is irregular. Furthermore, "starlike" crystals are precipitated (Zanotto and James, 1988). Hence, the primary crystals also contain a glassy phase of approximately 34 volume%. The stable low-temperature modification nucleates on the sperulites of the high-temperature modification and grows in needlelike form. The growth rates of the two forms of barium disilicate are quite different. The spherulites grow very slowly in their early stages. After a certain time, however, their crystallinity develops more rapidly.

2.1.2.2 Multicomponent Glass-Ceramics

Barium silicate glass-ceramics were developed in the $SiO_2–Al_2O_3–Na_2O–BaO–TiO_2$ system (Table 2-4). These glass-ceramics are of great technical significance. They were used to develop the Corning Code 9609® glass-ceramics. Corning 9609® glass-ceramics were produced according to the mechanism of heterogeneous nucleation rather than by the crystallization of base glasses demonstrating the near-stoichiometric composition of barium disilicate and controlled phase separation. The heterogeneous nucleation agent TiO_2 was added to the base glass. The primary crystal phases of celsian, $BaAl_2Si_2O_8$; nepheline, $NaAlSiO_4$; and anatase, TiO_2; were formed by controlled crystallization. This type of glass-ceramic material was predominantly

Table 2-4

Chemical Composition of Corning 9609® Glass-Ceramic		
Composition	wt%	mol%
SiO_2	43.3	53.0
Al_2O_3	29.8	21.4
Na_2O	14.0	16.0
BaO	5.5	2.6
TiO_2	6.5	6.0
As_2O_3	0.9	0.4

used to make Centura® tableware, but because of excessive heating from microwave absorption, it was eventually removed from the market.

The formation of $BaSi_2O_5$ crystals in multicomponent glass-ceramics with a variety of compositions is described by Stookey (1959). Nucleation is initiated by copper. That is, epitactic solid-state reactions between the copper and the growing primary nuclei initiate the formation of barium disilicate. Stookey (1959) demonstrated that different crystal phases could be produced in the different materials systems, depending on the heterogeneous nucleating agents selected. For example, Stookey determined that gold could be used as a heterogeneous nucleating agent for precipitating lithium disilicate.

2.2 ALUMINOSILICATES

2.2.1 SiO_2–Al_2O_3 (Mullite)

In the SiO_2–Al_2O_3 binary phase diagram, two crystalline compounds stand out: sillimanite, $SiO_2 \cdot Al_2O_3$, and mullite, $2SiO_2 \cdot 3Al_2O_3$ (Fig. 2-4). These crystalline compounds are very important in the ceramics industry since they are the main components of refractories. Moreover, these crystals are also contained in porcelain. Mullite is particularly suitable as a refractory material because of its high melting point and excellent chemical durability.

In the preparation of the SiO_2–Al_2O_3 phase diagram, differences of opinion occurred in the characterization of mullite crystals (Hinz 1970). According to Fig. 2-4, for example, mullite generally is an incongruent compound, that is, one that decomposes as it melts. Mixed crystal formation of mullite by the additional absorption of Al_2O_3 in the crystal lattice, however, was also observed. Subsequently, mullite should represent a congruently melting compound. The crystal structure is shown in Appendix Fig. 16.

The incorporation of foreign ions into the crystal lattice of the mullite for refractory applications has been examined in depth. These investigations established the content of ions of a specific size that can be incorporated into the mullite lattice without destroying it. For example, up to 9% ions with a radius of 0.5–0.7 Å can be incorporated into mullite, $2SiO_2 \cdot 3Al_2O_3$. Therefore, it is possible to incorporate up to 8% Fe_2O_3, 9% Cr_2O_3, or 1.5% TiO_2. These foreign ions

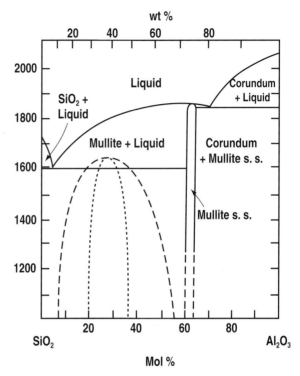

Figure 2-4 Phase diagram of SiO_2–Al_2O_3 (Levin et al., 1975, and MacDowell and Beall, 1969).

expand the crystal lattice of mullite. Ions with a radius greater than 0.7 Å, however, cannot be incorporated. Thus the following ions cannot be incorporated: Na^+, K^+, Mg^{2+}, Ca^{2+}, B^{3+}, or Fe^{2+} ions (Hinz 1970). This ability of the crystal to absorb ions into its lattice was used in the development of particular properties in glass-ceramics.

Mullite also exhibits a different morphology compared with other crystals. The double einer-chain silicate demonstrates preferred crystallization in a needle-like habit. If crystallization is uncontrolled, undesirable large mullite needles may occur in the glass-ceramic as a secondary reaction in the ceramic-forming process. This type of crystallization can be initiated by surface nucleation. The properties of such a material, its strength in particular, are negatively influenced.

To avoid uncontrolled crystallization in the development of a material exhibiting specific optical properties MacDowell and Beall (1969), and Beall (1992, 1993) produced a mullite glass-ceramic in the SiO_2–Al_2O_3–B_2O_3–ZnO–K_2O–Cr_2O_3 system. The composition of this optically optimized material is shown in Table 2-5.

Table 2-5

Composition (in wt%) of the Mullite Glass-Ceramic Used in Solar Cells (Beall and Pinckney, 1999)
Additives: 0.1 wt% Cr_2O_3, 0.4 wt% As_2O_5

SiO_2	48.0
B_2O_3	11.0
Al_2O_3	29.0
ZnO	10.0
K_2O	2.0

Nucleation in base glasses of the composition shown in Table 2-5 was initiated by controlled glass-in-glass phase separation. The microstructure of the base glass is characterized by typical microphase separation into two glass phases. Glass droplets measuring less than 100 nm in diameter are embedded in a glass matrix. These glass droplets are enriched in Al^{3+}. Thus, they exhibit a lower viscosity than the glass matrix during subsequent heat treatment. This permits nucleation and crystallization processes to occur in the droplets in the temperature range of 750–900°C.

Crystallization was carried out very carefully to ensure that mullite would only grow within the original droplet glass phase, without crossing the phase boundary into the glass matrix. In this way, the crystals remain smaller than 100 nm and large needles no longer occur. The resulting glass-ceramic demonstrated favorably high transmission in the visible spectrum. This special characteristic of microstructure formation in mullite glass-ceramics is discussed in detail in Section 3.2.5.

Furthermore, Andrews et al. (1986) were able to incorporate Cr^{3+} ions into these very small mullite crystals to produce luminescent properties. Reisfeld et al. (1984) and Kiselev et al. (1984) have reported about the luminescent properties produced by Cr^{3+}. Section 4.3.3 describes why luminescent mullite glass-ceramics, for example, are preferred in solar and laser technology. Selected properties of these glass-ceramics are listed in Table 2-6.

2.2.2 SiO_2–Al_2O_3–Li_2O (ß-Quartz Solid Solution, ß-Spodumene Solid Solution)

Glass-ceramics in the SiO_2–Al_2O_3–Li_2O system have gained a reputation for demonstrating very special properties, such as minimal or even zero thermal expansion in a large temperature range combined with desirable optical properties, such as high translucency or high transmissibility. These properties render this type of glass-ceramics suitable for a wide range of applications, particularly in technical fields.

Table 2-6

Selected Properties of Mullite Glass-Ceramics
(Beall and Pinckney, 1999)

Heat Treatment (T in °C – t in h)	Scattering Coefficient $\sigma(\times\ 10^3\ cm^{-1})$		Quantum Efficiency (0.1% Cr_2O_3)		
	458 nm	633 nm	458 nm	514 nm	633 nm
Precursor	9 ± 1	–	–	–	–
750–4/800–2	68 ± 5	36 ± 3	0.23	0.28	0.23
750–4/850–2	143 ± 5	55 ± 3	0.29	0.33	0.28
750–4/875–4	204 ± 9	76 ± 3	0.33	0.38	0.31

The unusual combination of properties was achieved by the formation of the crystals of β-quartz solid solution and β-spodumene solid solution in glass-ceramics. The aim of combining the aforementioned properties was achieved by precipitating a large volume percent of the desired crystal phases in the glass-ceramic in a controlled morphology ranging from very small crystals in the nanometer range to micrometer-sized crystals.

For a better understanding of the phase relationships, the part of the $SiO_2–Al_2O_3–Li_2O$ ternary phase diagram shown in Fig. 2-5 must be analyzed. A wide solid-solution range is evident. This solid solution and the different types of polymorphous phase transformations are described in Section 1.2. The particular nomenclature of stuffed derivatives of quartz is addressed in Section 1.2.1.

In the ternary system shown in Fig. 2-5, the main crystal phases β-quartz solid solution and β-spodumene solid solution were produced in glass-ceramics by controlling the crystallization of base glasses in the approximate composition range of 55–70 wt% SiO_2, 15–27 wt% Al_2O_3, 1–5 wt% Li_2O by incorporating specific additions. These additions are discussed in the following sub-sections.

2.2.2.1 β-Quartz Solid Solution Glass-Ceramics

As shown in Section 1.2.1, the β-quartz solid solution crystals are metastable and demonstrate a very low coefficient of thermal expansion. The basic chemical formula can be expressed as $(Li_2,R)O{\cdot}Al_2O_3{\cdot}nSiO_2$, where R represents either Mg^{2+} or Zn^{2+} ions and n ranges between 2 and 10. The limiting value of this composition would correspond to the thermodynamically stable β-eucryptite $(Li_2O{\cdot}Al_2O_3{\cdot}2SiO_2)$. This crystal can itself form SiO_2-rich solid solutions.

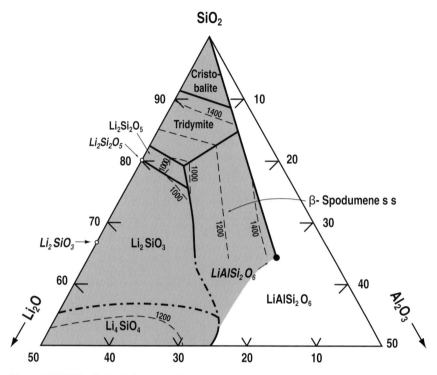

Figure 2-5 SiO_2–Al_2O_3–Li_2O phase diagram showing the liquidus field of a ß-spodumene solid solution (Levin et al., 1969). (wt%, T in °C.)

In the commercially important β-quartz solid solution glass-ceramics, values of 6–8 were achieved for n. These values improved the melting properties of the glasses and rendered the entire manufacturing process more economical (Beall 1986).

The basic process development for the production of β-quartz solid solution glass-ceramics was conducted by Corning Glass Works. The key to the controlled crystallization of β-quartz solid solution was soon shown to be controlled nucleation, in other words, the correct selection of the nucleating agents and their concentration.

Beall et al. (1967) developed an effective nucleation initiation method for the precipitation of very small crystals of the β-quartz solid solution type in the SiO_2–Al_2O_3–Li_2O–MgO–ZnO system. Nucleation for the subsequent crystallization of β-quartz solid solution was achieved with additions of TiO_2, ZrO_2, and occasionally Ta_2O_5 (Beall et al., 1967; Beall 1992). The desired crystallites of less than 100 nm were achieved in the crystallization of β-quartz solid solution which followed nucleation. Therefore, the fabrication of glass-ceramics demonstrating minimal expansion became possible. Given the small size of the precipitated crystals, the above property was also combined with high transmissibility.

What type of mechanism or controlling methods are available for crystallizing β-quartz solid solution in such small dimensions?

With the addition of TiO_2 to the base glasses of the SiO_2–Al_2O_3–Li_2O system, a nucleation mechanism was activated which had already been successfully used by Stookey (1959). Significant advances regarding the effects of TiO_2 as a nucleating agent were achieved by Doherty et al. (1967) and Beall (1967). They determined that phase separation processes in the glass state initiated nucleation. The base glass was composed of a glass matrix and a second phase consisting of approximately 50 Å droplets. A relatively high density of these droplets phase-separated from the matrix. This phenomenon was observed when 2 wt% TiO_2 was added to base glasses of the following system: 55–70 wt% SiO_2, 15–27 wt% Al_2O_3, 1–5 wt% Li_2O, 0.5–2 wt% MgO, 0.5–2 wt% ZnO.

Titanium ions are enriched in the droplet phase containing Al^{3+} and cause heterogeneous crystallization of β-quartz solid solution.

In-depth initial investigations about the nucleating effect of ZrO_2 were conducted by Tashiro and Wada (1967). Furthermore, Beall (1967) determined the role of ZrO_2 solid solution in the crystallization of β-quartz solid solution. The investigations by Tashiro and Wada (1967) showed that nucleation can be initiated by adding 4 wt% ZrO_2 to an SiO_2–Al_2O_3–Li_2O glass. The glass-ceramics in question demonstrate a composition that is very similar to the stoichiometric composition of β-spodumene. These glass-ceramics are discussed in detail in Section 2.2.2.2.

The initial β-spodumene glass-ceramic developed by Wada (1967) was white. In further developments involving ZrO_2 nucleation, he subsequently produced transparent glass-ceramics of the β-quartz solid solution type. Today, related glass-ceramics that are transparent in the visible spectrum are produced by Nippon Electric Glass. Their properties and applications are discussed in Section 4.1.5 on displays and Section 4.5 on electrical/electronic applications. The material in question is known as Neoceram™ N-0. Another material of this kind called Firelite™ is discussed in Section 4.6. on architectural applications. The combination of high transmissibility and near-zero expansion in a wide range of temperatures is a particularly favorable characteristic of these materials.

The combination of the two nucleating agents TiO_2 and ZrO_2 was also examined in-depth by Sack and Scheidler (1966), Beall et al. (1967), Stewart

(1971), Petzoldt (1967), and Müller (1972). An effective nucleating agent ratio of 2 wt% TiO_2 and 2 wt% ZrO_2 was established (Beall 1992) for the nucleation of highly dispersed $ZrTiO_4$ crystallites at 780°C. During subsequent heat treatment at 980°C, β-quartz solid solution measuring less than 0.1 μm grew. The microstructure of this type of β-quartz solid solution glass-ceramic is shown in Section 3.2.1.

The compositions of β-quartz solid solution glass-ceramics produced on a large scale for technical applications are summarized in Table 2-7. These glass-ceramics are called Vision®, Zerodur®, Narumi®, Neoceram™, Ceran®, and Keraglas®. Other products are Robax®, which represents a col-

Table 2-7

Composition of Commercial ß-Quartz Solid Solution Glass-Ceramics (wt%) According to Beall (1992), Beall and Pinckney (1999), and Sack and Scheidler (1974)

	Vision® (Corning/USA)	Zerodur® (Schott/ Germany)	Narumi® (Nippon Electric Glass/ Japan)	Neoceram™ N-0 (Nippon Electric Glass/Japan)	Ceran® (Schott/ Germany)
SiO_2	68.8	55.4	65.1	65.7	64.0
Al_2O_3	19.2	25.4	22.6	22.0	21.3
Li_2O	2.7	3.7	4.2	4.5	3.5
MgO	1.8	1.0	0.5	0.5	0.1
ZnO	1.0	1.6	–		1.5
P_2O_5	–	7.2	1.2	1.0	
F	–	–	0.1		
Na_2O	0.2	0.2	0.6	0.5	0.6
K_2O	0.1	0.6	0.3	0.3	0.5
BaO	0.8				2.5
CaO					0.2
TiO_2	2.7	2.3	2.0	2.0	2.3
ZrO_2	1.8	1.8	2.9	2.5	1.6
As_2O_3	0.8	0.5	1.1	1.0	
Sb_2O_3					0.85
Fe_2O_3	0.1	0.03	0.03		0.23
CoO	50 ppm				0.37
Cr_2O_3	50 ppm				
MnO_2	–		.		0.65
NiO				0.06	

orless version of Ceran® (Pannhorst 1993), and Eclair®, which is an analo-gous version of Keraglas®. The components of these glass-ceramics are divided into groups in this table to clarify their roles in these materials.

The first group of components, SiO_2, Al_2O_3, Li_2O, MgO, ZnO, P_2O_5, and F, is used for the formation of the main crystal phases of β-quartz solid solution glass-ceramics. The β-quartz solid solution crystals in the Vision® glass-ceramic can be described as stuffed derivatives of β-quartz because of the incorporation of Li^+, Mg^{2+}, Zn^{2+}, and minor Al^{3+} into interstitial sites in the structure with the bulk of the Al^{3+} partially replacing Si^{4+} for charge bal-ance. These structural relationships in quartz and the associated favorable thermal properties are discussed in Section 1.3.1.

The following formula is used by Petzoldt and Pannhorst (1991) and Petzoldt (1967) to describe the main crystal phase of the β-quartz solid solu-tion type found in the Zerodur® glass-ceramic:

$$Li_{2-2(v + w)}Mg_v Zn_wO \cdot Al_2O_3 \cdot xAlPO_4 \cdot (y - 2x)SiO_2$$

This formula demonstrates that Li^+, Mg^{2+}, Zn^{2+}, and Al^{3+} ions are incor-porated into the structure of β-quartz. As a result, the β-quartz is frozen in the structure of the glass-ceramic in a metastable state at room temperature. In addition, $AlPO_4$ structural units are incorporated. These units are iso-morphous with SiO_2 and replace the $[SiO_4]$ tetrahedra in the β-quartz solid solution structure. The microstructure of this glass-ceramic was studied by Maier and Müller (1987). It is evident that the β-quartz solid solution crys-tals measure approximately 100 nm. This microstructure was developed by heat treating the base glass at 700°C for 4 hours. Subsequently, the base glass was heat treated at 900°C for 10 minutes. Surprisingly, this relatively short heat treatment produced the fine crystals.

In the Narumi® glass-ceramic, Li^+, Mg^{2+}, F^-, and Al^{3+} ions are incorpo-rated into the structure of β-quartz and P^{5+} is incorporated in the network as $AlPO_4$ structural units.

In Table 2-7, the second group of components consists of the Na_2O, K_2O, CaO, and BaO oxides together with residual Al_2O_3 and SiO_2, which form the glass matrix of the glass-ceramic. The development of the residual glass phases in these glass-ceramics is significant for two reasons. The glass-ceramic manufacturing procedure (e.g., casting and pressing of the base glass) is ren-dered more economical through the additions of the ingredients and favorable properties such as high optical transparency are produced in the glass-ceramic partly as a result of grain-growth resistance caused by the glassy matrix.

The third group of components is composed of the nucleating agents TiO_2 and ZrO_2.

The fourth group is made up of refining agents. These components are mainly used as anti-foaming agents when manufacturing base glasses, particularly As_2O_3.

The fifth group of components is coloring agents from the transition metals (3-d). They produce the desired color in the glass of glass-ceramics by the familiar phenomenon of ion coloring.

These glass-ceramics combine the unusual properties of high translucency, near-zero thermal expansion in a wide temperature range, excellent optical polishability, and strength that exceeds the bending strength of glass by a wide margin. A detailed description of the properties of the Vision® and Zerodur® glass-ceramics is provided in Section 4.2 and Section 4.3.1, respectively.

Given these special properties, this high-performance material has been successfully used for a variety of applications such as telescope mirror blanks in precision optics and infrared-transmitting range tops (Section 4.3.1). This glass-ceramic has also been successfully mass produced for household applications such as cookware and woodstove windows (Section 4.2).

2.2.2.2 β-Spodumene Solid Solution Glass-Ceramics

The structure of β-spodumene is described in Section 1.2.1 in connection with the structurally related thermal properties. The solid solution β-spodumene ranges between $n = 4$ and $n = 10$ for $Li_2O \cdot Al_2O_3 \cdot n\ SiO_2$. The effect of the increasing SiO_2 content in the solid solution on the thermal expansion is shown in Section 1.3.1.

As in the case of β-quartz, the Li^+ ion can also be replaced by Mg^{2+} in β-spodumene, a stuffed derivative of keatite. The degree of substitution, however, is lower than that of β-quartz.

The transformation of β-quartz to β-spodumene is possible between 900°C and 1000°C. The irreversible procedure always entails the enlargement of the crystallites. As a result, the optical properties are often affected and the transmissibility decreases. Furthermore, if TiO_2 is used as a nucleating agent, crystalline TiO_2 forms as rutile. This crystal phase is known to demonstrate a high refractive index, which gives the material an opaque appearance.

The microstructure of a typical β-spodumene solid solution glass-ceramic in the form of Corning Ware® is presented in Fig. 2-6. The width of the main crystal phase of β-spodumene solid solution measures 1–2 μm.

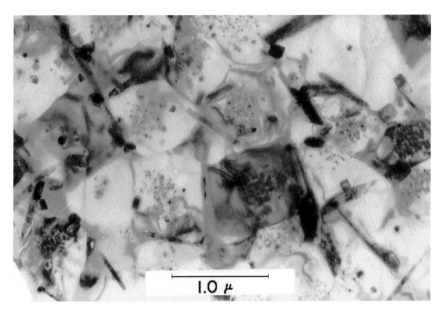

Figure 2-6 Microstructure of β-spodumene solid solution (Corning Ware®).

The compositions of commercial β-spodumene solid solution glass-ceramics are shown in Table 2-8. The composition of Corning Ware® differs from that of Cercor® in that it contains a number of additions that Cercor® does not contain.

Table 2-8

Commercial ß-Spodumene Solid Solution Glass-Ceramics (wt%)

	Corning Ware® (Corning/USA)	Cercor® (Corning/USA)	Neoceram™N-11 (Nippon Electric Glass/Japan)
SiO_2	69.7	72.5	65.7
Al_2O_3	19.2	22.5	22.0
Li_2O	2.8	5.0	4.5
MgO	2.6		0.5
ZnO	1.0		
Na_2O	0.4		0.5
K_2O	0.2		0.3
TiO_2	4.7		2.0
ZrO_2	0.1		2.5
As_2O_3	0.6		1.0
Fe_2O_3	0.1		
P_2O_5			1.0

In Corning Ware®, the additions and the low percentage of network-forming oxides modify the viscosity–temperature function in such a way to allow the base glass to be processed in automatic procedures such as pressing, blowing, or tube drawing. Base glasses processed in this way were net-shaped and required only a heat treatment at a very high temperature of 1125°C to effect the formation of the final glass-ceramic. The crystals grow from 1.8 µm to 3.2 µm in 200 h (Chyung 1969). A degree of crystallinity exceeding 93% was achieved for the β-spodumene main crystal phase, the secondary phases consisting of minor amounts of spinel and rutile crystals and the residual glass phase.

Given its composition, the Cercor® glass-ceramic is manufactured according to a different procedure. The material cannot be processed in the same way as Corning Ware® because the network-forming content exceeds 89 mol%, and there is no nucleation agent (e.g., nucleation proceeds only from the surface or ghost interfaces of the sintered glass particles).

The glass is ground into a frit, sintered, and heat treated. During the heat treatment, controlled crystallization takes place after deformation and sintering and a glass-ceramic with only slight internal porosity (<2 %) results. However, a honeycomb structure is produced by saturation of alternately crimped and plane paper with glass slurry and subsequent winding into a wheel.

Both of the mentioned glass-ceramics demonstrate a very low coefficient of thermal expansion.

Corning Ware®: 12×10^{-7} K^{-1} (0–500°C)
Cercor®: 5×10^{-7} K^{-1} (0–1000°C)

Given the favorable processing technology of Corning Ware®, this material is used to produce cookware and hot plates, while Cercor® is used in gas turbines.

In Japan, Tashiro and Wada (1963) conducted in-depth investigations into the nucleation of ZrO_2 for the crystallization of white spodumene glass-ceramics. Nucleation and crystallization as well as the favorable formation of the base glass were examined in glasses with the following general composition: 65 wt% SiO_2, 30 wt% Al_2O_3, 5 wt% Li_2O, 1 wt% K_2O, 3 wt% P_2O_5.

The percentages of the main components varied. The preferred compositions of Tashiro and Wada (1963) were close to the stoichiometric compound of spodumene, $Li_2O \cdot Al_2O_3 \cdot 4SiO_2$. To examine the formation of the glass for the fabrication of base glasses for glass-ceramics with low coefficients of thermal expansion, the effect of the P_2O_5 and Al_2O_3 content depending on that of ZrO_2 was evaluated. As a result, the solubility of ZrO_2 in the glass was shown to increase with the increase in the P_2O_5 content. Likewise, it was established that the Al_2O_3 content should not exceed 30 wt% to achieve a low coefficient of thermal expansion of less than $15 \times 10^{-7} K^{-1}(20–500°C)$. As a result, glass-ceramics that are highly thermal shock resistant were developed.

The most surprising effect observed by Wada (1998) during his examination of the catalytic effects of ZrO_2 on the nucleation and crystallization of lithium alumino silicate glasses was the synergistic relationship between ZrO_2 and TiO_2, also described by Stewart (1972). As a result, Wada managed to catalyze the nucleation of spodumene by incorporating 2.5 wt% ZrO_2 and 2 wt% TiO_2 as early as 1962. Neoceram™ N-11 spodumene glass-ceramic, produced by Nippon Electric Glass in Japan, was developed on this basis. The properties and applications of this glass-ceramic are discussed in Section 4.2 and Section 4.5.

2.2.3 SiO_2–Al_2O_2–Na_2O (Nepheline)

Glass-ceramics based on nepheline $(Na,K)AlSiO_4$ have been studied because of their ability to be easily strengthened by these two techniques: 1) the application of surface compression through glazing with glasses of lower coefficient of thermal expansion, and 2) ion exchange treatment involving potassium for sodium exchange.

The first glass-ceramic to be used as tableware was based upon a glazed nepheline-based formulation sold by Corning Glass Works under the brand-name Centura®. This glass-ceramic was optimized from the basic system SiO_2–Al_2O_3–Na_2O–BaO–TiO_2, yielding a crystal phase assemblage of nepheline $(NaAlSiO_4)$, celsian $(BaAl_2Si_2O_8)$, and rutile (TiO_2) (Duke et al., 1968). Some aluminosilicate residual glass was also present. The major reason for the incorporation of baria into this product was to lower the coefficient of thermal expansion from well over 100×10^{-7} to about 95×10^{-7}, thus ensuring the necessary resistance to thermal down-shock required in a tableware product. The barium feldspar, monoclinic celsian, has a coefficient of thermal expansion of nearly $40 \times 10^{-7} K^{-1}$, and could therefore be used to partially compensate for the overly high coefficient of thermal expansion of nepheline, roughly $115 \times 10^{-7} K^{-1}$. With a glass-ceramic body of coefficient of thermal expansion of $95 \times 10^{-7} K^{-1}$, it was possible to develop a durable glaze composition about 30 points lower in coefficient of thermal expansion. This composition, basically a complex calcium borosilicate, could be fired onto the glass-ceramic at a temperature below its point of deformation. A compressive stress resulting in an abraded flexural strength of 240 MPa resulted. This clearly compared favorably with the original strength of the glass-ceramic, some 91 MPa. Table 2-9 gives the oxide composition of both glass–ceramic and glaze, along with some key properties.

The highest flexural strength ever measured in a bulk glass-ceramic was from nepheline compositions that had been subjected to potassium for sodium ion exchange in a high-temperature salt bath (52 wt% KCl, 48% K_2SO_4 at 730°C) (Duke et al., 1967). Specifically, glasses of composition near the nepheline solid-solution range: $K_xNa_{8-x}Al_8Si_8O_{32}$, where x lies between 0 and 4.7, can thermally crystallize to fine-grained glass-ceramics with the aid of a nucleating agent, preferably TiO_2. Certain of these (per-potassic) nepheline glass-ceramics (x between 2 and 4.7) yielded the best strength, consistently above 1450 MPa after K^+ for Na^+ exchange for 8 h. These measurements were made on 6 mm rods after abrasion with 30-grit SiC for 15 min, and hence indicate reliability for articles of significant size.

The sequence of crystallization of a typical nepheline glass-ceramic of composition 42.0 wt% SiO_2, 31.2 wt% Al_2O_3, 12.5 wt% Na_2O, 6.2 wt% K_2O, 7.4 wt% TiO_2, and 0.7 wt% As_2O_3, was examined. Although TiO_2 is required to ensure fine-grained internal nucleation in this glass, the first phase identifiable by XRD is in fact metastable carnegieite, which appears at

Table 2-9

Composition (in wt%) and Properties of Commercial Nepheline Glass-Ceramic and Glaze (Centura®)

Glass-Ceramic		Glaze	
SiO_2	43.3	SiO_2	46.1
Al_2O_3	29.8	Al_2O_3	4.4
Na_2O	14.0	Na_2O	2.7
BaO	5.5	B_2O_3	18.2
TiO_2	6.5	CaO	16.5
As_2O_3	0.9	K_2O	1.0
		PbO	9.1
		ZrO_2	1.0
		CdO	1.0
Heat treatment	(4 h)	820°C	1140°C
Coefficient of thermal expansion	(0–300°C)	96.8×10^{-7} K^{-1}	65×10^{-7} K^{-1}
Modulus of rupture after abrasion	91.1 MPa	240 MPa for 0.8 mm thick glaze	

temperatures near 750°C. Surprisingly, this cubic crystal, a stuffed derivative of cristobalite, appears on the phase equilibria diagrams as a high-temperature phase at the nepheline stoichiometry. Upon heat treatment above 850°C, the carnegieite breaks down to the expected nepheline phase, and coincident with this transformation, the anatase form of TiO_2 also develops. Perhaps the TiO_2 was originally dissolved in solid solution in the cristobalite-structured carnegieite phase, since it is known to be soluble in cristobalite (SiO_2) itself. In any event, the nepheline glass-ceramic is best developed at a temperature of 1140°C, where the flexural strength (abraded) is 58 MPa and the CTE is 122.5×10^{-7} K^{-1}. The density at this point is 2.669, an increase of some 6% from that of the original glass.

After undergoing a K^+ for Na^+ ion exchange treatment for 8 h in the KCl-K_2SO_4 salt bath, the abraded flexural strength was increased from 58 MPa to 1300 MPa, a dramatic improvement of about 25-fold. A chemical analysis of a somewhat similar composition: 40.9 wt% SiO_2, 31.5 wt% Al_2O_3, 11.5 wt% Na_2O, 7.7 wt% K_2O, 7.2 wt% TiO_2, and minor As_2O_3 (not analyzed) was compared with a powdered version of the same material after ion exchange. The results were: 38.0 wt% SiO_2, 30.1 wt% Al_2O_3, 0.3 wt% Na_2O, 23.9 wt% K_2O, and 7.0 wt% TiO_2. Fig. 2-7 shows a secondary electron image of the ion-exchanged layer in this composition from a polished bar. The direction of ion exchange was from right to left. In the right side of the picture is the zinc cladding used to protect the sample edge during polishing. The thin dark layer is the surface of the glass-ceramic and the light-colored band corresponds to kalsilite. The kalsilite layer is about 20 µm thick following a 30 min exchange at 730°C. The dark left side illustrates the interior nepheline glass-ceramic. Evidently, the exchange proceeded in the nepheline until the K-solubility limit was reached, after which kalsilite formed, producing a sharp composition discontinuity, typical in interfacial reactions involving phase transformation (Fig. 2-8) (Thompson 1959).

It is interesting to note the effect of article thickness on strengthening by ion exchange. With thinner rods (1–2 mm), flexural strengths after ion exchange treatment followed by abrasion were measured as high as 2200 MPa. In fact these thin 10 cm rods could be bent more than 90° before breaking. The breaking phenomenon involved total and instantaneous pulverization to a fine powder with considerable and violent release of stored energy.

Strong microwave absorption is another unusual characteristic of some nepheline glass-ceramics. It is particularly obvious in pure soda nepheline-containing materials such as the commercial tableware composition listed previously. In fact, short periods (less than one minute) in a domestic microwave

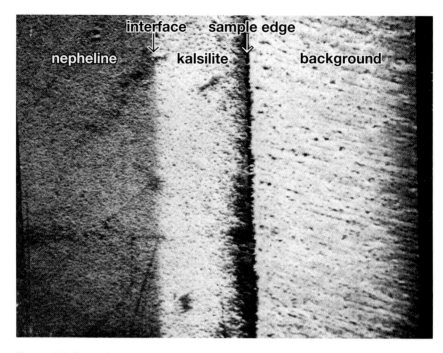

Figure 2-7 Secondary electron image of K^+/Na^+ exchanged glass-ceramic showing nepheline–kalsilite interface.

oven can raise the temperature of this glass-ceramic over 100°C. This was a key reason for the replacement of this tableware material by the potassium fluorrichterite glass-ceramic described in Section 2.3.4.

The reason for this microwave activity in soda nepheline is probably related to the crystal structure (see Appendix Fig. 15). Nepheline is a stuffed derivative of tridymite, the hexagonal form of SiO_2. There are vacant channels in tridymite, but in nepheline, where one-half of the network Si^{4+} is replaced by Al^{3+}, Na^+ ions are located in these vacancies in order to restore charge balance. These sodium ions are sufficiently small to oscillate in the channels along the c-axis of nepheline. When some of these holes are occupied by K ions, as in the case of natural nephelines where typically about one-quarter of the alkali positions contain K^+ ions, the thermal effects due to microwave absorption are reduced significantly. These larger ions apparently interfere with the capability of the Na^+ ions to freely oscillate and absorb energy in the microwave regime.

2.2.4 SiO_2–Al_2O_3–Cs_2O (Pollucite)

Following the discussion of lithium aluminosilicate and sodium aluminosilicate glass-ceramics, the presentation of glass-ceramics in the

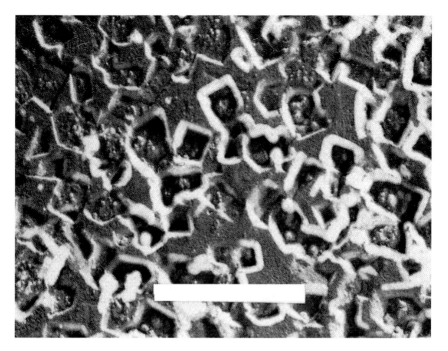

Figure 2-8 TEM/replica of nepheline–kalsilite interface showing growth of kalsilite around edge of each nepheline crystal. Direction of K⁺→Na⁺ is from lower right to upper left. (bar = 1 μm).

SiO_2–Al_2O_3–Cs_2O system may seem somewhat unusual. Intitially, it may seem unclear of what significance the very rare element cesium is to the fabrication of glass-ceramics. It must be noted, however, that the mineral pollucite, $CsAlSi_2O_6$, demonstrates a very high melting point above 1900°C. As a result, it exhibits better refractory qualities than any other silicate. Furthermore, pollucite is isostructurally related to the mineral leucite, $KAlSi_2O_6$, (see Section 1.3.1). On the basis of this structural similarity, pollucite, like leucite, might be expected to have a very high coefficient of thermal expansion of more than 200×10^{-7} K^{-1} (0°C–1000°C). However, pollucite, unlike leucite, does not invert upon cooling from cubic to tetragonal form, and therefore does not produce the large volume change.

An in-depth study of glass formation in the SiO_2–Al_2O_3–Cs_2O system and the development of new methods of controlling crystallization for the production of glass-ceramics was conducted by Beall and Rittler (1982). According to Fig. 2-9, the preferred composition ranges are in the triangle formed by mullite, pollucite, and SiO_2. It must be noted, however, that the formation of the base glass in the range of the stoichiometric composition of pollucite is not preferred because of the high temperatures above 1900°C

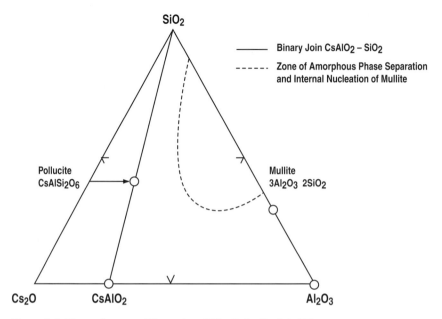

Figure 2-9 Phase diagram of the system SiO_2–Al_2O_3–Cs_2O (wt%).

that are required. Therefore, the following composition was considered to be favorable for manufacturing glass-ceramics:

25–70 wt% SiO_2, 20–50 Al_2O_3, 10–35 Cs_2O

To control the crystallization in glasses of this composition range, Beall and Rittler (1982) first had to examine the microstructure of the glasses. Furthermore, they had to analyze the crystallization tendency of the glass to determine whether volume or surface crystallization takes place. In the process, two types of glass were determined. The first type demonstrated spontaneous phase separation. This presented the possibility of self-nucleation by volume crystallization without the addition of a nucleating agent to the base glass composition. The second type of glass tended to demonstrate surface crystallization similar to glass-ceramics in the leucite system. Therefore, ZrO_2 was added as a heterogeneous nucleating agent to the base glass system to produce volume crystallization in the monolithic glass. The compositions selected for the two types of glass-ceramics are shown in Table 2-10. These base glasses were melted at temperatures ranging between 1800°C and 1900°C and the sequence of phase formation in the development of the associated glass-ceramics was examined.

Table 2-10

Composition A and B of Pollucite Glass-Ceramic (wt%)		
	A	B
SiO_2	35.0	29.6
Al_2O_3	40.0	29.6
Cs_2O	25.0	20.8
ZrO_2	–	20.0

Given its composition, Glass A is located behind the dashed line in the right part of the phase diagram shown in Fig. 2-9. This part of the phase diagram is poor in Cs_2O. Glass B is located on the dotted line in the Cs_2O-rich part of the diagram. Glass A demonstrates phase separation and permits self-nucleation. Glass B represents a transition to the other mechanism.

In Glass A, mullite develops as the primary crystal phase by self-nucleation when the glass is heat treated at temperatures above 920°C. This familiar reaction was also used by MacDowell and Beall (1969) to control crystallization in binary Al_2O_3–SiO_2 glasses. Additional heat treatment at 1000°C produces pollucite. Both phases are main crystal phases of the glass-ceramic. Further heat treatment at 1600°C forms the unusual coast-and-islands microstructure shown in Fig. 2-10. After etching the glass-ceramic with dilute hydroflouric acid, the mullite demonstrates higher chemical stability than the pollucite

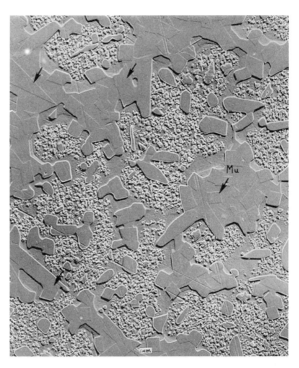

Figure 2-10 Microstructure of a coast-and-islands pollucite–mullite glass-ceramic A (Beall and Rittler, 1982). Bar = 1μm.

in SEM images. In Fig. 2-10, the chemically more-stable mullite within the islands of mullite and pollucite is marked with arrows. Pollucite exhibits some euhedral faces on the islands composed of both crystals.

In its basic ternary composition, Glass B is outside the field of phase separation. Therefore, ZrO_2 must be added to control crystallization. By adding ZrO_2, phase separation of droplets measuring approximately 250 nm occurs. Following heat treatment of the glass at 800°C, a primary crystal phase of tetragonal zirconia was produced. The crystallization of zirconia increased after 4 h of heat treatment at 1000°C. Furthermore, pollucite formed at 1200°C. The typical microstructure of a pollucite–mullite glass-ceramic, which also contains ZrO_2 crystals, developed after 4 h of treatment at 1400°C.

Experiments were conducted by adding various components to Glass B. In the process, it was discovered that the addition of La_2O_3, CeO_2, and P_2O_5 produced the simultaneous crystallization of the mineral monazite, $(La, Ce)PO_4$ (see structure in Appendix Fig. 20) in the glass-ceramic.

The most important property of the pollucite glass-ceramic is its refractoriness. It was determined that the viscosity–temperature function of this glass-ceramic is approximately 300°C higher than that of silica glass. The modulus of rupture of 15,000–20,000 psi (103–138 MPa) and the particularly high chemical stability are also favorable properties of this glass-ceramic. The chemical stability renders the material more resistant to autoclaving (at 300°C with water steam) than Pyrex® glass, for example. Pyrex® is a common borosilicate glass, which is used to manufacture equipment and installations, particularly in the chemical industry.

Because of these special properties, potential applications of the glass-ceramics are as refractories (thermal stablity at 1600°C), corrosion-resistant material for manufacturing equipment and installations, and as a pore-free material for radwaste containment. Electromelting is the preferred method for producing the base glasses.

2.2.5 SiO_2–Al_2O_3–MgO (Cordierite, Enstatite)

2.2.5.1 Cordierite Glass-Ceramics

Glass-ceramics with a primary crystal phase of cordierite, $Mg_2Al_4Si_5O_{18}$, are of great commercial importance. The crystal structure is shown in Appendix Fig. 11. The first glass-ceramics of this type were developed at the Corning Glass Works (Stookey 1959, Beall 1992). Subsequently, further glass-ceramics of this basic type were developed. The most important types in technical terms are presented in this section. Cordierite glass-ceramics are distinguished for their special properties such as high mechanical strength, excellent dielectric properties, good thermal stability, and thermal shock resistance.

The composition of cordierite glass-ceramics of the Corning 9606®-type is presented in Table 2-11. A comparison with the ternary phase diagram in Fig. 2-11 shows that the composition of the three basic components, SiO_2, Al_2O_3, MgO, of the bulk glass does not correspond to the stoichiometric composition of cordierite. This composition was selected to optimize the viscosity of the glass and to achieve advantageous processing properties. If the thermal stability and fatigue resistance of the glass-ceramic material had been the decisive criteria, however, a stoichiometric composition would have been selected.

A) Multicomponent System

In the multicomponent base glass system (Table 2-11), TiO_2 with a content of approximately 9 wt% acts as the heterogeneous nucleating agent. Magnesium dititanate, $MgTi_2O_5$, develops as the primary crystal phase during the heat treatment of the base glass (composition according to Table 2-11). During the crystallization process, cordierite develops as the primary crystal phase in hexagonal form. This phase is responsible for the properties mentioned above. Beall (1992) discovered that this type of cordierite, demonstrating a specific solid solution, grows toward magnesium beryl. That is, cordierite and a solid solution of cordierite, which are chemically modified toward beryl, are present. Consequently, there is an ion substitution of Mg^{2+} + Si^{4+} replacing 2 Al^{3+}. Apart from cordierite, as the primary crystal phase, and its solid solution, further crystals such as cristobalite and rutile

Table 2-11

	Composition of Glass-Ceramics (wt%)		
	Corning 9606® cordierite glass-ceramic	Corning E1 enstatite glass-ceramic	Corning E2 enstatite glass-ceramic
SiO_2	56.1	58.0	54.0
Al_2O_3	19.8	5.4	–
MgO	14.7	25.0	33.0
CaO	0.1	–	–
TiO_2	8.9	–	–
As_2O_3	0.3	–	–
Fe_2O_3	0.1	–	–
Li_2O	–	0.9	–
ZrO_2	–	10.7	13.0

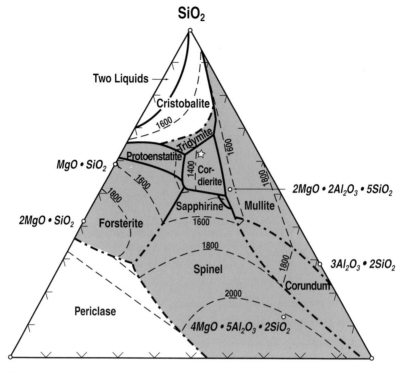

Figure 2-11 SiO_2–Al_2O_3–MgO phase diagram (Levin et al., 1964) indicating the base-glass composition of Corning 9606® cordierite glass-ceramic (☆). (wt%, t in °C).

occur. A small amount of residual glass matrix lies between the crystal phases. Phase formation dependent on the temperature at which the base glass is heat treated is shown in Table 2-12.

Table 2-12

Phase Assemblages during Crystallization of Corning 9606® Glass-Ceramic

Temperature, °C	Phase
700	glass
800	glass, $MgTi_2O_5$
900	$MgTi_2O_5$, β-quartz solid solution
1010	$MgTi_2O_5$, β-quartz, sapphirine, enstatite, rutile
1260	$MgTi_2O_5$, cordierite solid solution, rutile

A glass-ceramic material was produced with the formation of the primary crystal phase of cordierite. This glass-ceramic material is distinguished for its high fracture toughness of approximately 2.2 MPa·m$^{0.5}$, a high degree of hardness (Knoop hardness 700), and thermal conductivity of 37.7 W/(m·K). The linear thermal expansion coefficient is 45 × 10^{-7} K^{-1}.

B) Composites Using the Stoichiometric Composition of Cordierite

In contrast to the glass-ceramics mentioned in Section 2.2.5.1A, Semar and Pannhorst, et al. (1989) developed powder-pressed glass-ceramics by using a chemical composition that is similar to that of stoichiometric cordierite. In addition to 50 wt% of the cordierite glass-ceramic material, this composite also contained 50 wt% powdered ZrO_2 (3Y-TZP). In studying the process of rate-controlled sintering of the composite, it was possible to produce almost pore-free and homogeneous products by using a ZrO_2 grain size of approximately 1 μm. It was also possible to avoid the formation of $ZrSiO_4$ as a secondary crystal phase in the composite material by applying rate-controlled sintering. Moreover, it was possible to reduce the sintering temperature by approximately 50 K to approximately 1300°C by using this special type of processing.

The nucleation and crystallization process of this cordierite glass-ceramic material was the subject of an international research project of the Technical Committee 7 of ICG (International Commission on Glass) conducted by Pannhorst (2000), Müller et al. (1992), Mora et al. (1992), Szabo et al. (1992), Heide et al. (1992), and Höland (1995a). The results of this project in addition to those of Winter et al. (1995), and Schmelzer et al. (1995) demonstrated that the crystallization of a powdered base glass with the composition of $Mg_2Al_4Si_{11}O_{30}$ takes place to form a β-quartz solid solution as the primary crystal phase, according to the mechanism of surface nucleation and crystallization (see Sections 1.4.4.2 and 1.5.1.3). Nucleation takes place at the surface of the base-glass grains. Furthermore, special structural strains prefer the process of surface nucleation.

The growth of β-quartz solid solution is also preferred if the crystals impinge on each other. This process takes place at different temperatures, e.g., at 1150°C/2 min of heat treatment. During heat treatment at 1150°C for 60 min the conversion of the β-quartz solid solution into cordierite was observed by TEM (Winter et al., 1995). Cordierite grows dendritically into the cores of the β-quartz solid solution. The final glass produced with controlled crystallization features α-cordierite as the primary crystal phase.

It must be noted, however, that other crystal phases apart from the β-quartz solid solution are also involved in the crystal phase formation of primary crystal phases up to α-cordierite. These crystal phases render the solid-state reaction highly complex. Hence, several unknown crystal phases with flat, almost two-dimensional habits were observed and could not be identified as Mg-petalite or osumilite as shown by Schreyer and Schairer (1961).

According to Semar and Pannhorst (1991), this composite composed of cordierite glass-ceramics and 50 vol% powdered ZrO_2 demonstrated particularly high bending strength of approximately 350 MPa and fracture toughness of approximately 2–4 MPa·m$^{0.5}$. The thermal conductivity was 3.4–3.8 W/(m·K) and the average roughness of a polished sample was 0.1–0.2 μm.

2.2.5.2 Enstatite Glass-Ceramics

Beall (1991) successfully developed glass-ceramics featuring an enstatite $MgSiO_3$ (see the crystal structure in Appendix Fig. 7) primary crystal phase in the ternary SiO_2–Al_2O_3–MgO system with small amounts of Al_2O_3. This crystal phase demonstrated very interesting properties during cooling of the glass-ceramics. It also exhibited martensitic transformation, which contributed to the toughening of the final product. The stoichiometric composition of enstatite, however, did not permit the formation of a stable base glass. Therefore, it was necessary to develop special glass compositions that would allow controlled crystallization into enstatite. In a phase-separated base glass, tetragonal zirconia was formed as the primary crystal phase at temperatures between 800° and 900°C. Tetragonal zirconia subsequently nucleated protoenstatite in glasses with the compositions of E1 and E2 (Table 2-11) in heat treatments above 900°C. During slow cooling, e.g., 50 K/h, protoenstatite was transformed into fine crystals in the form of twinned clinoenstatite rather than orthoenstatite.

Since the proto-to-clino inversion was accompanied by a 4% volume shrinkage, toughening was not based on the metastable presence of a protoenstatite form. The toughening mechanism involved crack deflection and slippage along twinned crystals. Beall (1991) showed that splintering due to the intersection of cleavage (110) and twin (100) planes in clinoenstatite was a factor of toughening. The final enstatite glass-ceramic material was characterized by the following outstanding properties:

- Fracture toughness of approximately 5 MPa·m$^{0.5}$ was determined for glass-ceramic E2.
- The high content of zirconia was responsible for the excellent refractoriness of the glass-ceramics, that is, 1250°C for E1 and 1500°C for E2.

- In glass-ceramic E1, a useful secondary crystal phase of β-spodumene reduced the linear thermal expansion coefficient to approximately $68 \times 10^{-7} \text{ K}^{-1}$.

The microstructure of enstatite β-spodumene glass-ceramics (E2, Table 2-11) is shown in Fig. 2-12. The product was cerammed at 800°C/2h and 1200°C/4h. Fine polysynthetic twinning was observed in the enstatite grains parallel to the [100] axis. The β-spodumene crystals demonstrated conchoidal fracture surfaces (Echeverría and Beall, 1991). Echeverría and Beall also showed that the secondary phase could be changed. The addition of BaO produced either celsian, $BaAl_2Si_2O_8$, or Ba osumilite, $BaMg_2Al_6Si_9O_{30}$, depending on the Ba/Al atomic ratio.

Three different mechanisms and types of processing were used by Echeverría and Beall (1991) to develop enstatite-type glass-ceramics:

- bulk nucleation and crystallization of monolithic base glasses by using nucleating substances (see Table 2-11)
- surface nucleation mechanism with devitrified glass frits, which did not contain nucleating substances
- sol–gel route for developing ceramic specimens consisting of 100% enstatite with superior toughness of approximately 4 $MPa·m^{0.5}$

Another type of enstatite glass-ceramic material was developed by Partridge and Budd (1986) and Partridge et al. (1989). The base glasses had the following compositions: 30–50 wt% SiO_2, 10–40 wt% Al_2O_3, 10–30 wt% MgO, and 8–13 wt% ZrO_2.

After the heat treatment, the glass-ceramics contained enstatite and tetragonal zirconia as the primary crystal phases. Glass-ceramics with a composition of 43.7 wt% SiO_2, 22.2 wt% Al_2O_3, 22.3 wt% MgO, 11.8 wt% ZrO_2 crystallized according to the mechanism of surface crystallization The surface crystallization was so dominant that the final glass-ceramic material demonstated very high stress build-up in the surface area

Figure 2-12 Microstructure of enstatite-β-spodumene glass-ceramics showing lamellar twinning in enstatite.

due to difference in thermal expansion coefficient between cordierite at the surface and enstatite plus glass in the volume. Therefore, Partridge et al. (1989) determined a bending strength of 600–750 MPa. Since the surface stress caused the glass-ceramics to crack easily, they had to be handled very carefully. Hence, the main objective of Partridge et al. (1989) and Budd (1993) was to develop less-sensitive enstatite glass-ceramics. Enstatite-type glass-ceramics having ZrO_2 or TiO_2 as the nucleating substance were prepared from glass powders by using the surface nucleation and crystallization mechanisms. The following property values were registered for this processed powdered glass-ceramic (type NK2/3833 GEC Alsthom, Stafford, England):

Flexural strength: 250 MPa, Weibull modulus: 11.2, flexural modulus: 160 GPa, fracture toughness: 3.3 MPa·m$^{0.5}$, and fracture surface energy: 51 J/m^2

This type of glass-ceramic material is used to make circuits in the electronics industry.

2.2.6 SiO_2–Al_2O_3–CaO (Wollastonite)

The base glasses of the system discussed in this section are particularly suitable for producing glass-ceramics according to the mechanism of controlled surface crystallization. Conventional volume nucleation agents, such as TiO_2 and ZrO_2, are used in the process. Glass-ceramics with a wollastonite main crystal phase have been made from these base glasses according to the above mechanism. Glass-ceramics of this type demonstrate special optical effects and other favorable properties (see Section 4.6). These materials are produced on a large scale and used as cladding in the building industry.

Wada and Ninomiya (1995), Tashiro (1985), and Kawamura et al. (1974) showed that a base glass of the following composition must first be melted in a tank furnace at 1500°C to produce the wollastonite glass-ceramic: 59 wt% SiO_2, 7 wt% Al_2O_3, 17 wt% CaO, 6.5 wt% ZnO, 4 wt% BaO, 1 wt% B_2O_3, 3 wt% Na_2O, 2 wt% K_2O, 0.5 wt% Sb_2O_3.

Subsequently, this glass is poured into water to produce a frit. The glass grains required for controlling surface crystallization are thus produced. In a fully automated process, grains measuring 1–7 mm are used to fabricate glass-ceramic products for the building industry, such as sheets measuring $1000 \times 1000 \times 45$ mm^3. The glass grains are sintered to a dense monolithic glass by heat treatment. At temperatures above approximately 950°C, controlled surface crystallization of β-wollastonite, $CaO·SiO_2$ (see the crystal structure in Appendix Fig. 8), begins at the boundary of the former glass grains. At 1000°C, wollastonite grows in needlelike form from the surface of the glass toward the interior of the glass grain, without crossing the boundary to the neighboring glass particles (Fig. 2-13).

At heat treatment of 1100°C for 2 h, controlled surface crystallization of β-wollastonite for the production of finished glass-ceramic sheets ends. During this heat treatment phase, however, the crystallized boundaries join together to form large crystal needles. The glass-ceramic product demonstrates needle-like β-wollastonite with a crystallite length of 1–3 mm.

Figure 2-13 SEM image showing the microstructure of wollastonite glass-ceramic. (Courtesy M. Wada).

The content of large crystals is 40 wt% of the bulk while small crystals in the glassy matrix are a much smaller percentage. The boundaries of the base glass grains are virtually indiscernible in the glass-ceramic end product.

The optical characteristics of this glass-ceramic are produced by the formation of needle-shaped β-wollastonite crystals embedded in the glass matrix. The different light diffraction indices of β-wollastonite crystals and of the glassy matrix make the material look like marble or granite. If the β-wollastonite glass-ceramics are heated to 1200°C, rather than 1000°C, however, β-wollastonite of a granular crystallographic morphology is produced. Consequently, the particular appearance of the glass-ceramic provided by the β-wollastonite crystals is more opaque.

The most important properties of glass-ceramics with a β-wollastonite main crystal phase are shown in Table 2-13. The properties required by the building industry are shown in Section 4.6. This type of glass-ceramic is manufactured and sold by Nippon Electric Glass Co., Ltd, under the name of Neoparies®.

Table 2-13

Main Properties of Wollastonite Glass-Ceramic

	β-Wollastonite Glass-Ceramic
Expansion coefficient (30–380°C, × 10^{-7} K^{-1})	62.0
Density (g/cm³)	2.72
Modulus of rupture (MPa)	440
Hardness	
Mohs	6.0
Vickers (100g)	600
Acid durability*	
1% HCl	0.089
1% H_2SO_4	0.080

*Acid and alkali durability were determined by the weight losses of samples of 15x15x10mm after immersing in the solutions at room temperature for 650 h.

Maeda et al. (1992) reported on a bulk nucleation and crystallization mechanism of β-wollastonite in a glass of composition 55.5 wt% SiO_2, 11 wt% Al_2O_3, 3.8 wt% B_2O_3, 2.2 wt% MgO, 15.3 wt% CaO, 8.8 wt% Na_2O, 3.0 wt% K_2O, 0.4 wt% Sb_2O_3 by adding metallic species such as Ru-, Rh-, Pd-, Ir-, Pt-, Au-chlorides, or Ag-nitrate to the base glass. Sb_2O_3 and SnO were used as reducing agents. Thus, metal colloids were formed as heterogeneous nucleating particles following the equations:

$$Pt^{4+} + 2Sb^{3+} \rightarrow Pt^0 + 2Sb^{5+}$$
$$Pt^{4+} + 2Sn^{2+} \rightarrow Pt^0 + 2Sn^{4+}$$

A preferred content of platinum was characterized as 40 ppm. A β-wollastonite glass-ceramic of this type is produced by the Asahi Glass Company Ltd., Japan.

2.2.7 SiO_2–Al_2O_3–ZnO–MgO (Spinel, Gahnite)

2.2.7.1 Spinel Glass-Ceramic Without β-Quartz

The most important and economically most significant developments of glass-ceramics from the SiO_2–Al_2O_3–ZnO-MgO system were achieved by Pinckney and Beall (1997) and Beall and Pinckney (1999). The objective of their research involving the SiO_2–Al_2O_3–ZnO–MgO system was to develop a fine-grained glass-ceramic (Pinckney and Beall, 1997) and an ultrafine-grained glass-ceramic that is also transparent in the visible spectrum (Beall and Pinckney, 1999). In contrast with the development of β-quartz and β-spodumene glass-ceramics, therefore, Li_2O was not used. The composition range of spinel glass-ceramics is 40–50 wt% SiO_2, 22–30 wt% Al_2O_3, 7–18 wt% MgO, 0–12 wt% ZnO, and 7–13 wt% TiO_2 where the glass-ceramic can contain a number of additions (Beall and Pinckney, 1995). The chemical additions in question may serve to pigment the material, for example. The chemical composition of the glass-ceramic must meet a secondary requirement. That is, the sum of MgO and ZnO oxides must be at least 13 wt%. As a result, the formation of spinel as the main crystal phase is assumed. Spinel exhibits the general structural formula AB_2O_4. According to the above composition, spinel in glass-ceramics can range from traditional spinel of the $MgAl_2O_4$ type to gahnite, $ZnAl_2O_4$ (see the spinel structure in Appendix Fig. 17).

The preferred nucleation of Zn(II)-containing spinel crystals was achieved by Pinckney (1987) using ZrO_2 and/or TiO_2. A typical spinel glass-ceramic formed by the heterogeneous nucleation of ZrO_2 exhibits the following composition: 64.8 wt% SiO_2, 18.5 wt% Al_2O_3, 4.6 wt% ZnO, 4.6 wt% MgO, 7.5 wt% ZrO_2.

Primary crystals of the ZrO_2 type grow during heat treatment of base glasses with this composition. Prior to the formation of these ZrO_2 crystals, no glass-in-glass phase separation processes are observed. Thereafter, crystallization of spinel solid solution occurs by a heterogeneous solid-state reaction on the heterogeneous basis of ZrO_2 crystals at 1060°C. The microstructure of this type of spinel glass-ceramic, which is formed by heat treating the above base glass at 900°C for 6 h and 1060°C for 6 h, is shown in Fig. 2-14. The crystals are considerably smaller than 0.1 μm.

The nucleation process that occurs in the presence of TiO_2 is different from the heterogeneous process observed with ZrO_2 as the nucleating agent. TiO_2 favorably influences the nucleation of various crystal phases in the

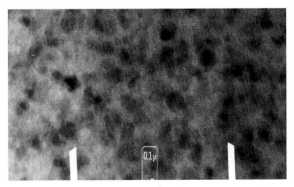

Figure 2-14 TEM image showing the microstructure of spinel solid solution glass-ceramic.

SiO_2–Al_2O_3–MgO system by fine-scale glass-in-glass phase separation. Beall and Pinckney (1999) proved that nucleation of TiO_2 by influencing phase-separation processes is also very effective for crystallizing spinel. At this stage of nucleation, the droplet glass phase is titania/alumina-rich; the glass matrix is silica-rich. The crystallization of spinel takes place in the droplet phase following phase separation.

This process of fine-scale phase formation is very interesting and can be compared to other glass-ceramic systems. Thus, in respect to a nanoscaled phase separation and additional nanophase crystallization, this reaction was also observed in the formation of SiO_2–Al_2O_3–MgO–TiO_2 glass-ceramics with quartz crystal phases (Höland et al., 1991a, 1991b).

The microstructure of spinel glass-ceramics provides special properties. Given the small dimensions of the crystals and their fine distribution in the glass matrix, the glass-ceramics are transparent in the visible spectrum and demonstrate minimal scattering. Since highly mobile ions collect in the spinel crystals, and for the most part become incorporated into the spinel structure with only additional oxides being the possible exception, a glass matrix that is very rich in SiO_2 develops. As a result, the chemical durability of the glass-ceramics is excellent. At the same time, the temperature resistance is increased. A strain point above 900°C has been recorded for the ZrO_2/TiO_2-nucleated glass-ceramic. Moreover, since the coefficient of linear thermal expansion of the spinel glass-ceramic is close to that of elementary silicon, the material can be used in the electronics industry for flat panel displays and a variety of photovoltaic substrates. Furthermore, rigid information discs for magnetic memory storage devices can also be produced with this glass-ceramic (Pinckney 1999).

2.2.7.2 β-Quartz-Spinel Glass-Ceramics

First, however, it must be noted that in glass-ceramics of the SiO_2–Al_2O_3–Li_2O system, a main crystal phase of β-quartz solid solution is formed by the addition of various components, ZnO among them (Beall et al., 1967). In this way, glass-ceramics with particularly low coefficients of thermal expansion can be produced (see Section 2.2.2). In a glass-ceramic of this system, Beall and Pinckney (1992) developed a β-quartz solid solution glass-ceramic that demonstrated controllable optical properties. The glass-ceramic was composed of 64–70 wt% SiO_2, 18–22 wt% Al_2O_3, 3.3–4 wt% Li_2O, 1.5–3.5 wt% ZrO_2, 0.5–2.5 wt% TiO_2, 0.5–1.5 wt% Al_2O_3. The ZnO content of this glass-ceramic was 2–5 wt%. Since the Al_2O_3 content was relatively high, gahnite, $ZnO \cdot Al_2O_3$, was formed as a secondary phase in the glass-ceramic.

Glass-ceramics in which gahnite, $ZnO \cdot Al_2O_3$, and β-quartz solid solution constitute the main crystal phase were introduced by Strnad (1986). The typical range of the chemical composition of β-quartz solid solution–gahnite glass-ceramics is shown below:

41 wt% SiO_2, 14–24 wt% Al_2O_3, 16–40 wt% ZnO, 0–5 wt% ZrO_2, 0–5 wt% TiO_2, 0–2 wt% P_2O_5, 0–0.01 wt% Pt, 0–0.5 wt% As_2O_3

By adding a nucleating agent such as ZrO_2, TiO_2, P_2O_5, or Pt to the material, β-quartz solid solution and/or gahnite ($ZnO \cdot Al_2O_3$), or willemite ($2ZnO \cdot SiO_2$) crystals containing Zn(II) were produced in the glass-ceramics. Volume crystallization was achieved in all cases using either metals or oxides as nucleating agents. Interesting differences in the phase formation processes were noted using ZrO_2 or TiO_2 as the nucleating agent.

When ZrO_2 was used as the nucleating agent, ZrO_2 (tetragonal modification) developed as the primary crystal phase in the phase-separated base glass, followed by β-quartz solid solution. Finally, gahnite, $ZnO \cdot Al_2O_3$, formed at elevated temperatures between 950°C and 1000°C.

When TiO_2 was used as the nucleating agent, gahnite, $ZnO \cdot Al_2O_3$, formed as the primary crystal phase. The crystallites ranged between 0.01 and 0.1 μm in size. This glass-ceramic containing gahnite was transparent in appearance.

The formation of β-quartz solid solution and gahnite crystals in glass-ceramics from the SiO_2–Al_2O_3–ZnO allowed specific properties, such as translucency and even transparency, to be combined with variable coefficients of thermal expansion of

−5.0 up to +31.0 × 10^{-7} K^{-1} (20°–500°C)

2.2.8 SiO_2–Al_2O_3–CaO (Slag Sital)

Pavluskin (1986) reported that slag sitals as crystallized glasses were produced from industrial slag, for example, from the steel and copper industries as well as from other metal industries, in the former USSR. Moreover, Pavluskin (1986), Hinz (1970), and Strnad (1986) reported that as early as the 1960s, the former USSR and Hungary produced many tons of slag sital a day. Glass melts that were converted into crystallizing glasses in a two-step heat treatment were produced by using inexpensive slag from the furnaces of the metal industry together with other additives.

This process can also proceed according to the mechanism of controlled crystallization, depending on the additives. The resulting products are called glass-ceramics, even though controlling of the main crystal phase is highly complex and controlling of the crystallization of all the phases, including the secondary crystal phases, is not possible because of the numerous secondary components and impurities. The bulk glasses for these slag sitals were produced according to the following method:

A raw material composed of 50–60 wt% slag and 20–40 wt% sand with additives of up to 11 wt% Al_2O_3, 4–6 wt% Na_2SO_4, 1–3 wt% carbon, and 0.5–10 wt% nucleating agents was produced. Predominantly sulfides, fluorides, Cr_2O_3, TiO_2, and P_2O_5 were used as nucleating agents. The compositions of Russian slag, depending on the procedure and the region in which it was produced, and the composition of the base glass are shown in Table 2-14. The phase diagram of the SiO_2–Al_2O_3–CaO base system is shown in Fig. 2-15.

On the basis of the compositions shown in Table 2-14, however, it must be noted that although the SiO_2–Al_2O_3–CaO glass system represents the base system, the amounts of the different additives such as P_2O_5, Fe_2O_3, or MgO are considerable. The large number of main components clearly shows

Table 2-14

Composition of Russian Furnace Slag and Slag Sitals (Pavluskin 1986)

	Russian furnace slag (wt%)	*slag sitals in wt%*
SiO_2	33–40	49–63
Al_2O_3	5–16	5.4–10.7
CaO	30–48	22.9–29.6
MgO	1–7	1.3–12
Fe_2O_3	0.1–5	0.1–10
MnO	0.5–3	1–3.5
Na_2O	–	2.6–5
Cr_2O_3	–	0.1–3
FeS, MnS	–	1.5–5
ZnS	–	2.2–4.5
F	–	1.6–2.5
TiO_2	–	3–6
P_2O_5	–	5–10

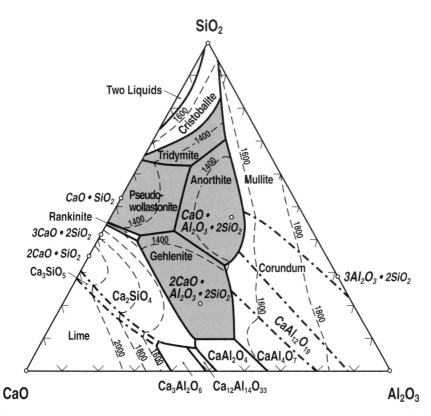

Figure 2-15 SiO_2–Al_2O_3–CaO phase diagram according to Levin et al. (1964). (wt%, T in °C).

that the solid-state reactions during the crystallization of the glasses are highly complex and characterized by numerous parallel and secondary reactions.

The phase separation process of slag sital is characterized by the fact that the base glasses demonstrate substantial liquid–liquid phase separation as a result of the high P_2O_5 content. This type of phase separation in the base glasses, which was determined by Pavluskin (1986) with scanning electron microscopy, is more strongly influenced by the phosphate additives than by the sulfides. Hence, the phospates are of particular importance for the entire nucleation procedure and for the formation of crystals.

Apart from the phase separation processes in base glasses, Pavluskin (1986) also attributes a great deal of importance to the process of heteroeogeneous nucleation (with sulfides). He states that the reactions of sulfides as nucleating agents are independent of the reactions of other nucleating agents such as TiO_2 and Cr_2O_3. Moreover, carbon does not have a nucleating influence. It functions only as a reduction agent.

According to Pavluskin (1986) the color of slag sital can be controlled with additives. By adding MnS + FeS slag sital acquires a grayish black color. If ZnO is added to the mixture, however, white ZnS is produced. Consequently, the slag sital also appears white in color. The formation of ZnS is explained with the following exchange reaction:

$$FeS + ZnO \rightarrow ZnS + FeO$$
$$MnS + ZnO \rightarrow ZnS + MnO$$

Additional colors can be achieved if other color components are added to the glass, for example, elements of 3 d-oxides.

The nucleation and crystallization of the bulk glasses were induced in heat treatments, generally conducted in a two-step procedure in the range of 800°–1000°C. Pavluskin (1986) showed that the process involved is one of controlled crystallization in a phase-separated glass. In his report, he wrote that wollastonite and anorthite developed as the main crystal phases. The secondary crystal phases developed as diopside, pyroxene, or gehlenite. Depending on the composition, however, the secondary crystal phases may represent the main crystal phases and vice versa.

Given the inexpensive manufacturing costs in the former USSR, the large amounts of slag produced by the metals industry (furnace slag), and the resulting properties of slag sital, this material was predominantly used in the construction industry for the facing of buildings. These products were manufactured by rolling the glass into sheets following the melting procedure. These sheets were subsequently converted into slag sital by heat treating.

Slag sitals are characterized by their good mechanical stability, hardness, and abrasion resistance (see Table 2-15). They can be processed with diamond tools. That is, they can be given a final shape, if necessary, after the rolling procedure.

Slag sital is also used in the construction of roads. Given the excellent chemical stability of slag sital, the material is also used in the construction of chemical apparatus, for example, as a foundation for absorption apparatus chambers or as a protection against corrosion in large metal apparatus.

Research activities are also carried out in different countries, such as Japan, Italy, Germany, and Switzerland to use industrial waste, slag ash, or fly ash as raw material for glass-ceramic production. Pelino et al. (1994) reported on the development of glass-ceramics obtained by recycling goethite industrial waste. The goethite waste contains high quantities of metals (Pb, Zn, Cu, Ni) from the Italian zinc production. Pelino et al. (1994) investigated the crystallization of iron silicate glasses and determined pyroxene crystals.

Table 2-15

Properties of Slag Sital (Pavluskin 1986)	
Coefficient of linear thermal expansion	$65–85 \times 10^{-7}$ K^{-1}
Softening point	950°C
Thermal shock resistance	200–300°C
Bending strength	90–130 MPa
Compressive strength	700–900 MPa
Young's modulus	93 GPa
Thermal conductivity	1.16–1.3 W·m^{-1}·K^{-1}
Chemical durability	
Compared with 96% H_2SO_4	99.8%
Compared with 20% HCl	98–99.8%
Compared with 35% NaOH	74.7–90%

Susuki et al. (1997) developed glass-ceramics from the SiO_2–Al_2O_3–CaO system using sewage slag ash as raw material. The nucleation of anorthite was generated by small amounts of FeS. The final anorthite-type glass-ceramic was characterized by a high mechanical strength and a good chemical durability.

A candidate for an application in industrial construction was developed by Boccaccini et al. (1997). They developed glass-ceramics from the SiO_2–Al_2O_3–CaO system. Fly ash was used as a special raw material. The investigation of the crystallization behavior showed the possibility of precipitating pyroxene-type phases.

Blume and Drummond III (2000) investigated synthetic basaltic glass-ceramics under high and low fugacity. Feldspar, spinel, and metasilicate phases represented the equilibrium main crystal phases under high oxygen fugacity, but feldspar was kinetically inhibited.

2.2.9 SiO_2–Al_2O_3–K_2O (Leucite)

This chemical system is of technical significance for sintered ceramic materials and traditional porcelains and also for glass-ceramics. As shown by the phase diagram by Schairer and Bowen (Kingery et al., 1976) in Fig. 2-16, potash feldspar ($K_2O·Al_2O_3·6SiO_2$), leucite ($K_2O·Al_2O_3·4SiO_2$) (see the structure in Appendix Fig. 14), mullite ($3Al_2O_3·2SiO_2$), potassium silicates, and SiO_2 modifications are important crystal phases of this system. Traditional porcelains are derived from the SiO_2–Al_2O_3–K_2O system. The formation of the crystal phases mentioned as well as the possibility of producing eutectoid melts determine the heat treatment of the raw-material mixture.

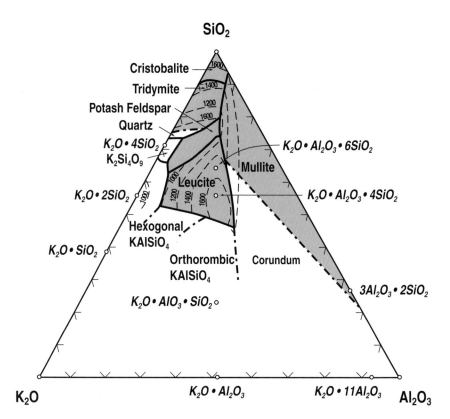

Figure 2-16 Ternary SiO₂–Al₂O₃–K₂O system. (Levin et al., 1964). (wt%, *T* in °C).

Natural substances such as feldspar, quartz, nepheline, and kaolin are usually used to produce this mixture.

Porcelain systems were also the basis for developing ceramics used in dental medicine. Ceramics were developed that could be fused to metal frameworks in restorative dentistry. Typical compositions for these types of sintering ceramics are given, for example, by O'Brien et al. (1978) as 52–62 wt% SiO_2, 11–16 wt% Al_2O_3, 9–11 wt% K_2O, 5–7 wt% Na_2O, and additives, such as Li_2O and B_2O_3. The formation of these types of ceramics, primarily dominated by the precipitation of the main crystal phase of leucite, is described by numerous authors (McLean 1972; Lindemann 1985; Schmid et al., 1992). Although there are many examples in which a glass or different glassy phases develop from the mixture of raw materials and crystals grow as a result of the subsequent heat treatment, as is the case in the manufacture of glass-ceramics, these products cannot be called glass-ceramics. Crystal growth does not occur in a controlled process with controlled nucleation and crystallization. Instead, it proceeds spontaneously in what is called "wild" crystallization. The resulting microstructure features crystallites of the leucite type in different sizes. Frequently, real growth fronts of crystals appear that look like a string of pearls.

These products, however, were the basis for the development of glass-ceramics of the SiO_2–Al_2O_3–K_2O system. A typical glass-ceramic that contains leucite as the main crystal phase, for example, is IPS Empress® glass-ceramic (see Section 4.4.2). It has the following composition: 59–63 wt% SiO_2, 19–23.5 wt% Al_2O_3, 10–14 wt% K_2O, 3.5–6.5 wt% Na_2O, 0–1 wt% B_2O_3, 0–1 wt% CeO_2, 0.5–3 wt% CaO, 0–1.5 wt% BaO, and 0–0.5 wt% TiO_2.

In addition to numerous additives such as pigments, stabilizers, and further silicate sintering products, the base glass in which the leucite is produced through controlled surface crystallization (see Sections 1.4, 1.5, and 3.2.3) is of particular importance. The base glass has the following chemical composition (Höland et al., 1995a): 63.0 wt% SiO_2, 17.7 wt% Al_2O_3, 11.2 wt% K_2O, 4.6 wt% Na_2O, 0.6 wt% B_2O_3, 0.4 wt% CeO_2, 1.6 wt% CaO, 0.7 wt% BaO, and 0.2 wt% TiO_2.

If the glass is heat treated in monolithic form, isolated leucite crystals grow inward from the surface of the glass specimen as a result of the low nuclei density on the surface. These leucite crystals demonstrate obvious direction of growth (dependent on direction: anisotropic growth) in which the apex of the polyhedron is located at the surface of the glass specimen (see Sections 1.4 and 3.2.4).

To increase the density of the nuclei and to achieve effective surface crystallization that proceeds in close connection with the components responsible for promoting nucleation, such as TiO_2, CeO_2, and B_2O_3, surface activation is significantly increased by finely grinding the base glass. A base glass thus produced demonstrates a significantly higher crystallization rate of leucite. For monolithic samples it increases 0.005 μm/min and for powdered samples to approximately 2 μm/min. At heat treatment between 920°C and approximately 1200°C, leucite initially grows dendritically from the nucleating center.

Thus within the first few minutes of heat treatment, a microstructure develops as shown in Fig. 2-17. At the end of heat treatment, that is, after approximately one hour, a significantly less disordered leucite is evident than that shown in Fig. 2-17.

Some of the leucite crystals differ considerably in their outer shape from the type of natural leucite crystals. The base glass-ceramic thus produced is distributed by the manufacturer (Ivoclar Vivadent AG) in the form of glass-ceramic ingots in the IPS Empress® dental restoration system for further processing in dental laboratories. To complete the formation of the final

x10000   2 µm   20kU   18mm
#0   leucite   ✳   IVOCLAR   ✳

Figure 2-17 SEM image after HF-etching of the crystallization of leucite in glass granulates. The growth process begins at the surface of the grains and proceeds inward. The figure shows a high resolution of the dendritic growth process.

microstructure of the glass-ceramic product, however, a second heat treatment of the raw glass-ceramic is required.

The second heat treatment that takes place in the dental laboratory is coupled with a viscous flow process (see Section 4.4.2). After approximately 35 min at a temperature range between 1050°C and 1180°C, the leucite crystals undergo a maturing process, resulting in the type of microstructure shown in Fig. 2-18. The microstructure is dense and crack-free. But after HF-etching, microcracks become visible at the surface of the glass-ceramic (compare Fig. 2-18a and Fig. 2-18b). The crystals possess the near-perfect outer shape of a tetragonal modification. As a result of their high coefficient of thermal expansion, they have a dispersion-strengthening effect on the glass matrix. Thus a glass-ceramic demonstrates a basic flexural strength of approximately 120 MPa. The strength of the material can be further increased to 170 MPa by applying glazes or layers of sintering ceramic (Dong et al., 1992).

As the crystal content of leucite of approximately 34 vol% in glass-ceramic of IPS Empress® (Ermrich et al., 2001) is quite high, linear coefficients of

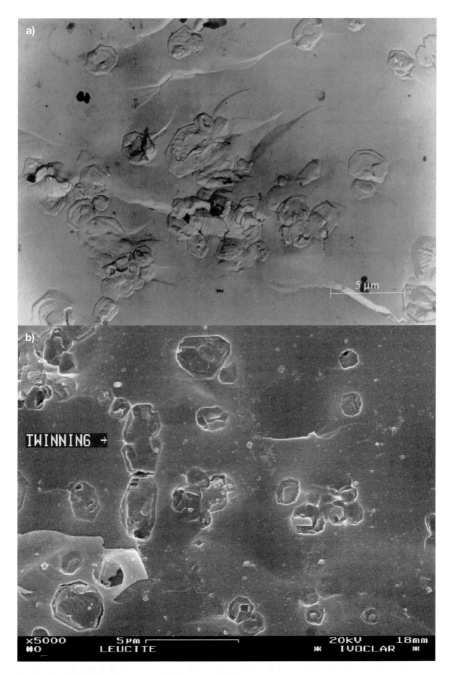

Figure 2-18 Microstructure of IPS Empress® glass-ceramic.
a) The microstructure is dense and crack-free. The leucite crystals demonstrate virtually uniform growth (TEM replica, fracture preparation).
b) Microstructure after HF-etching (2.5% HF, 3 sec). The arrows indicate the beginning of the etching procedure. At the same time, twinning of leucite crystals, typical of IPS Empress® glass-ceramic, becomes visible.

thermal expansion range from 14.9×10^{-6} K^{-1} to 18.25×10^{-6} K^{-1}, depending on the type of glass-ceramic.

The optical properties are particularly significant for the subsequent applications of the material. It is important that glass-ceramics be translucent, even though crystals measure between 2 and 5 μm. This characteristic is desirable in restorative dentistry.

Furthermore, glass-ceramics must also be capable of combining with other products such as sintering ceramics, glazes, and glasses with special optical properties in a sintering process. Thus it is possible to combine translucency with opalescence. This type of opalescent glass that adapts to a glass-ceramic and contains leucite crystals has the following composition (Höland et al., 1996b): 49–58 wt% SiO_2, 11–19 wt% Al_2O_3, 9–23 wt% K_2O, 2–12 wt% CaO, 0.5–6 wt% P_2O_5, 0.2–2.5 wt% F, and further additives up to 9 wt% of $R(I)_2O$, $R(II)O$, $R(III)_2O_3$, and $R(IV)O_2$.

The microstructure of this opal glass-ceramic shows the different phases of liquid–liquid phase separation responsible for producing opalescence. Initial crystallization of leucite is visible in the interface region between the individual glass grain boundaries (Section 3.2.3).

In comparison to the IPS Empress® microstructure, a different type of leucite glass-ceramic was developed by Schweiger et al. (1999). The leucite-reinforced glass-ceramic was developed as a biomaterial for dental restoration. The material can easily be machined by a CAD/CAM system. The mechanical properties are characterized by a flexural strength of 135–160 MPa and a fracture toughness as K_{IC} of 1.3 MPa·m$^{0.5}$ (section ProCAD® in Section 4.4.2 dental glass-ceramics).

2.3 FLUOROSILICATES

2.3.1 SiO_2–$R(III)_2O_3$–MgO–$R(II)O$–$R(I)_2O$–F (Mica)

The unique diversity of three-layer silicates that constitute the mineral mica is familiar to minerologists and crystallographers. The easy cleavage of mica crystals has been the subject of numerous technical papers, and the structure of mica is presented in Section 1.3.1 and Appendix Fig. 13 of this book. Not only is the crystallography of mica well documented, but its presence as a major component in rocks, such as granite, is also well known.

The controlled crystallization of mica in glass to produce glass-ceramics, however, presents a particular challenge for materials engineers. Mica alone, demonstrating the basic structural formula of $X_{0.5–1}Y_{2–3}Z_4O_{10}(OH,F)_2$, is a multicomponent system of alkaline or large alkaline earth ions X, small alkaline earth ions Y or aluminum in octahedral coordination (Y) and Z ions, representing the sum of Al (III) and Si (IV) tetrahedral building blocks, as well as of oxygen, hydroxyl, and/or fluoride anions. The list of elements, e.g., Na, K, Mg, Al, Si, O, H, and F, contained in mica indicates that model binary and ternary systems are inadequate for developing a mica glass-ceramic because mica would not form in these simple cases.

2.3.1.1 Alkaline Phlogopite Glass-Ceramics

Beall (1971) developed a glass-ceramic containing mica by controlling crystallization of the base glass. The objective of developing a material conducive to machining by methods such as lathing, drilling, milling, or thread cutting, using simple metal cutting tools, was fulfilled. In the glass-ceramic produced by Beall (1971), the following alkaline phlogopite mica was formed:

$$K_{1-x}Mg(Al_{1-x}Si_{3+x}O_{10})F_2$$

where $x = 0 - 0.5$ (see structure in Appendix Fig. 13). Various mechanisms were used to develop this material. The processes used and the intermediate stages of controlled microstructure formation achieved (Beall 1971; Chyung et al., 1974; Beall 1992) are described below:

- Base glasses of the following multicomponent system in the indicated composition ranges (in wt%) were melted: 30–50 SiO_2, 3–20 B_2O_3, 10–20 Al_2O_3, 4–12 K_2O, 15–25 MgO, and 4–10 F.

    The complexity of this system clearly shows that a simple comparison with the ternary system SiO_2–Al_2O_3–MgO would not provide accurate information regarding the glass formation potential of the above system and crystal phase formation, since the ternary system has been completely changed and expanded by the numerous additions.

    The microstructure of the glass is characterized by phase-separation phenomena (Fig. 2-19a). Phase separation was used as an essential process in nucleation. Without phase separation, nucleation would not have taken place in these base glasses. In this context, Chyung et al. (1974) established that Mg^{2+} has a specific function and that during phase separation, its coordination clearly changed from 4 to 6. This observation leads to a comparison with the role of Ti^{4+} in SiO_2–Al_2O_3–MgO glasses. In this case, an effect on the phase separation process was observed and the coordination also changed from 4 to 6 during phase separation. Chyung et al. (1974) determined that MgF_2 is

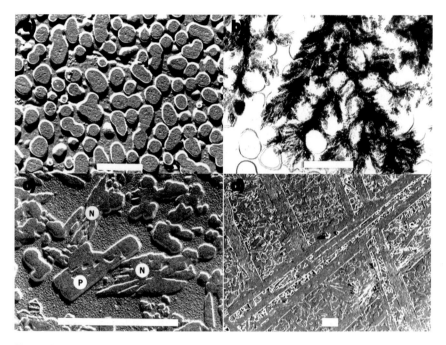

Figure 2-19 Microstructure formation in machinable glass-ceramics of the phlogopite-type (Chyung et al., 1974). Bar = 1 μm.

a) TEM replica image of the base glass that shows glass-in-glass phase separation of glass droplets of approximately 0.5 μm.

b) TEM image showing the crystallization of a primary crystal phase of the chondrodite-type in dendritic morphology at 700°C/1 h.

c) TEM replica image showing a subsequent crystallization process, in which norbergite develops at approximately 850°C/0.5 h.

d) Mica crystals after 950°C/4 h.

isostructural with TiO_2 in rutile (see the structure in Appendix Fig. 18). The droplet phase measuring approximately 0.5 μm was found to be composed of an F-rich potassium alumino borosilicate glass phase.

• Primary crystallization began at approximately 650°C in the form of dendritic chondrodite $[2Mg_{2-x}(Al,B)_{2x} Si_{1-x} O_4 \cdot MgF_2]$ crystals developing in the matrix from the phase-separation boundaries. Using X-ray diffraction analysis, these crystals were determined to be body-centered cubic chondrodite. The microstructure of this transition stage in the development of mica glass-ceramics is shown in Fig. 2-19b. It is interesting to note that this particular chondrodite clearly demonstrated a distorted structure compared with other orthosilicates or orthoborates, such as forsterite, Mg_2SiO_4, or sinhalite, $MgAlBO_4$.

- The next part of the crystallization process at temperatures above approximately 750°C was characterized by the development of the humite mineral norbergite, $Mg_2SiO_4 \cdot MgF_2$. These crystals were produced by means of a solid-state reaction between chondrodite and the droplet-shaped glass phase, in the process of which both were entirely dissolved. The microstructure containing norbergite crystals is shown in Fig. 2-19c.

  The desired main crystal phase of the phlogopite type was formed at temperatures above 850°C, entirely consuming the norbergite. Fluorborite, $Mg_3(BO_3)F_3$ developed as a secondary crystal phase. Optimal microstructure formation, however, occurred at 950°C.

The mechanism involved in the formation of alkaline phlogopite mica was studied from a crystallographic perspective by Chyung et al. (1974). These investigations revealed structural similarities between the (100) plane of norbergite and the (002) plane of phlogopite, specifically that lattice parameters deviated no more than 5% from one another, making epitaxal interaction possible. Therefore, phlogopite can grow heterogeneously on norbergite.

Fluorphlogopite is a solid solution with the following approximate formula:

$$K_{1-x}Mg_{3-y}Al_y[(Al,B)_{1+z}Si_{3+z}O_{10+w}]F_{2-w}$$

The total number of anions (oxygen and fluorine ions) must equal 12. The remaining parameters were determined: $x = 0.01 - 0.2$, $y = 0.1 - 0.2$, and $w = 0 - 0.1$.

The microstructure of the mica glass-ceramic of the phlogopite-type demonstrates direct interlocking of the mica crystals after heat treatment at 950°C for 4 h. This morphology is described as a house-of-cards in Section 3.2.6.

Beall (1971) determined that different microstructures can be produced, depending on the composition crystallization process used. Furthermore, the chemical stability of the respective glass-ceramics also changes. The machinability of this mica glass-ceramic proved to be very favorable. A ceramic containing one-third by volume mica was shown to demonstrate satisfactory machinability. Optimum machinability was obtained at two-thirds by volume.

The particularly favorable machinable properties of this glass-ceramic must be attributed to the fact that mica crystals are readily cleaved. The bonds between the three-layer units of the mineral (see Section 1.3.1.) can be broken relatively easily by external force. In addition, the cracks propagate along the crystal boundaries. Thus, machining of the material is made possible without causing the glass-ceramic to fracture. In addition to machinability, this material also exhibits the following favorable properties: heat resistance exceeding 800°C, electrical insulating properties, and other mechanical parameters, such as bending strength of approximately 120 MPa and compressive strength of 345 MPa in combination with a coefficient of linear thermal expansion of 93×10^{-7} K^{-1} (25°–300°C). These properties are superior to those of other mica-based materials, such as composites composed of sintered glass incorporating natural mica which are less uniform and prone to porosity (Beall 1971). This mica glass-ceramic of the phlogopite type is produced by Corning Incorporated, USA, under the MACOR® trademark. Its composition is listed in Table 2-16.

Additional mica glass-ceramics also containing an alkali phlogopite as the main crystal phase were developed by Vogel (1978) and Vogel and Höland (1987). Glass-ceramics produced from base glasses in the SiO_2–Al_2O_3–

Table 2-16

Composition and Crystal Phases of Commercial Mica-type Glass-Ceramics

(a) Compositions	MACOR® (Corning)	Fotovel™ (Japan)	Bioverit® II (Vitron)	DICOR® (Corning/ Dentsply)
	(wt%)	(wt%)	(wt%)	(wt%)
SiO_2	47.2	X	48.9	56–64
B_2O_3	8.5	X		
Al_2O_3	16.7	X	27.3	0–2
MgO	14.5	X	11.7	15–20
Na_2O		X	3.2	
K_2O	9.5	X	5.2	12–18
F	6.3	X	3.7	4–9
ZrO_2		X		0–5
CeO_2		X		0.05

(b) Crystal Phases	
Glass-Ceramic	Mica Type
MACOR®	$K_{1-x}Mg_3Al_{1-x}Si_{3+x}O_{10}F_2$ (flat crystals)
Fotovel™	$KMg_3AlSi_3O_{10}F_2$ (flat crystals)
Bioverit® II	$(Na_{0.18}K_{0.82})(Mg_{2.24}Al_{0.61})(Si_{2.78}Al_{1.22})O_{10.10}F_{1.90}$ (curved crystals)
DICOR®	$K_{1-x}Mg_{2.5+x/2}Si_4O_{10}F_2$ $x<0.2$ (flat crystals)

MgO–Na$_2$O–K$_2$O–F system demonstrated special characteristics depending on their composition and heat treatment. Phlogopite in the familiar plate-like morphology was precipitated as well as a new type of curved phlogopite.

The phase development sequence and the crystal growth mechanism of the machinable glass-ceramic composed of flat phlogopites are addressed in Section 1.4 on crystal growth. The glass-ceramic demonstrates a house-of-cards morphology. A detailed description of this microstructure is provided in Section 3.2.6. The formation of the phlogopite main crystal phase proceeds according to the same mechanism as the formation of the norbergite main crystal phase as established by Beall (1971) and Chyung et al. (1974) for the MACOR® glass-ceramic.

Unexpectedly, Vogel and Höland (1987) and Höland et al. (1981, 1991b) discovered a special glass-ceramic containing curved phlogopite. Table 2-16a shows the composition of Bioverit® II glass-ceramic, which demonstrates a typical composition of the SiO$_2$–Al$_2$O$_3$–MgO–Na$_2$O–K$_2$O–F system. Table 2-16 (a) and (b) also indicates that the morphology of this phlogopite crystal with its curved three-layer silicate of the phlogopite type differs from that of the other mica glass-ceramics. Even in nature, there is no evidence of a curved mica crystal phase. On the basis of the morphology of these crystals, a glass-ceramic with a particular microstructure was developed. The microstructure of this curved phlogopite is called a *cabbage head*. It is presented in detail in Section 3.2.7. Another special feature of this glass-ceramic concerns the solid-state reactions that lead to the formation of the main crystal phase. Prior to the formation of curved phlogopite, no other primary crystal phases have been established by X-ray diffraction. The properties and applications of Bioverit® II are covered in Section 4.4.1.

A fluorine mica glass-ceramic with a mica phase containing Au$^+$ ions, was developed in Japan. In addition, ZrO$_2$ crystals were formed in this glass-ceramic (Sumikin Photon Ceramic Co., 1998). The properties and applications of this material are discussed in Section 4.1.3. The trade name of this glass-ceramic is Fotovel™.

2.3.1.2 Alkali-Free Phlogopite Glass-Ceramics

Hoda and Beall (1982) developed a glass-ceramic free of alkaline ions, such as Na$^+$, K$^+$, or Li$^+$. The main crystal phase was formed of alkali-free phlogopite. The investigations of Hoda and Beall (1982) concentrated on the formation of the main crystal phases of

Ca-phlogopite $Ca_{0.5}Mg_3AlSi_3O_{10}F_2$
Sr-phlogopite $Sr_{0.5}Mg_3AlSi_3O_{10}F_2$
Ba-phlogopite $Ba_{0.5}Mg_3AlSi_3O_{10}F_2$, or mixed, e.g., Sr-Ba or Ca-Ba, phlogopite

The base glasses used to develop this material were in the following composition range: 35–40 wt% SiO_2, 10–15 wt% Al_2O_3, 20–30 wt% MgO. Alkaline earth oxides were incorporated systematically in the ranges 0–20 wt% BaO, 0–25 wt% SrO, and 0–6 wt% CaO.

Compared with alkaline phlogopite mica glass-ceramics, alkaline-free alkaline earth mica glass-ceramics demonstrated a different phase evolution in the crystallization process. Initially, the base glass exhibited similarly pronounced phase-separation processes as those found in glasses containing alkaline ions. Thereafter, however, the sequence of the phase evolution changed. Mica already formed as a primary crystal phase during heat treatment at 640°C for 4 h. These crystals were very small and measured less than 0.05 µm. In the temperature range of 700°–800°C, however, these crystals grew. Most of the glass-ceramics in the mentioned composition range demonstrated secondary crystal growth combined with increased growth rates at 1110°C. The Ca-phlogopites, which started to decompose at these temperatures, were the exception. A Ca-Ba phlogopite glass-ceramic with the following composition is given: 43.8 wt% SiO_2, 12.4 wt% Al_2O_3, 18.7 wt% MgO, 5.6 wt% CaO, 3.4 wt% BaO, and 16.6 wt% MgF_2.

A dense microstructure of mica crystals formed in the glass matrix on heat treatment of 625°C for 4 h and 1100°C for 6 h. The determination of the properties of this type of alkaline-free mica glass-ceramics was of particular interest.

As was expected with the removal of the alkali ions, the dielectric properties of the alkaline-free alkaline earth mica glass-ceramics were improved further. Glass-ceramics composed of mixed Ba and Sr mica with the following composition 39.5 wt% SiO_2, 14.0 wt% Al_2O_3, 15.0 wt% MgO, 6.6 wt% SrO, 10.4 wt% BaO, and 14.5 wt% MgF_2 exhibited a lower dielectric loss factor (tan δ) in the 0°–300°C range compared with the MACOR® glass-ceramic containing alkaline ions. Furthermore, the dielectric constant was higher than that of MACOR® in the 0°–200°C range.

A glass-ceramic with the following composition demonstrated particularly unexpected properties: 44.8 wt% SiO_2, 12.7 wt% Al_2O_3, 19.1 wt% MgO, 12.2 wt% SrO, and 17.0 wt% MgF_2.

This glass-ceramic, containing a Sr-phlogopite main crystal phase, exhibited water swelling behavior. Following the absorption of water by the glass-ceramic, a new glass-ceramic containing crystals free of Sr^{2+} and measuring 2.5 μm was produced. The crystals exhibited the following type of fluoro-vermiculite structure: $Mg_{0.5}Mg_3(AlSi_3O_{10})F_2 \cdot 4H_2O$. This reaction is reversible for the mineral vermiculite, and as a result, the glass-ceramic demonstrates an ion-exchange capacity.

2.3.1.3 Alkaline-Free Tetrasilicic Mica Glass-Ceramic

Another type of mica glass-ceramic was developed from the SiO_2–MgO–K_2O–F system by Grossman (1972). The main crystal phase was composed of tetrasilicic mica of the $KMg_{2.5}Si_4O_{10}F_2$ type. The crystals in this phase were smaller in size and possessed a smaller aspect ratio than in phlogopite glass-ceramics. As a result, the material demonstrated translucent properties. Furthermore, this crystal phase has no natural analog. As was the case for curved phlogopite, this crystal phase can be synthesized and stabilized in glass-ceramics by controlled crystallization utilizing well-known mechanisms that in nature, would not cause the phase to grow.

A mica glass-ceramic of the tetrasilicic type was developed in the following composition range: 56–60 wt% SiO_2, 15–20 wt% MgO, 12–18 wt% K_2O, 4–9 wt% F with additions of approximately 5 wt% ZrO_2 (Grossman 1972 and Pinckney 1993). Table 2-16 (a) and (b) lists the composition of a particular commercial glass-ceramic called DICOR® (Beall 1992). The basis for that development was given by Grossman (1972), who originally developed tetrasilicic mica glass-ceramics. His preferred composition range for the formation of this type of mica of the $Mg_{2.5}Si_4O_{10}F_2$ type was approximately 58–62 wt% SiO_2, 12–17 MgO, 10–11 MgF_2, and 11–16 K_2O, with additions of up to approximately 2 wt% ZrO_2 and As_2O_3.

Microcrystals of the mica type measuring approximately 0.25 μm in size were formed when a glass melt of base glasses of this composition area was slowly cooled. This crystal phase, however, did not form if the melt was cooled quickly.

If the rapidly cooled glass is heat treated at temperatures between 650°C and 950°C, the phase formation processes of nucleation and crystallization can be studied in detail. At 650°C, for example, fine-scaled glass-in-glass phase separation containing initial quasispherical mica grains was observed. These crystallites measure approximately 400 Å in diameter. The crystallization of mica is progressively more complete at temperatures of 910°, 940°, and 960°C. At 960°C, the microstructure exhibits blocklike mica crystals of the $KMg_{2.5}Si_4O_{10}F_2$ type of relatively low aspect ratio. These crystals are structurally comparable to tetrasilicic mica containing OH^- (Seifert and

Schreyer, 1965). At high temperatures, such as 980°C, an additional crystal phase develops. This crystal phase has been identified as enstatite, $MgSiO_3$.

The kinetics of mica crystallization in DICOR® glass-ceramic was studied by Bapta et al. (1996). The formal kinetic parameters were established according to the JMAK equation (Eq. 1-6 in Section 1.4.3). The activation energy of crystallization was determined as 203 kJ mol^{-1}, the formal reaction order n as 3.4 ± 0.2 and the pre-exponential factor as 2.88 × 10^{11} s^{-1}.

Grossman (1972) found that this tetrasilicic mica glass-ceramic exhibits certain favorable properties, such as high mechanical strength (bending strength of 157 MPa), high electrical-isolation capacity (dielectric constant at 25°C/10 kHz: 6.61, loss tangent at 25°C/10 kHz: 0.0054) and excellent chemical durability (weight loss in mg/cm^2 after treating in 5% HCl for 24 h at 95°C: 0.69). In addition, the coefficient of thermal expansion is approximately 64–71 × 10^{-7} K^{-1}. Another important property of the material is its translucency, which is achieved by the formation of relatively small mica crystals measuring less than 1 μm.

This development, which dates back to 1972, forms the basis for the fabrication of DICOR® glass-ceramics. The properties of this material were modified to develop a biomaterial (see Section 4.4.2). For example, the translucency of the material had to be adjusted to the translucency of natural dental enamel. This objective was fulfilled by reducing the crystallite size during the development of DICOR® MGC glass-ceramics (Grossman 1989 and Beall 1992).

2.3.2 SiO_2–Al_2O_3–MgO–CaO–ZrO_2–F (Mica, Zirconia)

As stated at the beginning of this book and supported by a number of examples, controlled crystallization of several main crystal phases has been shown to permit the combination of different properties in the development of glass-ceramics, according to special requirements.

The development of SiO_2–Al_2O_3–MgO (cordierite, enstatite) glass-ceramic by Beall (1989) has already been addressed in Section 2.2.5. He found that the precipitation of ZrO_2 crystals in the glass matrix increased the mechanical strength of glass-ceramics. Furthermore, ZrO_2 crystal precipitation also plays an important part in increasing the strength of the glass-ceramic in the SiO_2–P_2O_5–Li_2O–ZrO_2 (Section 2.6.5) glass system.

Various investigations have been conducted in an effort to combine two specific properties in the fluorosilicate glass-ceramic system and in the mica glass-ceramic system in particular, that is, high strength and machinability as a result of mica precipitation.

Bürke et al. (2000) demonstrated that controlled crystallization takes place in two steps in the SiO_2–Al_2O_3–MgO–ZrO_2–K_2O–F system. They initiated mica precipitation in base glasses. Once the material had cooled, it became machinable, for example, by drilling. In an additional heat-treatment step, secondary crystallization resulted in nucleation and crystallization of ZrO_2 phases. These phases increased the strength of the material.

Uno et al. (1993) were able to combine the controlled crystallization of mica and ZrO_2 crystals. It was their aim to develop a machinable glass-ceramic that demonstrates high strength as a result of ZrO_2 crystal precipitation. This type of glass-ceramic would not require a second heat treatment and would therefore be easier to manufacture. Uno et al. (1993) based their work on the mica glass-ceramic developments of Grossman (1972), Chyung et al. (1974), and Höland et al. (1983a). Uno et al. (1991) developed a CaO- and BaO-rich glass-ceramic in the SiO_2–Al_2O_3–MgO–BaO–CaO–P_2O_5–F system. By heat treating the base glass at 1000°C, they achieved a main crystal phase of barium-rich mica. The crystals measured between 0.2 and 0.5 μm. Furthermore, they demonstrated an interlocking microstructure embedded in a glassy phase, as well as bending strength of 350 MPa and fracture toughness K_{IC} of 2.3 MPa·$m^{0.5}$.

In the development of a machinable, high-strength glass-ceramic, Uno et al. (1993) had to produce this type of microstructure containing an interlocking mica crystal phase to assure the machinable properties of the glass-ceramic. In the process, they also added ZrO_2 to the base glass. They successfully added ZrO_2 to a glass-ceramic in the SiO_2–Al_2O_3–MgO–CaO–K_2O–F system composed of 91.75 wt% [$^{XII}Ca_{0.43}$ $^{XII}K_{0.14}$ $^{VI}Mg_3$ ($^{IV}Si_3$ ^{IV}Al O_{10})F_2] and 8.25 [ZrO_2].

The base glass of this composition demonstrated phase separation on a very small scale. By heat treating the glass at 700°C, a crystal phase measuring approximately 1–2 nm in diameter was produced. This phase was shown to be ZrO_2 when it was analyzed with TEM-EDS. The content of mica of the phlogopite type was still very low at this temperature. At 900°C, clusters (20–50 nm) of ZrO_2 crystals were formed. Furthermore, the mica crystals grew to approximately 0.5 μm in length. Although the mica crystals continued to grow to a crystallite size of approximately 1 μm at 1050°C, the ZrO_2 crystals remained unchanged. These crystals were unable to grow, since they were embedded in the glassy matrix phase and surrounded by mica crystals.

Since the crystal phase of ZrO_2 has been identified as being both monoclinic and tetragonal ZrO_2, it is thought to contribute to the increase in strength of the glass-ceramic by transformation toughening. The most favorable machinable properties and strength were produced in a glass-ceramic of the above composition, which was heat treated at 950°C. Its microstructure is shown in Fig. 2-20. The glass-ceramic with this type of microstructure was machinable, very strong, and tough. Uno et al. (1993) determined the strength of the material to be 500 MPa and K_{IC} 3.2 MPa·m$^{0.5}$.

2.3.3 SiO_2–CaO–R_2O–F (Canasite)

The primary objective of a number of research projects devoted to developing new glasses and glass-

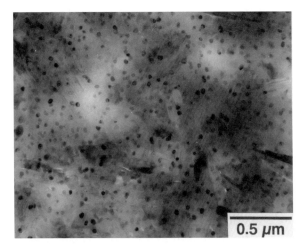

Figure 2-20 Microstructure of mica ZrO_2 glass-ceramic (containing 7.5 wt% F) after heat treatment at 950°C.

ceramics from the middle to the end of the twentieth century has been to increase the strength of these materials. In glass technology, "thermal hardening" is often used to develop stronger glasses. By rapidly cooling the surface of glass products, e.g., vehicle windscreens, compressive strain is produced on the surface, while tensile stress develops within the volume of the glass. The high compressive strain on the surface considerably strengthens the glass product. Therefore, serrated cracks and fractures are prevented.

A further principle developed in glass technology is reinforcement by ion exchange on the surface of a glass object. Using both procedures, high compressive strain is produced to a small depth measuring only a few µm. As a result, the glass object demonstrates overall high strength. Within the object, however, tensile strain is produced. This type of material is used to produce reinforced flat glass for the construction industry, for safety glass in the automotive industry, and for encapsulation glasses in the electrical industry.

This effective glass-forming principle was also used in the development of glass-ceramics. The objective of achieving high-strength products has been pursued in the development of glass-ceramics from the beginning. Based on the above experiences regarding the reinforcement of surfaces in glass technology, high-strength glass-ceramics were produced according to two principles: surface reinforcement by ion exchange processes and surface reinforcement by surface crystallization.

By exhanging Mg^{2+} ions for 2 Li^+ ions (from Li_2SO_4 bath at 800°–900°C) in the SiO_2–Al_2O_3–MgO system, Beall et al. (1967), Beall (1971), and Beall and Duke (1983) managed to increase the fracture strength of glass-ceramic surfaces by approximately 300 MPa. Stewart (1971, 1972) produced similar results by exchanging Li^+ ions for Na^+/K^+ ions in the SiO_2–Al_2O_3–Li_2O system. Pirooz (1973) used controlled surface crystallization of β-eucryptite and β-spodumene to increase the strength of materials from the SiO_2–Al_2O_3–Li_2O system. The combination of a range of potential mechanisms, that is, ion exchange and surface crystallization, has also been shown by Stookey (1959). In a number of examples, the β-eucryptite crystal phase was precipitated in a surface layer of approximately 100 μm thickness. Furthermore, a bending strength of up to 600 MPa was also achieved. Suggestions have been made to use the material in the construction of technical installations and laboratories.

The main problem in using glass-ceramics with reinforced surfaces, however, is the high risk of damaging the surface. If the relatively thin, warped surface layer were damaged or destroyed, the glass-ceramic could break even if a small amount of force were exerted upon it.

In response to this problem, Beall (1991) developed a glass-ceramic with the objective of not only reinforcing the surface, but also the entire glass-ceramic product. The glass-ceramic he developed contained crystals with a chain-silicate structure. Section 2.2.5 also addresses the fabrication of a glass-ceramic with chain-silicate crystals, that is, enstatite glass-ceramics, $MgSiO_3$. Beall established that an increase in strength could be achieved.

Significant progress was also made in the development of chain-silicate glass-ceramics in the SiO_2–CaO–$R(I)_2O$–F and SiO_2–MgO–CaO–$R(I)_2O$–F systems, where $R(I)_2O$ represents the alkali oxides Na_2O and K_2O. These important developments in the fabrication of high-strength glass-ceramics are described in this section as well as in Section 2.3.4.

Beall (1986, 1991) systematically developed glass-ceramics in the SiO_2–CaO–$R(I)_2O$–F system with the objective of controlling the crystallization of the fluorocanasite chain silicate, $Ca_5Na_{3-4}K_{2-3}Si_{12}O_{30}F_4$. The pronounced chain structure of fluorocanasite is shown in Section 1.3.1. This section also addresses the structurally dependent anisotropic properties in connection with the chain structure. Anisotropy of the coefficient of thermal expansion in particular is important for developing a strengthening mechanism in glass-ceramics.

In a first approximation, glasses of this multicomponent system can be represented by the classical SiO_2–CaO–Na_2O/K_2O system (Fig. 2-21). This glass system of the soda-lime-silicate type is often used to manufacture bulk glasses for the packaging industry. Major crystal phases of sodium

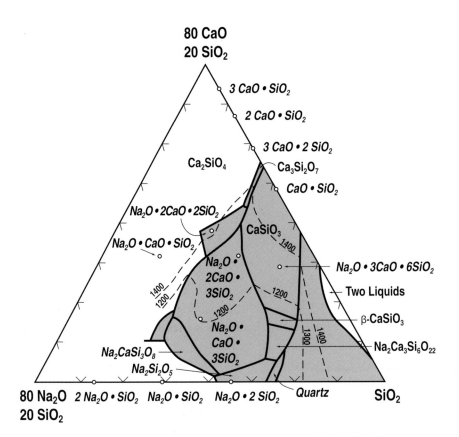

Figure 2-21 SiO_2–CaO–Na_2O phase diagram (Levin et al., 1974) (wt%, *T* in °C).

disilicate, $Na_2Si_2O_5$, or devitrite, $Na_2Ca_3Si_6O_{16}$, can be formed in this system under conditions of equilibrium. Furthermore, the examination of the ternary phase diagram also reveals liquid–liquid phase separation in the SiO_2-rich area.

By incorporating fluorine, Beall (1991) managed to develop new crystal phases that could not be formed in the pure ternary system. In addition, it was relatively easy to melt glasses of this composition, since the viscosity curve was approximately 150°C lower than that of technical soda-lime silicate bulk glass.

Canasite, $Na_{4-3}K_{2-3}Ca_5(Si_{12}O_{30})F_4$, represents the most important crystal phase in this system used to produce high-strength glass-ceramics. Other chain silicates, however, may also be formed, such as agrellite, $NaCa_2Si_4O_{10}F$, or fedorite with the assumed, but not yet accurately determined formula $(K,Na)_{2.5}(CaNa)_7[Si_8O_{19}]$ (Sokolova1983).

While searching for a suitable base-glass composition, which would permit the controlled crystallization of canasite, Beall (1986) discovered that glasses with a practically stoichiometric composition with small alumina additions are the most favorable. A typical glass contains 57 wt% SiO_2, 8 wt% Na_2O, 9 wt% K_2O, 15 wt% CaO, and 13 wt% CaF_2.

Compared with the mineral canasite, $Na_4K_2Ca_5Si_{12}O_{30}(OH,F)_4$, which possesses a stoichiometric composition, the glasses demonstrated a very small amount of excess CaF_2.

Glasses with this composition are relatively easy to melt. The Tg ranges between 470°C and 530°C (Likivanichkul and Lacourse, 1995). Furthermore, the low viscosity allows the base glass to be cast, pressed, or rolled at 950°C. As a result, the base glass can be produced economically and used for a wide range of applications. After cooling, the glasses can also be machined, for example, by diamond-machining or polishing.

Following the fabrication of the base glass, controlled crystallization takes place between 700°C and 850°C during approximately 1 h of heat treatment. A heterogeneous mechanism was used to initiate the nucleation of the fluorocanasite crystals. Beall (1991) managed to form microcrystals of CaF_2 (fluorite) measuring approximately 0.5 μm as heterogeneous nuclei. Furthermore, he was able to grow canasite on these crystals at approximately 800°C. This heterogeneous nucleation mechanism is shown in Fig. 2-22. Following the nucleation process, fluorocanasite crystals rapidly grow to form a dense micro-structure. A matrix composed of crystals and a residual glass phase is formed. Beall (1991) described the microstructure of the matrix to be interlocking. As the crystals also appear bladelike, the entire micro-structure is described as being acicular interlocking (see Section 3.2.8).

A typical microstructure of the canasite glass-ceramic is shown in Fig. 3-19 of Section 3.2.8. This morphology exhibits by lathlike

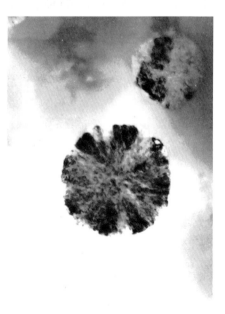

Figure 2-22 TEM image showing heterogeneous nucleation of fluorocanasite on CaF_2 crystals at approximately 800°C.

canasite crystals grown *in situ* in accordance with a schedule of 500°–900°C for 1–4 h. The length of the crystals ranging between 1 and 25 μm, the width 0.25–2 μm and the thickness less than 1 μm (Beall 1983). Morphologically, this microstructure can be compared with that of the amphibole mineral nephrite jade (Bradt et al., 1973). Nephrite jade also exhibits a very high fracture toughness. It must be noted that as a result of the selected fracture preparation of the sample prior to the electron microscopic investigation, cleavage splintering of the microstructure can be observed. This image of the microstructure is the result of energy absorption in the sample producing crack branching and deflection.

The special microstructure with a high percentage of anisotropic crystals helps produce very tough glass-ceramic products. Depending on the selected test procedures, fracture toughnesses of approximately 4.8–5.2 MPa·m$^{0.5}$ can be attained (5.1 ± 0.2 MPa·m$^{0.5}$ for the single-edge notched beam technique, 4.8 MPa·m$^{0.5}$ for the chevron-notched technique). No other glass-ceramic material demonstrates a high fracture toughness of this kind. This degree of fracture toughness exceeds that of very strong Al_2O_3 ceramics that demonstrate a K_{IC} of 4–5 MPa·m$^{0.5}$ (Norton 1998). They almost attain the fracture toughness of 7 MPa·m$^{0.5}$, which is achieved by densely sintered yttrium-stabilized ZrO_2 ceramics (Metoxit 1998). This material, however, must be produced in a high-temperature sintering process at approximately 1500°C. Furthermore, it cannot be shaped by pressing, drawing, rolling, or casting, like glass-ceramics. In addition, during the sintering process, ZrO_2 ceramics linearly shrink approximately 18%–25%.

During the crystallization process of canasite glass-ceramics, Beall (1991) determined minimal shrinkage of approximately 1%. This percentage is below that determined in the shrinkage of other sintered ceramics.

The high fracture toughness of canasite glass-ceramics is produced by a new mechanism, which differs from that in ZrO_2 and other glass-ceramics. It must be noted that the high fracture toughness of ZrO_2 ceramics is produced by the mechanism of transformation toughening involving a phase transformation of ZrO_2 from tetragonal to monoclinic. In various glass-ceramics and in the densely sintered Al_2O_3, the mechanism of dispersion toughening is used. With this mechanism, a fine distribution of crystals in the matrix reinforces the material. In canasite glass-ceramics, however, a mechanism of anisotropic thermal expansion microcracking comes into play. This mechanism is based on the high anisotropy of the coefficient of thermal expansion of the crystals [$(\alpha)a = 159 \times 10^{-7}$ K^{-1}, $(\alpha)b = 82 \times 10^{-7}$ K^{-1}, $\alpha(c) = 248 \times 10^{-7}$ K^{-1} for 0°–700°C]. This large anisotropy gives rise to microcracks in the microstructure of the glass-ceramic.

Moreover, it must be noted that the fracture toughness of the glass-ceramic is a function of the temperature. This functional relationship is presented in Fig. 2-23. It shows that the K_{IC} value would drop to 1 MPa·m$^{0.5}$ if the glass-ceramic were exposed to a temperature of 600°C. As a result, canasite glass-ceramics are better used at room temperature or other low temperatures.

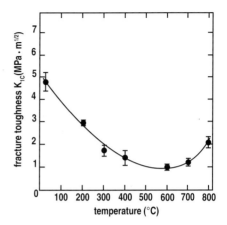

Figure 2-23 Fracture toughness of canasite glass-ceramic as a function of temperature.

In addition to the high fracture toughness at room temperature, canasite glass-ceramics also demonstrate high flexural strength of 300 MPa and a low Young's modulus of 82 GPa compared with sintered ZrO_2 or Al_2O_3 ceramics. These properties render this glass-ceramic particularly suitable for a variety of applications. The high fracture toughness compared with soda-lime silicate glasses permits the material to be used in the construction industry. Furthermore, the material has also been successfully used in the electronic industry (see Section 4.1.4).

The canasite glass-ceramic family was extended (Wolcott 1994) by the controlled co-precipitation of two phases from an appropriate base glass. In this process, a canasite-apatite glass-ceramic was developed with the composition 42–70 wt% SiO_2, 6–12 wt% Na_2O, 3–10 wt% K_2O, 20–30 wt% CaO, 3–11 wt% F, and 2–13 wt% P_2O_5, with additions of B_2O_3, Al_2O_3, and ZrO_2 up to 6 wt%.

The controlled crystallization was carried out in the preferred temperature range of approximately 580°–640°C for nucleation and 900°–950°C for crystallization. Canasite is formed according to the same mechanism of heterogeneous nucleation on the CaF_2 primary crystals. The crystallization of apatite, however, takes place during the nucleation of the droplet glass phase. At the end of the crystallization process, the apatite is dispersed in the glass-ceramic in the form of small hexagonal crystallites.

Apatite-canasite glass-ceramics demonstrate high fracture toughness of up to 3.9 MPa·m$^{0.5}$ combined with special bioactive properties. Therefore, the material has been suggested for use as a biomaterial for replacing bone tissue.

2.3.4 SiO₂–MgO–CaO–R(I)₂O–F (Amphibole)

Like the canasite crystals described in Section 2.3.3, minerals of the amphibole type also demonstrate a chain structure. Given the double chain structure $(Si_4O_{11})^{-6}$ of the tetrahedra, the crystals exhibit a needlelike shape and a high aspect ratio (length:width) of up to 20:1. These characteristics are typical of fluorrichterite amphibole crystals, $KNaCaMg_5Si_8O_{22}F_2$ (see the structure in Appendix Fig. 10). This crystalline phase often demonstrates a random acicular microstructure in nature. By developing this crystal morphology in glass-ceramics, materials with favorable properties, such as high fracture toughness, can be produced.

In the course of his comprehensive research on mica glass-ceramics, Beall (1991) examined a similar glass formation system, that is, SiO_2–MgO–CaO–R(I)$_2$O–F, where R(I)$_2$O represents the alkali oxides Na_2O and K_2O. Small amounts of Al_2O_3, P_2O_5, Li_2O, and BaO were added. As a first approximation, this multicomponent system can be represented by the simple ternary glass formation system of SiO_2–MgO–CaO. Figure 2-24 shows two-liquid melts in the SiO_2-rich region of the phase diagram.

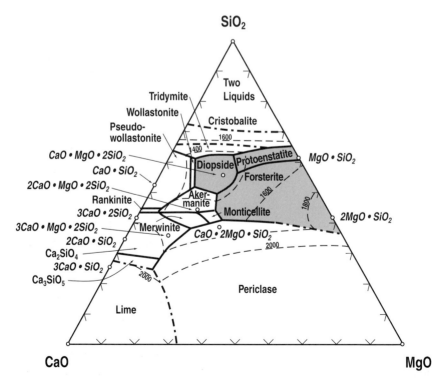

Figure 2-24 Phase diagram of the SiO_2–MgO–CaO system (Levin et al., 1964) (wt%, T in °C).

These liquid melts provide clear proof of the separation processes in this system. Thus, liquid–liquid separation takes place in the SiO_2-rich part of the ternary system. Furthermore, subliquidus separation can be expected in the immediately surrounding compositions. The phase diagram also indicates crystal phases that can be formed under a thermodynamic equilibrium. These phases include diopside, $CaMgSi_2O_6$, akermanite, $Ca_2MgSi_2O_7$, monticellite, $CaMgSiO_4$, and merwinite, $Ca_3MgSi_2O_8$.

The incorporation of alkali ions and most importantly of fluorine allowed Beall (1991) to change and expand the ternary system significantly. Furthermore, new crystal phases, which would not be expected in a simple ternary SiO_2–MgO–CaO system, were produced in the base glasses. In the process, certain reactions were achieved that were similar to the ones that had already been successfully tested in the development of mica glass-ceramics. Furthermore, Beall (1991) demonstrated that controlled crystallization of base glasses in the composition range 55–70 wt% SiO_2, 10–25 wt% MgO, 2–6 wt% CaO, 2–6 wt% Na_2O, 2–7 wt% K_2O, 2–5 wt% F with additions of up to 10 wt% combined Li_2O, Sb_2O_3, Al_2O_3, BaO, P_2O_5 produces the chain silicate richterite, $KNaCaMg_5Si_8O_{22}F_2$. This crystal phase represents the most important chain silicate in this system. It plays a significant role in the glass-ceramic product that will be discussed at a later stage.

Glass formation in this composition range is favorable, although extra SiO_2, Al_2O_3, and P_2O_5 beyond the composition of stoichiometric K–F-richterite is required for best results. A viscosity that would allow a pressing or casting production process was expected on the basis of these results.

An optimized composition relatively close to the stoichiometric composition of fluorrichterite is shown in Table 2-17. It demonstrates a higher SiO_2 content compared with the stoichiometric composition. With this base glass, Beall (1991) was able to produce a fluorrichterite glass-ceramic using various solid-state reactions. This glass-ceramic is characterized by high fracture toughness (approximately 3 $MPa \cdot m^{0.5}$).

Nucleation represents the first step of the complex solid-state reaction process. In this process, phase separation occurs in the base glass after the melt has cooled. During heat treatment at approximately 600°C, a tetrasilicic mica, taeniolite, $KMg_2LiSi_4O_{10}F_2$, forms as the primary crystal phase. As the temperature rises, the solid-state reactions continue. Diopside, $CaMgSi_2O_6$, which also occurs in the ternary SiO_2–MgO–CaO system, forms at 700°C. Both the primary mica and the diopside react with the glass matrix at approximately 950°C and form fluorrichterite and SiO_2 solid solution crystals of the cristobalite type. The needlelike morphology of the fluorrichterite phase is

Table 2-17

Composition of a Commercial Fluorrichterite Glass-Ceramic (wt%)

SiO_2	67.1
MgO	14.3
CaO	4.7
Na_2O	3.0
K_2O	4.8
F	3.5
Li_2O	0.75
BaO	0.3
P_2O_5	1.0
Sb_2O_3	0.2
Al_2O_3	1.8
	101.5

That is, the fluorine equivalent oxygen is 1.5%

even more developed at 980°C. Fluorrichterite is formed according to the following equation:

$$KMg_2LiSi_4O_{10}F_2 + CaMgSi_2O_6 + [(Na/K)_2O + 7MgO + \tfrac{1}{2}Al_2O_3 + 11SiO_2 + CaF_2] \rightarrow 2KNaCaMg_5Si_8O_{22}F_2 + SiO_2 \cdot (LiAlO_2)/\text{stuffed cristobalite}$$

This phase development sequence and the associated solid-state reactions can be described using Figs. 2-25 (a)–(c). Primary mica crystals and a high content of phase-separated glass matrix are discernible in Fig. 2-25(a). Surprisingly, the phase-separated base glass remains intact to a large degree during mica crystallization. Fig. 2-25(b) shows that mica with a relatively high aspect ratio continues to grow at 700°C. Fig. 2-25(c) shows the microstructure of the final glass-ceramic of the fluorrichterite type, which is formed at 980°C. In this glass-ceramic, fluorrichterite forms the main crystal phase; residual mica and cristobalite form the secondary crystal phases. The aspect ratio of fluorrichterite crystals is approximately 10:1. It tends toward the ratio of 20:1 often found in the natural mineral.

The solid-state reactions shown in the microstructure formation were examined as a function of viscosity by Beall (1991). He determined that mica crystals form at a high viscosity exceeding 10^{11} dPa·s. The viscosity increases during crystallization, and decreases at higher temperatures. These fluctuations

Figure 2-25 Phase formation in glasses of the SiO_2–MgO–CaO–R_2O–F system (composition according to Table 2-17)
a) after heat treatment at 650°C
b) after heat treatment at 700°C
c) after heat treatment at 980°C

in the viscosity in connection with the solid-state reactions during the formation of crystal phases are shown in Fig. 2-26. Similarly, the crystallization of mica and diopside at approximately 700°C causes the viscosity to increase. At temperatures above 950°C, fluorrichterite forms and the viscosity also increases. Moreover, a slight increase in viscosity was also established in connection with the crystallization of the secondary cristobalite phase.

Fluorrichterite glass-ceramics exhibit outstanding mechanical properties, such as their fracture toughness of 3.2 $MPa \cdot m^{0.5}$ and bending strength of 150 ± 15 MPa. The application of a glaze increases the strength to 200 MPa.

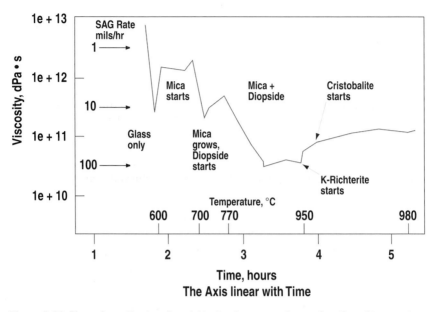

Figure 2-26 Phase formation in a fluorrichterite glass-ceramic as a function of temperature and viscosity.

It is relatively easy to apply a glaze to these materials, as richterite glass-ceramics exhibit a high coefficient of thermal expansion of 115×10^{-7} K^{-1} (0°C–300°C) due to the formation of a cristobalite secondary phase. Therefore, a glaze with a lower coefficient of thermal expansion can be applied to produce compressive strain in the surface of the material. As a result, the material is strengthened as described in Section 2.3.3.

Other favorable properties of fluorrichterite glass-ceramics are optical translucency and high resistance to thermal shock. The glass-ceramic resists temperature fluctuations between 170°C and 8°C without demonstrating any defects. Therefore, the resistance to thermal shocks of this material is comparable to that of the popular borosilicate Pyrex® glass, which is used to fabricate technical equipment and domestic glassware. Pyrex® glass, however, demonstrates lower fracture toughness than fluorrichterite glass-ceramics. As a result, richterite glass-ceramics are more thermomechanically versatile than Pyrex® glass.

Because of its special properties and the associated manufacturing methods, this material is intended for use in the production of domestic tableware, both commercial (hotels, restaurants, etc.) and retail (Corelle® mug). Furthermore, it may also be used for a variety of technical applications.

2.4 SILICOPHOSPHATES

2.4.1 SiO_2–CaO–Na_2O–P_2O_5 (Apatite)

The starting point for the development of apatite-glass-ceramics in this material system was the successful production of the first bioactive glasses for the replacement of bone in human medicine. As early as the end of the 1960s and beginning of the 1970s, Pantano et al. (1974) and Hench (1991) succeeded in developing bioactive glasses in this system. As regards bioactivity, that is, the rapid formation of a direct bond, without connective tissue, between the biomaterial and the living bone, Hench (1991) achieved better results with glasses of the BIOGLASS®-type than with glass-ceramics of the same composition as the glasses. Nevertheless, glass-ceramics of this system should be mentioned at this point. Glass-ceramics composed of 45 wt% SiO_2, 24.5 wt% CaO, 6 wt% P_2O_5, and 24.5 wt% Na_2O were produced by volume crystallization of apatite. Hydroxyapatite and apatite that does not contain hydroxyl groups were the main crystal phases of the glass-ceramic (Hench et al., 1972).

Hench (1991), and Cao and Hench (1996) described the bioactive properties of BIOGLASS® in 11 stages (Table 2-18). It is important to note that they attributed particular significance to SiO_2 and its silicate structure, such as the $[\equiv Si\text{-}OH]$ groups. Their significance carries through from the initial reaction process to the entire reaction for the formation of bone structure. Subsequently, an SiO_2 gel layer forms on the surface of the glass. After this layer has been formed, an amorphous calcium phosphate can be precipitated on it. Furthermore, these phosphates are transformed into hydroxy-carbonate apatite. This type of apatite bonds to living bone. It is, however, important to note that up to the formation of new, mature bone structure, biological cells must be active in the kinetic stages up to stage 11 (Table 2-18).

2.4.2 SiO_2–MgO–CaO–P_2O_5–F (Apatite, Wollastonite)

The above glass-ceramic system was selected with the intention of producing biomaterials to replace bone in human medicine. It was particularly important to achieve a high content of apatite in the glass-ceramic as well as a Ca^{2+} reservoir to assure the subsequent ion release of the implant by adding CaO and P_2O_5 to the glass-ceramic.

Kokubo (1993) succeeded in developing an apatite-wollastonite glass-ceramic (CERABONE® A-W) according to the principles of controlled surface crystallization of powdered glass. This glass-ceramic had the following composition 34 wt% SiO_2, 44.7 wt% CaO, 4.6 wt% MgO, 16.2 wt% P_2O_5, 0.5 wt% CaF_2. The glass powder was fully densified at approximately 830°C.

Table 2-18

Sequence of Interfacial Reactions Involved in Forming a Bond Between Tissue and BIOGLASS® (Hench 1991)

Stage	Process
1 and 2	Initiation and formation of $\equiv$Si-OH bonds on the surface of bioactive glass
3	Polycondensation of $\equiv$Si-OH + $\equiv$Si-OH $\rightarrow$ $\equiv$Si-O-Si$\equiv$
4	Adsorption of amorphous Ca^{2+} + PO_4^{3-} + CO_3^{2-} + OH^-
5	Crystallization of HCA (hydroxy–carbonate apatite)
6	Adsorption of biological moities in HCA layer
7	Action of macrophages
8	Attachment of stem cells
9	Differentiation of stem cells
10	Generation of matrix
11	Crystallization of matrix

Furthermore, oxyfluoroapatite, $Ca_{10}(PO_4)_6(O, F_2)$ and wollastonite, $CaSiO_3$, were precipitated with heat treatment at approximately 870°–900°C (Kokubo 1993). The final glass-ceramic was crack- and pore-free. The crystals were homogeneously distributed in the glassy matrix and demonstrated a size of approximately 50–100 nm (Fig. 2-27). It was possible to crystallize approximately 38 wt% apatite and 24 wt% wollastonite according to X-ray diffraction measurements. Thus, the residual glassy matrix demonstrated approximately 28 wt%.

A glass-ceramic with this type of microstructure is characterized by a bending strength of 215 MPa, compressive strength of 1080 MPa, and fracture toughness of 2.0 MPa·m$^{0.5}$. Clearly, glass-ceramics demonstrate more favorable mechanical properties than does pure glass. Therefore, they are suitable for load-bearing implants in human medicine.

In the reaction mechanism that characterizes the bioactivity of these glass-ceramics, Kokubo (1993) discovered that a dissolution process of Ca^{2+} ions takes place during immersion in simulated body fluid because of the high CaO content of the glass-ceramics. Hence, the concentration of Ca^{2+} in the body fluid rises. $\equiv$Si-OH groups on the surface of the glass-ceramic are favorable sites for apatite nucleation (Fig. 2-28). After the nucleation of apatite, additional requirements of Ca^{2+} ions and phosphate groups are covered by mass transport from the surrounding body fluid. In these glass-ceramics, however, Kokubo (1993) did not observe the formation of an SiO_2 gel layer as was the case for BIOGLASS®.

2.4.3 SiO_2–MgO–Na_2O–K_2O–CaO–P_2O_5 (Apatite)

Bioactive glass-ceramics from this system were developed by Brömer et al. (1973), at the beginning of the 1970s, immediately following the development of BIOGLASS®. Various glass-ceramics of the SiO_2–CaO–MgO–Na_2O–K_2O–P_2O_5 base system and the main crystal phase of apatite were produced by the Leitz, Wetzlar company under the collective brand name of CERAVITAL®. The composition range of CERAVITAL®-type glass ceramics was characterized by Brömer et al. (1997) as 40–50 wt% SiO_2, 30–35 wt% CaO, 2.5–5.0 wt% MgO, 5–10 wt% Na_2O, 0.5–3 wt% K_2O, and 10–15 wt% P_2O_5. Nucleation of the base glass was carried out at 600°C for 24 h and apatite crystallization took place at 750°C after 24 h. It was possible to add 5–15 wt% Al_2O_3, 0–5 wt% Ta_2O_5, and 5–15 wt% TiO_2 to the base glass-ceramic composition. This glass-ceramic was tested as dental root implants (Brömer et al., 1977).

The bioactive behavior of various glass-ceramics of this type, for example, 46.2 wt% SiO_2, 25.5 wt% $Ca(PO_3)_2$, 20.2 wt% CaO, 4.8 wt% Na_2O, 2.9 wt% MgO, 0.4 wt% K_2O has been successfully tested by Gross (1981). CERA-VITAL® is character-ized by a crystal size of 40–50 nm, a bending strength of 150 MPa, and compressive strength of 500 MPa (Brömer et al., 1977).

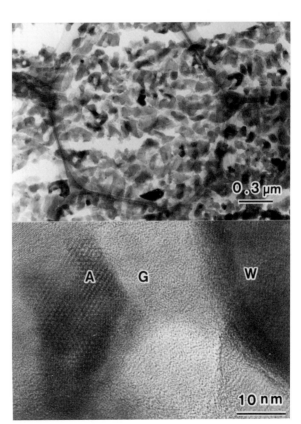

Figure 2-27 TEM image showing the microstructure of apatite (A)–wollastonite (W) glass-ceramic. (G): glassy phase. (Courtesy T. Kokubo).

Body fluid

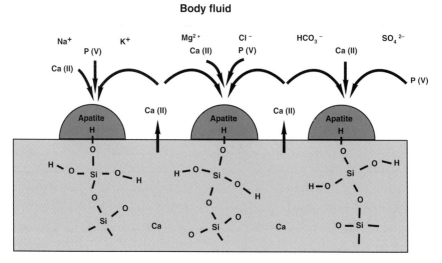

Figure 2-28 Schematic representation of the mechanism of apatite formation on the surface of CaO–SiO$_2$-based glass-ceramics in the human body (Kokubo 1993).

2.4.4 SiO$_2$–Al$_2$O$_3$–MgO–CaO–Na$_2$O–K$_2$O–P$_2$O$_5$–F (Mica, Apatite)

The base glasses of this multicomponent system are derived from the ternary system SiO$_2$–Al$_2$O$_3$–MgO (Section 2.2.5). The starting point of materials development is barely recognizable, as many additions are incorporated and Al$_2$O$_3$ is greatly reduced or even replaced entirely in the process. For a better understanding of this process, it is useful to describe and discuss the compositions of the simpler subsystems from a chemical point of view before describing the systems with eight components (seven oxides and one fluoride).

Glasses of the ternary base glass system SiO$_2$–Al$_2$O$_3$–MgO whose chemical composition belongs in the cordierite and mullite liquidus phase field (see Section 2.2.5) demonstrate only a minimal tendency toward liquid–liquid phase separation in their microstructure. During subsequent heat treatment, thermodynamically stable phases such as cordierite and mullite develop at a very early stage. If the additional components Na$_2$O, K$_2$O, F from 3 to 10 wt% are incorporated in this type of glass, controlled crystallization takes place to form mica crystals when the glasses are heat treated. Mica precipitation occurs through the enhancement of phase separation and primary precipitation of other crystal phases (see Section 2.3).

However, if Na$_2$O, K$_2$O, F, and further additions under 3 wt%, such as CaO and P$_2$O$_5$ are selected as shown in the following example, then phase separation of the base glass is greatly reduced, preventing nucleation of the primary phase. As a result, glasses with the composition shown in Table 2-19 permit

Table 2-19

BIOVERIT® II (Mica Glass-Ceramic) Composition (wt%)

	Composition range	Example
SiO_2	43–50	44.5
Al_2O_3	26–30	29.9
MgO	11–15	11.8
Na_2O/K_2O	7–10.5	4.4/4.9
F	3.3–4.8	4.2
Cl	0.01–0.6	0.1
CaO	0.1–3	0.1
P_2O_5	0.1–5	0.1

mica formation without another primary phase (Vogel and Höland, 1987; Höland and Vogel, 1992).

A new type of mica with a special curved form develops in this BIOVERIT® II-type glass-ceramics (Fig. 3-15, see also Section 3.2.7). From a materials development point of view, this special appearance of the mica crystals of phlogopite are of great interest. A detailed discussion of this topic will follow in Section 3.2.7. At the same time, however, it must be stressed that new or improved properties compared with mica glass-ceramics with plane or petal-shaped crystals result from the new, curved mica crystals. Specifically, it was demonstrated that the new glass-ceramics can be machined more rapidly than glass-ceramics with plane mica crystals.

All these materials developments, connected in part with the formation of actual products, are prerequisites for the development

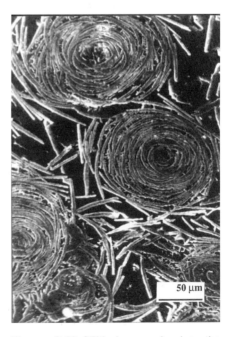

Figure 2-29 SEM image showing the advanced stage of the precipitation of curved phlogopite crystals in a mica glass-ceramic. Polished sample and etched 2.5 sec in 2% HF.

of materials in the eight-component system. Thus if the low CaO and P_2O_5 content in the aforementioned mica glass-ceramics is considerably increased, a new system of materials with new properties develops. The compositions of base glasses and mica–apatite glass-ceramics are shown in Table 2-20 (Höland et al., 1983b; Höland et al., 1985).

The base glass possesses a characteristic three-phase glass microstructure. This microstructure is characterized by two droplet phases and a glassy matrix (Fig. 2-30). The smaller droplet phase is comparable to the alkali-fluorine rich droplet phase of the mica system (Section 3.2.6). The larger droplet phase corresponds to a CaO-, P_2O_5-, and F-rich glassy phase. The reason for the formation of two droplet phases might be the differences in the field strength (defined as z/a^2, where a is the distance between Si–O or P–O, and z is the valency) of silicon to oxygen in comparison to phosphorous to oxygen (2.1 for P–O versus 1.57 for Si–O). After heat treatment of this glass, mica as well as apatite crystals develop. Thus twofold controlled crystallization of this base glass is possible when it is heat treated between 750°C and 1100°C. The precipitation of mica corresponds to that mechanism reported in Section 2.3.1. The formation of fluorapatite follows a homogeneous nucleaction mechanism within the CaO-, P_2O_5-, and F-rich droplet phase in a temperature range of 750°–1000°C. This is demonstrated in Fig. 2-31 (Vogel and Höland, 1987; Höland and Vogel, 1992). The growth of fluorapatite in the droplet phase stops at the boundary of the glassy droplet. Figure 2-31 shows one single apatite crystal per droplet phase. It is doubtful whether a nucleus formed a single crystal within the glassy droplet phase. The final result of the nucleation and crystallization is fluorapatite in a typical

Table 2-20

BIOVERIT® I (Mica–Apatite Glass-Ceramics) Composition (wt%)

	Composition range	Example 1	Example 2
SiO_2	29.5–50	30.5	38.7
MgO	6–28	14.8	27.7
CaO	13–28	14.4	10.4
Na_2O/K_2O	5.5–9.5	2.3/5.8	0/6.8
Al_2O_3	0–19.5	15.9	1.4
F	2.5–7	4.9	4.9
P_2O_5	8–18	11.4	8.2
TiO_2	additions		1.9

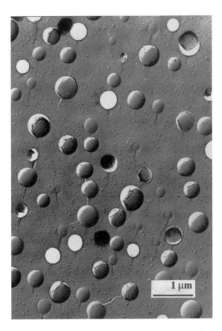 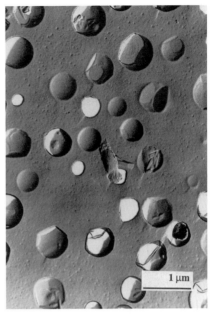

Figure 2-30 Three glassy phases in the base glass for mica–apatite glass-ceramic. SEM. TEM/replica. Fractured surface etched (HCl, 5 sec).

Figure 2-31 Precipitation of fluorapatite within the large CaO–P_2O_5–F-rich droplet phase of mica–apatite glass-ceramic. TEM/replica. Fractured surface etched (HCl, 5 sec).

hexagonal morphology (Höland 1997). A quite different mechanism of the formation of needlelike apatite is reported in Section 2.4.6.

Fluorapatite and mica crystallization are examples of a twofold controlled mechanism to develop a glass-ceramic having a combination of different properties, such as machinability and bioactivity. The microstructure of the final glass-ceramic comprising mica and apatite crystals embedded in a glass matrix is shown in Fig. 2-32.

Although the mica crystals are no longer curved and their content is reduced compared with pure phlogopite glass-ceramics, the glass-ceramic can still be machined. The process, however, is more time consuming. At the same time, new properties are achieved as a result of apatite precipitation. Because of its high biocompatibility, the new material can be used as a bone substitute.

The reactive behavior of the mica–apatite glass-ceramic was thoroughly examined in *in vitro* and *in vivo* studies to establish its biocompatibility. Also, bone implants were studied to establish the bioactivity of BIOVERIT® I glass-ceramics (Section 4.4.1). The two examples of mica–apatite glass-ceramics (Table 2-20) have both demonstrated bioactivity, to a different degree. Example 2 shows a higher bioactivity than example 1.

Basic research on bioactivity of glasses and glass-ceramics has demonstrated how to control the bioactivity of implants. Hench (1993) and Kokubo (1993) established that ≡Si-OH structural elements decisively contribute to the bioactivity of glass-ceramic surface (Section 4.4.1). Thus it can be inferred that the preferred formation of reactive OH⁻-groups in ≡Si-OH structural elements and layer thickness of more than 10 μm on implants surfaces play a significant role in achieving high bioactivity.

Figure 2-32 Mica–apatite glass-ceramic after one week in Ringer's solution. TEM/replica.

Hench and Andersson (1993) showed that high contents of Al_2O_3 in Bioglass® reduce its bioactivity by reducing the activity of the ≡Si-OH-rich layer.

Hence, to obtain high bioactivity in a glass-ceramic that should be both bioactive and machinable, that is, containing both mica and apatite as major crystal phases, a tetrasilicic mica–apatite glass ceramic with low Al_2O_3 content (Table 2-20, example 2) must be used instead of an Al_2O_3-rich glass-ceramic (Table 2-20, example 1). Glass-ceramic example 1 contains a high content of alumina in the phlogopite crystal and also a small quantity in the glassy phase. Therefore, its bioactivity should be lower than that of example 2 (Section 4.4.1). The bioactive reactions of phlogopite–mica glass-ceramics were also investigated by Chen et al. (1998).

2.4.5 SiO_2–MgO–CaO–TiO_2–P_2O_5 (Apatite, Magnesium Titanate)

Most apatite glass-ceramics have been developed as biomaterials. This section, however, addresses biocompatible glass-ceramics without bioactivity. Volume nucleation and crystallization are used to produce these glass-ceramics. In addition, phase separation processes in the base glass are important for nucleation.

Hobo and Takoe (1985) developed apatite glass-ceramics containing large amounts of CaO and P_2O_5 (more than 15 wt% CaO and more than 9 wt%

P_2O_5) in the SiO_2–MgO–CaO–P_2O_5 and Wakasa et al. (1992) in the SiO_2–MgO–CaO–P_2O_5–Al_2O_3 system. Glass-ceramics with the main crystal phase of apatite $Ca_{10}(PO_4)_6(O, F_2)$ (Hobo and Takoe, 1985) and apatite and diopside (MgO·CaO·2SiO_2) were produced by the controlled crystallization of base glasses. These glass-ceramics were developed with the objective of producing biomaterials for restorative dental applications. Special optical properties, such as high translucency, as well as excellent chemical resistance and strength were required. These goals, however, were not always met. The glass-ceramics of Hobo and Takoe (1985), for example, did not entirely fulfill the requirements for a dental biomaterial. Consequently, this glass-ceramic was produced for only a limited time by Kyocera, Japan, under the CeraPearl® product name.

A glass-ceramic of the SiO_2–MgO–CaO–TiO_2–P_2O_5 system with an apatite crystal phase demonstrating the favorable properties of a biomaterial for restorative dentistry was presented by Kokubo et al. (1989). In the development of this glass-ceramic, they examined glass formation in the pseudoternary CaO·MgO·2SiO_2 (diopside), MgO·TiO_2 (geikielite), 3CaO·P_2O_5 (whitlockite) system. Moreover, they studied glass formation and controlled crystallization in the composition range of 30–60 wt% CaO·MgO·2SiO_2, 0–35 wt% MgO·TiO_2, and 30–50 wt% 3CaO·P_2O_5 with the addition of 8 wt% F. Since the transformation range of the glasses (Tg) was located between 680° and 720°C, controlled crystallization of the glasses was initiated during heat treatment at approximately 760°C. It continued up to 1150°C. The main crystal phases of apatite, $Ca_{10}(PO_4)_6(O, F_2)$, geikielite, $MgTiO_3$, and diopside, $CaMgSi_2O_6$ (see the structure in Appendix Fig. 9), formed in the process of controlled crystallization. Apatite formation, however, was only possible in glasses containing flourine. Glasses without flourine formed a whitlockite crystal phase, β-3CaO·P_2O_5, instead of apatite.

Since the base glass had to be subjected to centrifugal casting before controlled crystallization, the glass with the following composition was selected to examine controlled crystallization more closely and to determine the properties of the resulting glass-ceramic more accurately: 16.3 wt% SiO_2, 22.8 wt% TiO_2, 16.9 wt% MgO, 24.8 wt% CaO, 15.7 wt% P_2O_5, 2.0 wt% CaF_2, with additions of 1.0 wt% Al_2O_3, 0.5 wt% ZrO_2, and 0.01 wt% MnO.

The technical process of forming the glass-ceramic and of initiating controlled crystallization is described below.

The base glass of this composition is transformed into a biomaterial (e.g., for a dental crown) following centrifugal casting. Subsequently, the glass is heated from room temperature to 940°C at a rate of 5 K/min. It is transformed into a glass-ceramic in a 5-min heat treatment at 940°C.

Since the main crystal phases of fine crystalline apatite and magnesium titanate were formed, the glass-ceramic demonstrated translucency in layers of 1-mm thickness. The glass-ceramic also demonstrated bending strength of approximately 205 MPa and fracture toughness of K_{IC} 1.46 MPa·m$^{0.5}$. As a result, the material is used as a biomaterial for dental applications (Kokubo et al., 1989). The material is produced as Casmic® glass-ceramic by Yata Dental MFG Co., Ltd., Japan.

2.4.6 SiO_2–Al_2O_3–CaO–Na_2O–K_2O–P_2O_5 (Apatite, Leucite)

Based on the development of a leucite glass-ceramic of the IPS Empress® (see Section 2.2.9) a new material, a chemical durable, translucent apatite–leucite glass-ceramic, has been developed by extending the chemical composition range and by using an additional nucleation mechanism (Höland et al., 1994). Apatite was precipitated in the glass-ceramics as flourapatite. Its crystal structure is shown in Section 1.3.2 and Appendix Fig 19. In recent publications other needlelike apatites have been observed in white opaque glass-ceramics, for example by Moisescou et al. (1999) or apatite/mullite needles (Clifford and Hill, 1996).

In contrast to the leucite system, the new material contains CaO, P_2O_5, and F and several additives. The composition is: 49–58 wt% SiO_2, 11–21 wt% Al_2O_3, 9–23 wt% K_2O, 1–10 wt% Na_2O, 2–12 wt% CaO, 0.5–6 wt% P_2O_5, 0.2–2.5 wt% F, with additives of up to 6 wt%, but preferrably CeO_2, B_2O_3, Li_2O. A CaO/P_2O_5 molar ratio of approximately 5.8 and a fluorine content of 0.6 wt% of the base glass composition were preferred.

The controlled crystallization of this base glass was carried out with glass powder of an average grain size of 20–40 μm. Nucleation and the sintering processes led to a monolithic body and proceeded almost simultaneously at temperatures between 800° and 1100°C. When the nucleation and crystallization processes were studied, it was revealed that two nucleation processes proceeded that led to two crystalline phases (Höland 1997). Leucite was produced by surface nucleation, while apatite was produced by volume nucleation combined with glass immiscibility. In contrast to apatite formation in the mica–apatite glass-ceramic (Section 2.4.4), apatite in leucite glass-ceramics is formed by a different mechanism. In the apatite glass-ceramic discussed in this section, the stage of glass immiscibility is rapidly surpassed, and the apatite quickly grows past the phase boundary of the amorphous droplet. The apatite in Section 2.4.4, however, grew up to the phase boundary of the droplet, beyond which without demonstrating anisotropic growth processes, it grew only slightly.

After nucleation in the apatite–leucite glass-ceramic, however, the apatite grows anisotropically in a preferred orientation into needlelike apatite (Höland et al., 1995). Previously, these crystals had only been produced under hydrothermal conditions (Newesely 1972; Jaha et al., 1997). Their morphology corresponds to that of the hydroxyapatite in natural teeth (enamel).

Experiments were performed at 1050°C to study the crystal growth processes of apatite explicitly. It was determined that as early as 10 min into the heat treatment (Fig. 2-33) a relatively large number (approximately 4–5 crystals per μm^2) of longish crystals measuring approx. 1–2 μm in length had been formed. The number of crystals decreased as the length of the heat treatment at 1050°C was increased. The remaining crystals, however, grew in size until they reached an average size of 7 μm after 500 min. Figure 2-33 shows the entire crystallization process at 1050°C. Figure 2-34 shows an intermediate stage. These initial kinetic studies (Höland 1996) have shown that this process is a case of Ostwald ripening. In short, this means that the mass transport to the location of the crystal growth, that is, the diffusion process from the glass matrix, is the rate-determining step for crystal growth. Non-steady-state

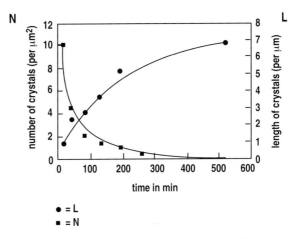

Figure 2-33 Number (N) and length (L) of needlelike apatite crystals as a function of time for apatite-leucite glass-ceramics.

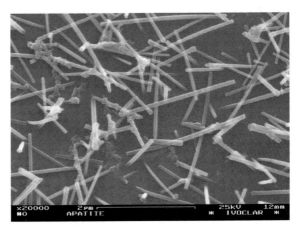

Figure 2-34 SEM image of the microstructure of apatite–leucite glass-ceramics with heat treatment at 1050°C/20 min.

processes were not observed in any of the cases. The mechanism of diffusion-controlled Ostwald ripening for needlelike apatite in a apatite–leucite glass-ceramic was proven by Müller et al. (1999). The mean diameter of the apatite needles and the length increases with $t^{1/3}$. The number of crystals decreases with t^{-1} due to a constant volume fraction of fluorapatite.

As reported the mechanisms of the nucleation and crystallization of needlelike apatite in apatite–leucite glass-ceramics is very difficult. Therefore, the mechanism was investigated in detail and the main results can be explained by the following two investigations.

To investigate the nucleation and crystallization mechanisms in apatite–leucite glass-ceramics, two kinds of samples were prepared: monolithic bulk and powdered samples. The chemical composition of these two samples was 54.8 wt% SiO_2, 14.1 wt% Al_2O_3, 8.4 wt% Na_2O, 10.6 wt% K_2O, 4.9 wt% CaO, 1.0 wt% ZrO_2, 0.3 wt% TiO_2, 3.9 wt% P_2O_5, 0.8 wt% CeO_2, 0.2 wt% Li_2O, 0.3 wt% B_2O_3, and 0.7 wt% F (Höland et al., 1999, 2000a).

2.4.6.1 Monolithic Glass-Ceramics

To discuss and to understand the findings, it is necessary to explain the experimental procedure selected. A monolithic base glass was cast as a block of approximately 10 mm × 10 mm × 50 mm and cooled slowly from T_g to room temperature (3–4 K/min). DSC measurements revealed a very interesting phenomenon (Fig. 2-35). The glass demonstrates two endothermic reactions. The endothermic reaction at 565°C corresponds to the T_g-range of the glassy matrix. The second peak at 634°C (reversible reaction) corresponds to the transformation or decomposition of a crystalline phase. The base glass was cooled and heat treated. After this procedure, crystalline phases were determined by X-ray diffraction at room temperature (Fig. 2-36). It was very surprising that primary crystals were determined already at room temperature up to 580°C. Fig. 2-36 (A) shows the diffraction peaks of $NaCaPO_4$ in a sample heated to 580°C. The

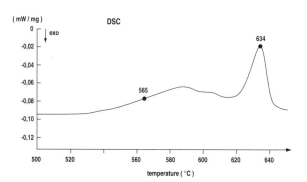

Figure 2-35 DSC functions of the base glass heated at 580°C for 15 min.

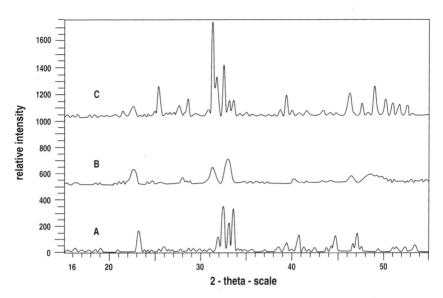

Figure 2-36 X-ray diffraction pattern of the base glass for apatite–leucite glass-ceramic after treatment at A) 580°C, 15 min; B) 640°C, 15 min; C) 700°C, 8 h and 1050°C, 2 h. A) represents $NaCaPO_4$ crystals, B) unknown phase, and C) fluorapatite formation.

microstructure (Fig. 2-37) of the corresponding sample (580°C) shows rounded crystals comparable in shape to phase separation of the glassy droplet phase. The lattice of $NaCaPO_4$ is orthorhombic corresponding to the space group *Pnam*. The lattice constants are a = 6.797Å, b = 9.165Å, c = 5.406Å on the basis of ICDD (International Center of Diffraction Data) pattern 29-1193.

At higher temperatures, above 610°C, a new interesting phenomenon was discovered. XRD investigations (Fig. 2-36 (B)) showed the precipitation of an additional crystalline phase at 640°C and no $NaCaPO_4$ crystals were present any longer.

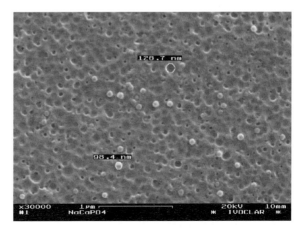

Figure 2-37 Primary crystal formation in apatite–leucite glass-ceramics. Crystals of $NaCaPO_4$ precipitate as very tiny species at 580°C after 15 min. SEM (10s, 2.5% HF).

But the XRD diagram (Fig. 2-36 (B)) could not be interpreted on the basis of ICDD (International Center of Diffraction Data) (Höland et al., 1999).

The primary crystals of apatite were determined at 700°C for 8 h heat treatment (Fig. 2-38). But these crystals did not show a needlelike habit. Needlelike fluoroapatite was precipitated at 700°C for 8 h and an additional heat treatment of 1050°C for 2 h. It is interesting to note in Fig. 2-39 that ball-like phases or areas are visible in this microstructure after the sample has been etched. This droplet-like area might be a diffusion zone responsible for the ion transport to the fast-growing apatite needles. This phenomenon is also visible at 800°C at the beginning of the apatite precipitation.

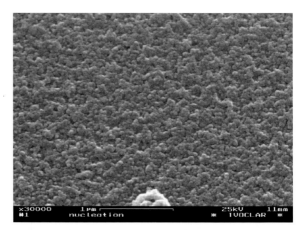

Figure 2-38 Fluoroapatite and X-crystal formation at 700°C/8 h. SEM (10 sec, 2.5% HF).

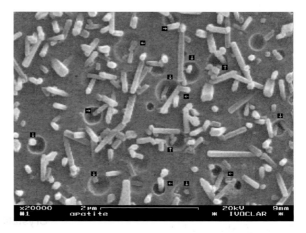

Figure 2-39 Crystallization of needlelike apatite in the volume of the glass-ceramic after heat treatment at 700°C/8h and additionally at 1050°C/2 h. SEM (10 sec, 2.5% HF). SEM micrograph of a representative microstructure; arrows indicate the needlelike apatite and its special circled environment (diffusion area).

In addition to the cast and slowly cooled samples, quenched base glass samples were investigated (Höland et al., 2000a). The glass was quenched (800 K/min) by pressing the glass melt to form a disc with a thickness of approximately 2 mm and a diameter of 25 mm. Then, the glass disc was cooled slowly from Tg to room temperature (3–4 K/min). The base glass was investigated by high-temperature X-ray diffraction, applying a heating rate of 1 K/min. Crystalline phases were detected like those shown in Fig. 2-40(a). NaCaPO$_4$ was analyzed as the primary crystal phase in the quenched base glass. This phase is transformed into another phase—most probably a phosphosilicate phase—at approximately 600°C. Apatite was not detected by this method because of the surface crystallization of leucite. Therefore, additional heat treatment of the sample at 700°C and higher was necessary. The surfaces of the cooled samples were ground and XRD measurements conducted at room temperature. Based on this method it was possible to ana-

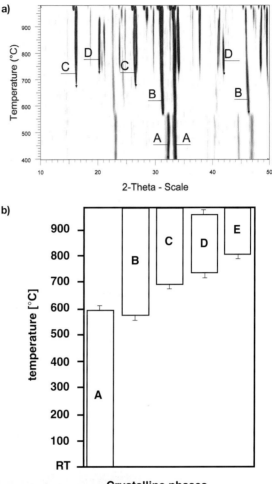

Figure 2-40 Crystal formation in apatite–leucite glass-ceramic (A: NaCaPO$_4$, B: Na$_2$Ca$_4$(PO$_4$)$_2$SiO$_4$, C: KAlSi$_2$O$_6$, D: unknown phase, E: Ca$_5$(PO$_4$)$_3$F.
a) HT-XRD
b) reaction scheme of phase formation

lyze the beginning of apatite formation at approximately 800°C (Fig. 2-40(b)). This method enabled the determination of the sequence of phase formation in a multicomponent system. A more detailed analysis of the structure can be conducted by NMR investigations in addition to room temperature XRD and high temperature XRD investigations. Initial results have been published by Schmedt auf der Günne et al. (2000).

2.4.6.2 Sintered Glass-Ceramics

Heat treatment of glass powders results in the precipitation of leucite from the surface of the glass-ceramics and the volume crystallization of needlelike apatite. Fig. 2-41 shows that leucite and needlelike apatite are formed in glass powders after heat treatment at 850°C for 1 hour and 1050°C for 1 hour. The formation

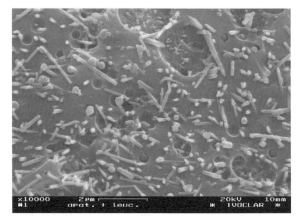

Figure 2-41 SEM image (etching 10 sec, 3% HF) of the microstructure of apatite–leucite glass-ceramic for dental restorations. Heat treatment of the glass powder at 850°C/1 h and 1050°C/1 h. The apatite crystals measure approximately 0.1–0.5 and 1–2 μm in diameter. The leucite crystals measure approx. 2 μm.

of needles at less than 1000°C was studied by Völksch et al. (2000) by TEM.

The leucite crystals measure approximately 1–2 μm in diameter. The considerably smaller needlelike apatite crystals are located between the leucite crystals. With regard to the morphology of these apatite crystals, it should be mentioned that the needlelike apatite formed in the glass-ceramic is comparable to the needlelike apatite in natural teeth. For comparison, a special high

resolution SEM micrograph was carried out to investigate the needlelike apatite of a tooth in the area of dental enamel. The result of this investigation is shown in Fig. 2-42. The needlelike apatite in natural teeth measures approximately 200–500 nm in length.

The crystallization process of the two

Figure 2-42 High-resolution SEM of natural tooth (etching 30s, 1 N HCl).

main phases in the glass-ceramic microstructure was conducted to the point where approximately 10–25 vol% leucite and 5–10 vol% apatite were determined by XRD. Main properties of this type of glass-ceramic were investigated by Szabo et al. (2000). These properties and the application of apatite–leucite glass-ceramic as a main component of a biomaterial in restorative dentistry is demonstrated in Section 4.4.2.

2.4.7 SiO_2–Al_2O_3–CaO–Na_2O–P_2O_5–F (Needlelike Apatite)

Needlelike apatite represents the main crystal phase in glass-ceramics derived from the SiO_2–Al_2O_3–CaO–Na_2O–P_2O_5–F system. The needlelike morphology of fluoroapatite has been shown in apatite–leucite glass-ceramics (Section 2.4.6). Two mechanisms, surface and volume nucleation, were applied to develop this type of glass-ceramic.

The development of a chemically durable, translucent glass-ceramic containing needlelike fluoroapatite was possible in the following composition range: 45–70 wt% SiO_2, 5–22 wt% Al_2O_3, 1.5–11 wt% CaO, 4.5–13 wt% Na_2O, 0.5–6.5 wt% P_2O_5, 0.1–2.5 wt% F.

Additives of K_2O up to 8.5 wt% and R(II)O up to 5 wt% are also possible. A typical microstructure of this type of glass-ceramic is shown in Fig. 4-45a of Section 4.4.2.

The glass-ceramic can be made to demonstrate a grain size of less than 100 µm. Following sintering of the glass-ceramic to a compact product at approximately 800°C, secondary crystallization was not observed. Needlelike fluoroapatite is the main crystal phase. The mechanism of nucleation and crystallization of apatite may be compared to that reported in Section 2.4.6. The application of the sintered glass-ceramic on the surface of a lithium disilicate substrate is reported in Section 4.4.2 (Schweiger et al., 1999; Höland et al., 2000b).

2.5 IRON SILICATES

2.5.1 SiO_2–Fe_2O_3–CaO

Glass-ceramics of this materials system were developed with the objective of obtaining improved magnetic properties over those found in products achieved with the precipitation from aqueous solutions. Ebisawa et al. (1991), for example, produced glass-ceramics of the SiO_2–Fe_2O_3–CaO materials system with the composition of 40 wt% Fe_2O_3 and 60 wt% $CaO·SiO_2$. The selection of this system was primarily directed toward the development of magnetic properties by controlled crystallization of iron-oxide crystals. At the same time, however, special biological properties were achieved with the insertion of Ca^{2+} ions.

Controlled crystallization of the base glass took place at temperatures between 700°C and 950°C. The formation of magnetite crystals (Fe_3O_4) was determined as the primary phase. The first primary crystals measured 6–30 nm. At heat treatment above 1000°C, the magnetite was converted to hematite (α-Fe_2O_3).

The glass-ceramic showed a maximum in saturation magnetization of 32A·m^2/kg and coercitive force of 39.8 kA/m when the base glass was heat treated at 950°C. The magnetic properties of the end product were accurately adjusted in relation to the content of magnetite in the glass-ceramic. The coercitive force, however, can only be determined qualitatively in terms of the size of the magnetite crystals. The most important result regarding the magnetic properties, however, was that the magnetic values achieved in the glass-ceramics were higher than those obtained from powders that had been produced from aqueous solutions.

This glass-ceramic contains no P_2O_5, but it may possess bioactive properties. As is shown in a comparison with bioactive apatite–wollastonite glass-ceramics (Section 2.4.2), the formation of apatite crystals on the surface of bioactive glass-ceramics is necessary when these materials come in contact and react with body fluids. Therefore, apatite forms on the surface of the P_2O_5-free SiO_2–Fe_2O_3–CaO glass-ceramic and generates a bond with human bone. This bond is free from connective tissue. Unlike the other glass-ceramic systems, no P_2O_5 is added here. This feature is new in the SiO_2–Fe_2O_3–CaO glass-ceramic system. Hence, bioactivity is achieved with reactive surface groups of the $\equiv$Si-OH-type and by the release of Ca^{2+} ions.

Given this special combination of bioactivity and magnetism, this type of glass-ceramic can be used as thermoseeds for the hyperthermia of cancers, particularly in the case of bone tumors. When the glass-ceramic, for example, is implanted in the human body in powder form, a direct bond to the bone is generated as a result of its bioactivity. For cancer treatment by magnetic hysteresis loss, tumors are locally heated to temperatures above 43°C under an alternating magnetic field. Consequently, a bioactive and at the same time ferromagnetic glass-ceramic implant can contribute to the destruction of a bone tumor.

2.5.2 SiO_2–Al_2O_3–FeO–Fe_2O_3–K_2O (Mica, Ferrite)

Glass-ceramics, having ferrimagnetic properties, were developed by Le Bras (1976). The magnetic property was achieved on the basis of the precipitation of calcium ferrite in glasses. Therefore, iron oxides were incorporated into silicate glasses. The base glass was characterized by the composition in:

34–40 wt% SiO_2, 25–38 wt% CaO, 18–27 wt% Fe_2O_3, 2–6 wt% Al_2O_3, 0–10 wt% MgO, and 0.7–2 wt% Cr_2O_3.

Beall and Reade (1979) developed hematite–magnetite glass-ceramics with ferrimagnetic properties. The glass-ceramic was characterized by high permeability and low resistivity (approximately 10^{-4} Ω/cm). The base composition of these materials is shown in Table 2-21.

Reade (1978) reported on a ferrimagnetic glass-ceramic, which spontaneously developed crystallites of $NiFe_2O_4$ and $CoFe_2O_4$ in the surface. Its composition was 40–75 wt% SiO_2, 16–27 wt% Al_2O_3, 1–3 wt% Fe_2O_3, 2.5–5.5 wt% Li_2O, 1.7–6 wt% TiO_2, 0.5–3 wt% NiO, 1–3 wt% CoO.

Smith (1984) showed that a ferrimagnetic glass-ceramic can be applied in medicine for use in hyperthermia. This concept included a local heating (42–45°C) that causes tumors to shrink, or in some cases, to disappear. A glass-ceramic was produced having a large hysteresis effect. The major crystal phases were lithium ferrite ($LiFe_5O_8$) and hematite (Fe_2O_3). Coercitive force, H_c, was 14.5 kA/m. After injection of the glass-ceramic in the tissue of animals and the application of 10 kHz and 39.8 kA/m the temperature rises approximately 9°C. But Smith (1984) reported that much more testing with animals will be necessary before this can be considered a method for treatment of human tumors.

Glass-ceramics that may be both machinable and ferrimagnetic should contain ferrite crytals and mica crystals. The reason for the good machinability of mica–type glass-ceramics is demonstrated in Section 4.1.3. The base glass system for machinable ferrimagnetic glass-ceramics is derived from the SiO_2–Al_2O_3–MgO–K_2O–F system and additives of iron oxides.

Multiple additions of FeO and Fe_2O_3 to SiO_2–Al_2O_3–MgO–K_2O base glasses increase the phase separation and favor the processes of nucleation

Table 2-21

Hematite–Magnetite Glass-Ceramic (Beall and Reade, 1979) Chemical Composition (wt%)

SiO_2	45–66
Al_2O_3	10–20
FeO	10–40
Li_2O	1.5–6
TiO_2/ZrO_2	0–5
B_2O_3	1

and crystallization. It was demonstrated that additions of FeO and Fe_2O_3 to the base glass composition of 20.7 mol% MgO, 19.6 mol% Al_2O_3, and 59.8 mol% SiO_2 increased phase separation in the base glass (Höland et al., 1982b). The addition of FeO and Fe_2O_3 was carried out by substituting these substances for MgO and Al_2O_3. The content of Fe(II)/Fe(III) within the molten base glass was determined in a Mossbauer study (Höland et al., 1982b).

Different microstructures of the glass were formed, depending upon the chemical composition of the base glass. Figure 2-43 shows the phase formation processes of glasses dependent upon the chemical composition. Status A, B, and C represent the microstructures of glasses having an increasing content of FeO and Fe_2O_3. On the basis of these quite-different starting conditions of the base glasses, nucleation and controlled crystallization were initiated by heat treating the three types (A, B, C) of base glasses. As a net result, it could be shown that a special chemical composition and heat treatment of the glass are necessary to develop a material demonstrating the microstructure of a machinable ferrimagnetic glass-ceramic material with magnesium ferrite and mica of the biotite-type (status F in Fig. 2-43). It is necessary to have a relatively high content of both iron and magnesium oxide to precipitate magnesium ferrite and biotite.

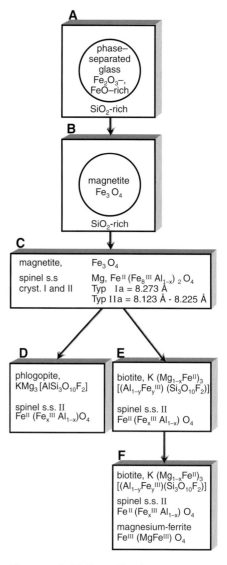

Figure 2-43 Crystallization sequence (schematic) of SiO_2–Al_2O_3–MgO–K_2O–F–FeO–Fe_2O_3 glasses. A, B: phase-separated base glasses, B: base glass having primary crystals after cooling, D, E, F: glass-ceramics after heat treatment of the base glasses.

Therefore, if the iron-oxide content (sum of FeO and Fe_2O_3) is higher than 6 mol% and MgO is higher than 15 mol%, a ferrimagnetic and machinable glass-ceramic material may develop following an additional heat treatment of 1180°C for 60 min. The chemical composition is shown in Table 2-22 and the corresponding microstructure is demonstrated in Fig. 2-44.

These ferrimagnetic machinable glass-ceramics have specific magnetic moments of 4–20×10^{-4} T cm^3g^{-1}. The magnetic hysteresis loop yields a coercitive force of 200 kA/m. These magnetic properties are lower than those of high-density sintered ferrite materials, such as magnesium or Ni, Co, Mn ferrite. Nevertheless, because of its machinability, the material presents potential for future applications.

2.5.3 SiO_2–Al_2O_3–Fe_2O_3–$R(I)_2O$-$R(II)O$ (Basalt)

Basalt is a gray to black, fine-grained volcanic rock that forms large lava flows on oceanic islands and is also common at many continental sites. Chemically it is composed of major oxides: SiO_2, Al_2O_3, FeO, CaO, MgO, and to a lesser degree Fe_2O_3, Na_2O, K_2O, TiO_2, MnO, and P_2O_5, with many other trace ingredients. The two major minerals that are always present are monoclinic pyroxene and plagioclase feldspar, but magnetite, olivine, and glass are often present as well as other accessory minerals.

Basalts are easily melted at 1400C° or above and are readily cooled to a glass (Beall and Rittler, 1976). Upon reheating, they can form fine-grained glass-ceramics if the Fe_2O_3/FeO weight ratio is sufficiently high, normally above 0.5. The major crystalline phase in these glass-ceramics, and sometimes the only crystal species is monoclinic pyroxene: a complex solid solution of $CaMgSi_2O_6$ (diopside), $CaFeSi_2O_6$ (hedenbergite), $MgSiO_3$ (enstatite), $FeSiO_3$, $NaFeSi_2O_6$ (acmite), $CaAl_2SiO_6$, $CaTiAl_2O_6$, and $MnSiO_3$.

Table 2-22

Ferrimagnetic Mica Glass-Ceramic (Höland et al., 1982b) Chemical Composition (in mol%)	
SiO_2	48.7
Al_2O_3	10.8
MgO	15.8
K_2O	5.6
F	12.5
Fe_2O_3	2.8
FeO	3.8

The feldspathic ingredients, for the most part, remain in the residual glass as a complex alkali–alkaline earth aluminosilicate.

To control internal nucleation, the basalt glass is typically melted under oxidizing conditions (Beall and Rittler, 1976). Since the original rock normally has a low Fe_2O_3/FeO ratio, usually below 0.5, the final glass must acquire some oxygen to produce a well-nucleated glass-ceramic. This can often be achieved simply by grinding the rock to a powder and melting it in air. However, to minimize deformation during crystallization and to achieve a very fine grain size, oxidizing agents like ammo-

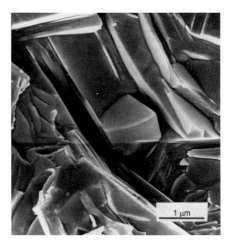

Figure 2-44 Microstructure of machinable ferrimagnetic glass-ceramic (see the composition in Table 2-22) after heat treatment (1180°C/1 h). Cubic spinel crystal of the ferrite-type and mica crystals characterize the morphology of the glass-ceramic (SEM).

nium nitrate (NH_4NO_3: Beall and Rittler, 1976) or MnO_2 (El-Shennawi et al., 1999) have been found effective. Indeed, nanocrystalline glass-ceramics with clinopyroxene crystals well below 100 nm in size have been reported (Beall and Rittler, 1976). The role of iron oxide in the nucleation has been attributed to a clustering of Fe^{3+} in the glass that upon heating yields magnetite (Fe_3O_4) as the nucleating agent, typically from 650° to 800°C, depending upon the composition. Sometimes the magnetite persists and can be seen as nuclei within clinopyroxene crystals or clusters; in other cases it is resorbed by soda to produce acmite ($NaFeSi_2O_6$) and other species capable of solid solution in pyroxene.

A series of physical properties was measured on an oxidized basalt glass-ceramic made entirely from the Holyoke basalt quarried at Westfield, MA, U.S.A. The composition of this raw basalt was analyzed as follows: 51.6 wt% SiO_2, 14.1 wt% Al_2O_3, 9.3 wt% CaO, 8.4 wt% FeO, 6.4 wt% MgO, 4.4 wt% Fe_2O_3, 3.2 wt% Na_2O, 1.2 wt% K_2O, 1.0 wt% TiO_2, 0.2 wt% P_2O_5, and 0.2 wt% MnO. The glass-ceramic, after melting with a 4% NH_4NO_3 addition and subsequent crystallization at 650°C for 4 h followed by 1 h at 880°C, had a more oxidized iron condition, namely FeO 4.8 and Fe_2O_3 10.0. Its properties were as follows: coefficient of thermal expansion MPa $72\cdot10^{-7}$ K^{-1}, Knoop hardness 900 MPa, modulus of rupture after abrasion

100 MPa, and alkali durability similar to soda–lime glass. Further increases in strength of basalt glass-ceramics may be possible with additives such as limestone and soda, which have been reported to increase the percent crystallinity (El-Shennawi et al., 1999). Because of the good strength and abrasion resistance, as well as the ease of pressing and centrifugal casting of the parent glasses, basalt glass-ceramics have been considered for roofing tile and liners for steel pipe.

2.6 PHOSPHATES

2.6.1 P_2O_5–CaO (Metaphosphates)

Since phosphate glass-ceramics exhibit some properties that are more favorable and others that are less favorable than those of silicate glass-ceramics, their special characteristics should be addressed at the beginning of this section. Most of these features are attributed to the different nucleation and crystallization conditions of phosphate glass-ceramics compared with those of silicate glass-ceramics. The glass structure provides the basis for these characteristics. Therefore, the structures of the two glass-ceramics families must be addressed.

The network-forming structure of silicate glasses, such as alkali or alkaline-earth silicate glasses, is composed of (SiO_4) tetrahedra (Kreidl 1983). Because of the tetravalence of silicon, (SiO_4) tetrahedra can link together on four corners by means of oxygen ions to form a fully three-dimensional network. In alkali or alkaline-earth silicate glasses, the alkali and alkaline-earth ions break Si–O–Si bonds, loosening the (SiO_4) tetrahedra network.

Phosphate glasses are also made up of short-range structures of tetrahedral units. The (PO_4) and (SiO_4) tetrahedra, however, demonstrate one major difference. That is, the (PO_4) tetrahedra cannot link together on four corners to form a complete network as is the case for silicate glasses because of the pentavalence of phosphorous. The (PO_4) tetrahedra can only link with other tetrahedra on three corners. Figure 2-45 shows the nomenclature for and examples of (PO_4) units with different degrees of polymerization. The nomenclature for the different tetrahedra is also shown. Tetrahedra are differentiated according to their degree of linkage: Q^0, Q^1, Q^2, or Q^3. In addition, Fig. 2-45 shows that Q^4 tetrahedra are formed by (SiO_4) tetrahedra in contrast to (PO_4) tetrahedra.

In Fig. 2-45, the degrees of polymerization of (PO_4) units are shown as the number of linked tetrahedra. Therefore, different structural units result. These units are produced by adding network-modifier ions to the phosphate glasses. First, however, it must be ensured that the addition of alkali or alkaline-earth

a) **[PO₄] – structure units**

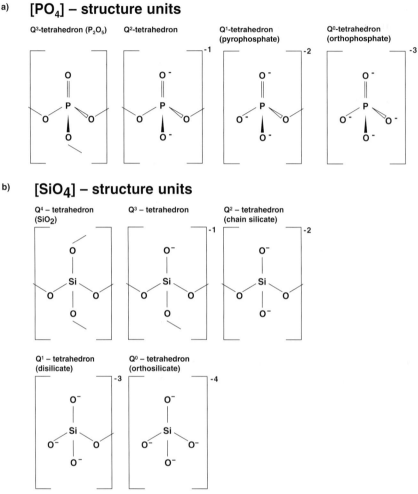

b) **[SiO₄] – structure units**

Figure 2-45 Short-range structure of phosphate (a) (Kirkpatrick and Brow, 1995; Brow et al., 1995) and silicate glasses (b).

oxides to pure P_2O_5 will strengthen rather than weaken the P_2O_5 network, as was shown by Kreidl (1983) (Fig. 2-46). Increasing amounts of alkali or alkaline-earth ions, however, systematically break down the long-range order of the networks until only chains (polyphosphates, metaphosphates) or even diphosphate or orthophosphate structural units are evident (Kreidl 1983).

By increasing the amount of network-modifier ions, a glass containing a network was modified into an invert glass, composed of diphospate or orthophospate units, containing more than 50 mol% network modifier ions. The phosphate structural units of the glass that became increasingly smaller in the process, were identified with chromatographic methods (Vogel 1978) and nuclear magnetic resonance (Kirkpatrick and Brow, 1995; Brow et al., 1995).

Metaphosphate glasses are of particular interest in the development of glass-ceramics and glass-ceramic fibers (Griffith 1995) by the controlled crystallization of phosphate base glasses. These glasses are composed of (PO_4) tetrahedra chains. In chemical terms, these structural units must be called condensed phosphates.

A $NaPO_3$-type metaphosphate glass was used to investigate con-

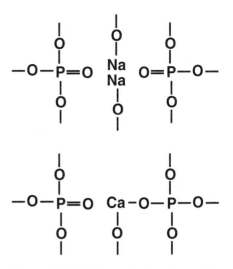

Figure 2-46 Addition of network modifier components to phosphate glasses (Kreidl 1983)

trolled crystallization and controlled surface crystallization in particular (Gutzow 1987). The stress–strain relations that influence surface crystallization of glasses were also examined in this glass (see Section 1.4).

Calcium metaphosphate glass-ceramics, however, are of practical significance in the production of materials based on phosphate glass-ceramics. Abe (1979) developed a Ca-metaphosphate glass-ceramic by successfully using the special nucleation and crystallization characteristics of Ca-metaphospate glasses. Because of the Q^2 chain structure, these glasses demonstrated high-crystal-growth rates of up to 10^7 units. That is, the mass-transport processes in the form of ion mobility in phosphate glasses were considerably higher than in silicate glasses. Abe et al. (1984) used this phenomenon to develop a mechanism of unidirectional crystallization. A detailed description of this crystallization process is given below.

A calcium metaphosphate glass rod composed of $CaO:P_2O_5 = 0.95$ (molar ratio) was produced. One area of this rod was heated to the softening point in a special furnace at 600°–650°C. In the process, nucleation and controlled crystallization of β-$Ca(PO_3)_2$ needlelike crystals occurred. It was interesting to note that crystals formed along the glass rod in one direction only, from the hotter to the colder areas. As the glass rod was pushed farther into the furnace, crystallization continued in the hot areas until unidirectional Ca-metaphosphate was produced as the main crystal phase. That is, the crystals were precipitated anisotropically and with preferred orientation.

β-Ca(PO$_3$)$_2$ always formed the main crystal phase and small amounts of 2CaO·3P$_2$O$_5$ the secondary crystal phase.

In addition to this anisotropic crystal growth, Abe et. al (1982) determined a further difference in the formation of phosphate glass-ceramics compared with that of silicate glass-ceramics. In silicate glasses, it is generally accepted that nucleation and primary crystallization is not possible below the transformation point (T$_g$). In other words, the transformation point of the silicate glass-ceramic is never below that of the base glass. In phosphate glasses, however, nucleation and primary crystallization can be initiated below the transformation point of the base glass.

The high flexural strength along the vertical axis of the crystal represents a special characteristic of oriented Ca-metaphosphate glass-ceramics. As a result, bending strengths of up to 800 MPa have been determined with a Young's modulus of 85 GPa. This material, therefore, achieves clearly lower Young's moduli compared with the 300–400 GPa exhibited by high-quality sintered Al$_2$O$_3$ ceramic, a material of similar strength.

Although there are many advantages of phosphate glass-ceramics over silicate glass-ceramics, there are also some critical disadvantages. The low chemical durability of phosphate glass-ceramics compared with that of silicate glass-ceramics presents the main disadvantage. Therefore, considerable research efforts were required to produce chemically durable glass-ceramic products according to the principles used in the development of Ca-metaphosphate glass-ceramics. The low chemical durability was most evident in the residual glass matrix and the glass-crystal interface. Kasuga et al. (1993) improved the chemical durability of P$_2$O$_5$–CaO glass-ceramics by adding small amounts of Al$_2$O$_3$, Y$_2$O$_3$, and ZrO$_2$ to the base glass. The composition 46 mol% P$_2$O$_5$, 52 mol% CaO, 2 mol% Al$_2$O$_3$ is typical for this type of glass-ceramic.

In contrast to the binary glass (CaO:P$_2$O$_5$ = 0.95, molar ratio), in which unidirectional growth is initiated by gradient crystallization, controlled crystallization of the durable glass (46 mol% P$_2$O$_5$, 52 mol% CaO, 2 mol% Al$_2$O$_3$) is conducted as volume crystallization of the monolithic glass at a constant temperature. Calcium metaphosphate crystals as fine dendritic δ-Ca(PO$_3$)$_2$ phases (1–5 μm in length) are produced with nucleating agents that produce volume crystallization. The crystals do not exhibit a preferred direction in the glass matrix. The glass-ceramic produced in this way demonstrates a bending strength of 150 MPa and fracture toughness of 1.8 MPa·m$^{0.5}$. This biomaterial is used for restorative dental applications. The product name is Crys-Cera®, produced by Kyushu Dentceram Co., Ltd., Okayama-ken, Japan.

Although the low chemical durability of the binary Ca-metaphospate glass-ceramic is unfavorable for the fabrication of monolithic glass-ceramic products, this property has been helpful in the development of fibers for composites of various ceramics. Kasuga et al. (1992, 1996), for example, took advantage of the high solubility of the glass matrix of Ca-metaphosphate glass-ceramics to produce β-Ca(PO$_3$)$_2$ fibers.

A base glass composed of CaO:P$_2$O$_5$ = 0.85 (molar ratio) was used to produce these fibers. Crystallization took place at 600°C for 48 h. Unidirectional crystallization techniques were not conducted. Crystallization of β-Ca(PO$_3$)$_2$ proceeded in needle form, with random orientation yielding an isotropic body. An ultraphosphate of the substantially lower chemical durability of the glass matrix and the ultraphosphate crystals compared with that of the β-Ca(PO$_3$)$_2$, the matrix was easily extracted by aqueous leaching. These β-Ca(PO$_3$)$_2$ fibers were subsequently used in a composite material with a very low modulus of elasticity (Kasuga 1996).

2.6.2 P$_2$O$_5$–CaO–TiO$_2$

Abe et al. (1995) and Hosono et al. (1994) developed glass-ceramics in the P$_2$O$_5$–CaO–TiO$_2$–Li$_2$O system by combining the P$_2$O$_5$–CaO and P$_2$O$_5$–Li$_2$O–TiO$_2$ systems. The base glass with the exact equimolar composition of Ca$_3$(PO$_4$)$_2$ and LiTi$_2$(PO$_4$)$_3$ was melted. Subsequently, the monolithic glass was converted into a uniform glass-ceramic by heat treatment. Nucleation took place at 620°C for 20 h and controlled crystallization at 730°C for 12 h. The monolithic glass-ceramic contained two main crystal phases: calcium orthophosphate, Ca$_3$(PO$_4$)$_2$, and lithium-titanium orthophosphate, LiTi$_2$(PO$_4$)$_3$. These two crystal phases possessed different degrees of chemical durability when treated with HCl. Since Ca$_3$(PO$_4$)$_2$ crystals dissolve in HCl, and LiTi$_2$(PO$_4$)$_3$ crystals are resistant, an open-pore glass-ceramic with an LiTi$_2$(PO$_4$)$_3$ main crystal phase was produced by treating the glass-ceramic with HCl. A small amount of Al$_2$O$_3$ was added to the glass-ceramic to form solid-solution crystals of composition Li$_{1.4}$Al$_{0.4}$Ti$_{1.6}$(PO$_4$)$_3$ type (Fig. 2-47). Bulk samples demonstrated continuous pores of approximately 30–50 nm in diameter (Hosono et al., 1993).

Furthermore, since the Li ion is highly mobile in the glossy matrix, ion exchange may take place with some Ag$^+$ ions, even in the presence of Na$^+$, replacing Li$^+$ to form an open-pore glass-ceramic with AgTi$_2$(PO$_4$)$_3$ (Hosono et al., 1994). This type of glass-ceramic is biocompatible and demonstrates favorable bactericidal effects (Abe et al., 1995).

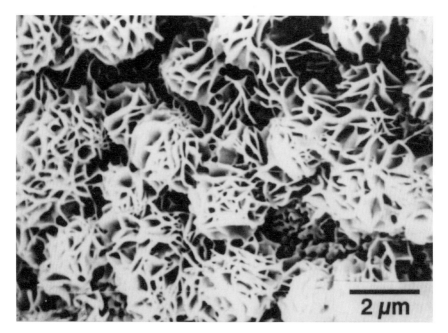

Figure 2-47 Microstructure of a porous glass-ceramic with a skeleton of $Li_{1.4}Al_{0.4}Ti_{1.6}(PO_4)_3$ crystals (Hosono et al., 1993).

2.6.3 P_2O_5–Na_2O–BaO and P_2O_5–TiO_2–WO_3

2.6.3.1 P_2O_5–Na_2O–BaO System

Wilder et al. (1982) developed glass-ceramics derived from the P_2O_5–Na_2O–BaO system. The microstructure and properties of the glass-ceramics were controlled by adding up to 3 mol% Al_2O_3 to base glasses of the composition 40 mol% Na_2O, 10 mol% BaO, 50 mol% P_2O_5.

The glasses were crystallized during a 2-h heat treatment cycle at 400°C and an additional 2-h cycle at 450°C. Main crystal phases of $NaPO_3$ (sodium metaphosphate), $(NaPO_3)_3$ (sodium trimetaphosphate), and $NaBa(PO_3)_3$ (sodium barium phosphate) were analyzed by X-ray diffraction. Effective nucleation was achieved by adding 0.01 mol% Pt to the base composition. The morphology of the microstructure varied from spherulitic to circular. Because of their high coefficients of thermal expansion ranging from 14.0×10^{-6} K^{-1} to 22.5×10^{-6} K^{-1}, these glass-ceramics can be joined to metals, e.g., alumina and copper. This application requires good chemical durability, which was achieved by this type of glass-ceramic.

2.6.3.2 P_2O_5–TiO_2–WO_3 System

Aitken (1992) was able to develop glass-ceramics in the P_2O_5–TiO_2–WO_3 system. A typical composition was 42 wt% P_2O_5, 29 wt% TiO_2, 29 wt% WO_3.

The base glass crystallized when it was heat treated at approximately 900°–950°C for 2 h. The main crystal phases of WO_{3-x} and TiP_2O_7 were precipitated in the glass matrix. Analysis showed WO_{3-x} to be present in its cubic form at temperatures below 900°C and as its prismatic polymorph at 950°C. WO_{3-x} glass-ceramics demonstrate special electronic properties. Materials of this type exhibit electrical conductivity and display temperature insensitive resistivity as low as 10^{-2} Ω·cm. The glass-ceramics show also another special property. The material is semi-conductive, because the WO_3 lathlike crystals are connected forming a continuous conduction path. This was revealed by micrographs.

2.6.4 P_2O_5–Al_2O_3–CaO (Apatite), P_2O_5–CaO (Metaphosphate)

2.6.4.1 P_2O_5–Al_2O_3–CaO (Apatite)

Glass-ceramics of the P_2O_5–Al_2O_3–CaO-type are produced as monolithic bulk glass-ceramics and as composites of glass-ceramics and metals. Both types are used as biomaterials for bone substitution in human medicine.

Glass-ceramics of the BIOVERIT® III-type represent monolithic phosphate glass-ceramics that do not contain silica. These glass-ceramic materials contain the following main crystal phases: apatite and $AlPO_4$ crystals of the berlinite-, tridymite-, and cristobalite-type. Complex phosphate structures, such as $Na_5Ca_2Al(PO_4)_4$, $Na_{27}Ca_3Al_5(P_2O_7)_{12}$ (see Section 1.3.2) are precipitated as secondary phases (Wange et al., 1990; Höland et al., 1991a).

Within the scope of the development of BIOVERIT® III as a phosphate glass-ceramic biomaterial for bone substitution in human medicine, a new way of controlling crystallization in phosphate glasses was discovered (Vogel et al., 1987). The processing of this type of glass-ceramics begins with the formation of a phosphate invert glass in the P_2O_5–Al_2O_3–CaO–Na_2O system. The structure of this glass is formed by mono- and diphosphate units. The base glass demonstrates an invert glass structure. Phase-separation processes were not observed. Therefore, nucleation via phase-separation processes was not possible. Doping with a suitable nucleation agent, e.g., iron oxide or ZrO_2, however, led to supersaturation of the glass with these oxides in a certain concentration range. This supersaturation was reduced by subsequent heat treatment, which resulted in the precipitation of a primary crystal phase according to the mechanism of bulk nucleation. In the present case, this phase is an Na–Zr phosphate. Thus crystal nuclei have developed, which initiate the precipitation of the main crystal phases of apatite and $AlPO_4$ and specific complex phosphates of $Na_5Ca_2Al(PO_4)_4$, $Na_{27}Ca_3Al_5(P_2O_7)_{12}$ (see Section 1.3.2). The chemical composition of BIOVERIT® III glass-ceramic is shown in Table 2-23.

The glass-ceramic material shows good chemical properties (hydrolytic durability). Furthermore, its thermal properties make the material well suited for the preparation of composites with certain metals, especially with the Co–Cr alloys, which are widely used in implantology. Jana et al. (1996) prepared a suspension of phosphate glass-ceramic BIOVERIT® III in water or alcohol and coated a Co–Cr alloy. The subsequent thermal treatment at approximately 700°C for a few minutes resulted in a dense and almost glassy coating of a thickness of 150–400 mm. Thermal treatment at 500°–600°C formed the crystal phases in the glassy matrix. Long-term *in vitro* tests in a trisbuffer solution demonstrated that a protective layer was formed during the corrosion process. This was a prerequisite for long-term stable implants of Co–Cr implants coated with BIOVERIT® III. Vogel et al. (1995) reported on clinical tests of phosphate glass-ceramics coated on metal implants for orthopedic surgery.

Composites of phosphate glass-ceramics and metals were developed by Pernot and Rogier (1992). The base glass was characterized by the composition 69.0 wt% P_2O_5, 8.3 wt% Al_2O_3, and 22.7 wt% CaO. The glass was prepared by melting at 1300°C for 2 h. After cooling, it was milled to a particle size of

Table 2-23

BIOVERIT® III		
Composition	**Range (wt%)**	**Example (wt%)**
P_2O_5	45–55	51.0
Al_2O_3	6–18	10.0
CaO	13–19	14.0
Na_2O	11–18	15.0
$MeO/Me_2O_3/MeO_2$ (MnO, CoO, NiO, FeO, Fe_2O_3, Cr_2O_3, ZrO_2)	1.5–10	
F		2.0
ZrO_2		5.0
TiO_2		3.0

less than 50 μm. A Co–Cr alloy was milled to form particles measuring 22, 40, and 60 μm. The fraction with the smallest particle size was oxidized at 800°C for 1 to 100 h. The composites were prepared by hot pressing. A dense material was formed at a sintering temperature of 700°C for 1 h. Crystallization took place during this process of heat treatment. $AlPO_4$ of the cristobalite-type was the main crystal phase that was precipitated in the glass matrix. The result was a composite material that permitted the controlling of its coefficient of thermal expansion, dependent upon the glass-ceramic–to–metal ratio. The composite was characterized by Young's modulus of 68–124 GPa, dependent upon the composition and fracture toughness as K_{IC} of approximately 2.2 MPa·m$^{0.5}$. The biomaterials were developed for clinical tests as implants in orthopedic surgery.

Rogier and Pernot (1991) also succeeded in developing composites of phosphate glass-ceramics and titanium, and composites of phosphate glass-ceramics and stainless steel (Pernot and Rogier, 1993).

2.6.4.2 P_2O_5–CaO (Metaphosphate)

Abe (1979) and Abe et al. (1984) developed phosphate glass-ceramics by unidirectional crystallization. The final glass-ceramic material demonstrated improved mechanical properties. Abe et al. prepared $Ca(PO_3)_2$ glasses by using a melting process at 1200°–1400°C. After remelting, rods 1–5 mm in diameter were pulled from the melt at approximately 800°C. To carry out the unidirectional crystallization, one end of the glass rod was heated until it almost reached its softening point at approximately 600°C. Subsequently, the rod was placed in a furnace having a temperature gradient of 30 K/cm. Crystals of $β$-$Ca(PO_3)_2$ grew from the high-temperature end to the low-temperature end and became aligned parallel with the length of the rod. The crystal growth rate at 560°C was determined as 20 μm/min for a glass having a CaO/P_2O_5 ratio of 0.95. Whitlockite ($2CaO·3P_2O_5$) was precipitated as a secondary crystal phase during the process of unidirectional crystallization of $Ca(PO_3)_2$ precipitation.

The final glass-ceramic demonstrated very special mechanical properties. Bending strength was determined at approximately 640 MPa and Young's modulus at approximately 85 GPa. The results also demonstrated that toughness had been improved.

2.6.5 P_2O_5–B_2O_3–SiO_2

At first glance, the development of useful glass-ceramics seems unlikely, since borophosphate-rich glasses are known to exhibit relatively poor chemical durability. Glass-ceramics from this system, however, also illustrate how large a change in a key property can be achieved by crystallization.

MacDowell (1989) developed glass-ceramics from the $P_2O_5-B_2O_3-SiO_2$ system, in the composition range 35–57 wt% P_2O_5, 16–35 wt% B_2O_3, and 14–46 wt% SiO_2.

Glass and glass-ceramic compositions are shown in Table 2-24 and Fig. 2-48. In the development of the glass-ceramic, the composition was established according to the stoichiometric molar oxide ratios of B_2O_3:P_2O_5:SiO_2 in Table 2-24. The base glasses were melted at 1600°C. Following cooling, the microstructure of the glasses was analyzed using the TEM/replica technique. The base glass composition 6 (Table 2-24), for example, exhibits a homogeneous microstructure, which is typical of a variety of other phosphate glasses (see Sections 2.6.1–2.6.3). Therefore, no phase-separation processes occur, despite the fact that considerable SiO_2 is contained in the base glass. As a result, nucleation proceeds as a homogeneous process in glasses with the stoichiometric composition of the base glasses from the $P_2O_5-B_2O_3-SiO_2$ system.

During heat treatment at 900°C, homogeneous crystallization of the BPO_4 (boroorthophosphate) main crystal phase is shown to begin in the absence of energy reducing interfaces. The crystallites measure approximately 0.05 µm. During further heat treatment at 1100°C, secondary crystal growth begins and the boroorthophosphate crystals grow to approximately 1 µm.

In the crystallization process of these base glasses, special properties are developed in the glass-ceramic that cannot be produced in glasses. As an example, the aforementioned poor chemical durability, particularly the base

Table 2-24

Composition of a Glass-Ceramic of the $P_2O_5-B_2O_3-SiO_2$ System (after MacDowell and Beall, 1990)

Composition	Representative composition (wt%)			
	SiO_2	P_2O_5	B_2O_3	B_2O_3:P_2O_5:SiO_2 mole ratio
1	46.0	36.3	17.8	1:1:3
2	36.2	42.8	21.0	1:1:2
3	22.1	52.3	25.6	1:1:1
4	29.9	35.4	34.7	2:1:2
5	14.5	68.7	16.8	1:2:1
6	24.2	57.2	18.7	2:3:3

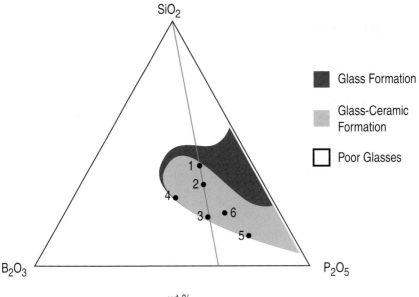

Figure 2-48 Glass and glass-ceramic formation in the P_2O_5–B_2O_3–SiO_2 system.

glasses in the P_2O_5–B_2O_3–SiO_2 system, is improved three orders of magnitude. The hydrolytic durability, the durability to 5% HCl and to alkali are in particular affected (MacDowell 1989). The improvement of the chemical durability must be attributed to the formation of crystals, which become embedded in a practically pure SiO_2 glass matrix that exhibits high chemical durability. This glass matrix protects the crystals from further chemical attack.

BPO$_4$ glass-ceramics are also characterized by a low dielectric constant between 3.8 and 4.5. As a result, the material is suitable for microelectronic applications, where circuit speed is inversely related to the square root of the dielectric constant. Since phosphate-based glass-ceramics are known to exhibit a coefficient of thermal expansion exceeding 100×10^{-7} K^{-1}, this coefficient had to be lowered before the material could be used for microelectronic packaging of silicon circuitry. This objective was fulfilled using SiO_2-rich glass-ceramics, for example, with composition 1 in Table 2-24. A coefficient of thermal expansion of almost 40×10^{-7} K^{-1} (20°C–300°C) was achieved in this glass-ceramic, making the material suitable for the above application (Section 4.5.2).

Using another process, MacDowell and Beall (1990) developed a further type of microstructure in BPO$_4$ glass-ceramics. This unique and effective process involves the development of a very porous "gas-ceramic" with low density from the dense BPO_4 glass-ceramic using ammonium phosphate. During the melting process of the glass, and according to one hypothesis, phosphorous is believed to be reduced from the $P5^+$ to the $P3^+$ state. Thereafter, the material is heat treated to conduct controlled crystallization, and the $P3^+$ is oxidized to the $P5^+$ state. Hydrogen is produced according to the following reaction:

$$P^{3+} + 2OH^- \rightarrow P5^+ + 2O^{2-} + H_2$$

Another and perhaps a simpler hypothesis involves the breakdown of ammonia (NH_3) from the ammonium phosphate batch material to N_2 plus H_2. The molecular hydrogen is partially dissolved in the hot glass and trapped in the glass structure on cooling. Then, on reheating, it is exsolved as bubbles, perhaps aided by the nucleation of BPO_4 crystals.

Pores are produced by the hydrogen and a "gas-ceramic" with the microstructure in Fig. 2-49 is formed. This reaction has been proven with the help of structural analyses of the glasses. NMR spectroscopy was successfully used for this purpose (Dickinson 1988). The size of the pores can be controlled to measure between 1 and 100 μm, depending on the hydrogen content. At the same time, the dielectric constant can be reduced to approximately 2. The properties of this glass-ceramic are discussed in detail in Section 4.5.2.

2.6.6 P_2O_5–SiO_2–Li_2O–ZrO_2

To produce a high-strength glass-ceramic by viscous flow during forming under pressure, glass-ceramics in the P_2O_5–SiO_2–Li_2O–ZrO_2 system were developed (Höland et al., 1994).

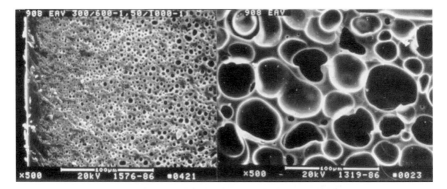

Figure 2-49 Microstructure of a "gas-ceramic" from the P_2O_5–B_2O_3–SiO_2 system.

Glass-ceramics containing ZrO_2 are known in other glass systems, for example, according to Beall (1991) and Budd (1993) in the enstatite system, and according to Uno et al. (1993) and Bürke et al. (2000) in the mica system. The new P_2O_5–SiO_2–Li_2O–ZrO_2 system should demonstrate formability under pressure. At the same time, it should be possible to incorporate high contents of more than 15 wt% ZrO_2 in the base glass to increase the strength of the material. Fundamental research on this glass system, however, has not yet been conducted.

The formation of specific main crystalline phases was observed in the SiO_2–Li_2O–ZrO_2 subsystem (Polezhave and Chukklantsew, 1965). As a result, glass formation was thoroughly studied. The formulation had to be considerably expanded with new composition ranges beyond the previous level of knowledge. Suitable viscosity ratios and nucleation conditions were examined, in the course of which the glass-ceramic system was discovered in the composition range of 4–15 wt% P_2O_5, 42–59 wt% SiO_2, 7–15 wt% Li_2O, 15–28 wt% ZrO_2 with additions of K_2O, Na_2O, Al_2O_3, and F to the amount of 11 wt%. Controlled crystallization for the production of two types of glass-ceramic was preferrably conducted with two different ZrO_2 contents. These glass-ceramics contained 16 wt% or 20 wt% ZrO_2. Both types of glass-ceramics were produced by pressing.

It was possible to produce glass-ceramics according to Table 2-25. In the first step, the glass-ceramic ingots are manufactured from a powdered glass. Nucleation occurs concurrently with this step. In the second step, viscous flow and further crystal growth take place. The glass-ceramic is thus already formed, eliminating the need for further heat treatment. Viscous flow was conducted in the apparatus described in Section 4.4.2.4.3.

Table 2-25

Processing of ZrO_2 Glass-Ceramics

	Process
1	Glass melt → frit → glass powder → formation of glass-ceramic ingot at 790°C (glass-ceramic A containing 16 wt% ZrO_2) or 950°C (glass-ceramic B containing 20 wt% ZrO_2)
2	Pressing (viscous flow at 900°C for glass-ceramic A or 1000°C for glass-ceramic B) → monolithic glass-ceramic

2.6.6.1 Glass-Ceramics Containing 16 wt% ZrO₂

The microstructure of the glass-ceramic after pressing at 900°C (Table 2-25) is characterized by the precipitation of ZrO_2-rich crystals of $Li_2ZrSi_6O_{15}$ type. These crystals could be determined by X-ray diffraction and SEM investigations (Fig. 2-50). The glass-ceramic demonstrates a bending strength of up to 160 MPa (Schweiger et al., 1998). The material shows very good optical properties, especially a high translucency. It is produced as IPS Empress® Cosmo glass-ceramic by Ivoclar Ltd., Schaan, Liechtenstein (see Section 4.4.2.4)

2.6.6.2 Glass-Ceramics Containing 20 wt% ZrO₂

Controlled crystallization was conducted on powdered glass, whereby sintering and nucleation were coordinated. Three crystal phases were precipitated in the base glass: Li_3PO_4, ZrO_2 microcrystals, and ZrO_2 macrocrystals. The formation of the Li_3PO_4 main crystal phase and the different modifications of the ZrO_2 crystals proceeded by volume nucleation. Nucleation and growth had to be carefully controlled. Given the sintering process and the subsequent viscous flow phenomenon, they were not allowed to proceed too rapidly since the viscosity would increase too dramatically. Nucleation in the phase-separated base glass is initiated by numerous local composition gradients. The steep compositional gradient becomes visible when the sample is etched. In SEM micrograph of the base glass, P_2O_5-rich areas thus appear as spherical cavities (Höland et al., 2000a). The primary crystal phase Li_3PO_4 becomes visible on the edges of the spherical etched areas. The subsequent crystallization process was characterized by the precipitation of other crystal phases (740°C: ZrO_2, 940°C: $ZrSiO_4$ by surface crystallization).

Various modifications of ZrO_2 crystals were determined with X-ray diffraction measurements. ZrO_2 crystals of baddeleyite-type and ZrO_2 crystals

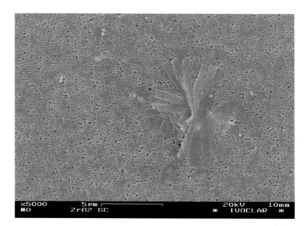

Figure 2-50 Microstructure of a glass-ceramic (16 wt% ZrO₂) after pressing at 900°C. SEM, etched sample.

of the tetragonal modification were primarily formed as microcrystals of 200–300 nm at 940°C. These crystals, however, grew into ZrO_2 macrocrystals of 1–20 μm in length during extended heat treatments. It was also determined that large crystals were formed by clusters of smaller crystals. If the main phases were growing at 1050°C

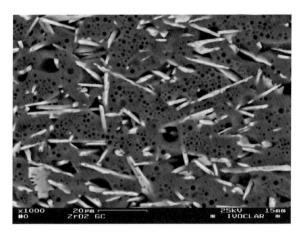

Figure 2-51 Precipitation of ZrO_2 macrocrystals in a monolithic sample at 1050°C for 6 h. SEM, etched sample.

(Fig. 2-51), the growth rate of the ZrO_2 macrocrystals was the fastest at 3.5–5.0 μm/h (Table 2-26) (Höland et al., 1996). Non-steady-state time lags were not observed for any of the growing crystal phases as they had been in other glass-ceramic systems. This fact indicated that nucleation proceeds rapidly and does not determine the rate of subsequent crystal growth.

Glass-ceramic microstructures like those depicted contrast in Fig. 2-52 (a) and (b) were produced in the viscous flow process at 1000°C (Table 2-25). The special SEM-composition technique (Fig. 2-52 (b)) revealed small, lightly colored crystals, shown by X-ray diffraction to be ZrO_2 crystals. The rod-shaped cavities visible in Fig. 2-52 can be attributed to Li_3PO_4 crystals that were dissolved by the etching conditions selected.

Table 2-26

Crystal Growth Rates (μm/h) of the Individual Crystal Phases at Different Temperatures of a Glass-Ceramic Containing 20 wt% ZrO_2

	Crystal growth rate	
	at 950°C	at 1050°C
rod-shaped lithium phosphate phase	0.25	4.0
ZrO_2 microcrystals	0.1	0.2
ZrO_2 macrocrystals	3.5	5.0

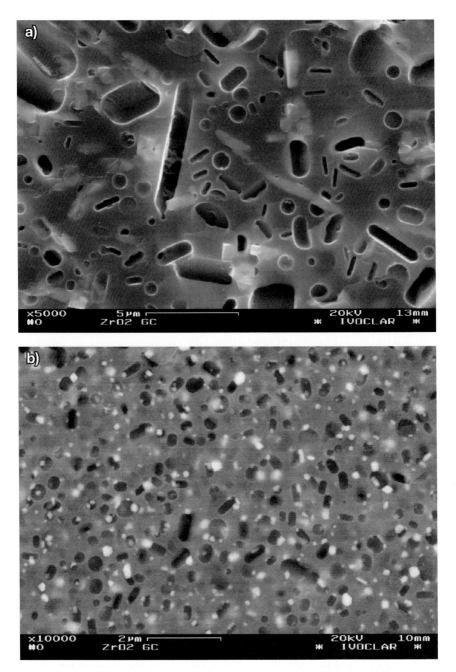

Figure 2-52 SEM image showing a glass-ceramic (20 wt% ZrO$_2$) containing ZrO$_2$ microcrystals and ß-Li$_3$PO$_4$ crystals after pressing.

a) SE-technique

b) composition contrast technique

The glass-ceramic with 20 wt% ZrO_2 is white opaque and shows small translucency. The material demonstrates bending strengths of 280 MPa and tougher with a K_{IC} of 2.0 MPa·m$^{0.5}$. In comparison to Sridharan and Tomozawa (1995) and Sarno and Tomozawa (1995), the toughening of the P_2O_5–SiO_2–Li_2O–ZrO_2 glass-ceramic is based on transformable tetragonal ZrO_2 and transformed monoclinic ZrO_2. Microcracking toughening was not observed.

2.7 OTHER SYSTEMS

2.7.1 Perovskite-Type Glass-Ceramics

The glass-ceramics of these systems were developed with the objective of producing materials for the electronics industry with high dielectric constants or optoelectronic effects where transparency was achieved.

2.7.1.1 SiO_2–Nb_2O_5–Na_2O–(BaO)

The first transparent glass-ceramics containing sufficient nanocrystalline perovskite to achieve dielectric constants well above 100 were made by heat treating glasses in the SiO_2–Nb_2O_5–Na_2O system, sometimes including optional BaO (Allen and Herczog, 1962).

Glasses of composition 5–25 wt% SiO_2, 50–80 wt% Nb_2O_5, 0–20 wt% Na_2O and 0–31 wt% BaO, with the total amount of Na_2O and BaO between 5% and 35% by weight, were melted and quenched into ribbon or thin (approximately 1–2 mm) patties. These were heat treated between 750°–950°C to achieve efficient nucleation and resulting transparency.

It was found that small additions of CdO raised the dielectric constant to over 400. The most stable glasses contained considerable BaO. Some compositions and electric properties are shown in Table 2-27.

Although $NaNbO_3$, the major perovskite phase, is antiferrielectric, small additions of solid-solution-forming impurities like cadmium render it ferroelectric. Borrelli et al. (1965) measured the electro-optic effect on a composition similar to composition 4 of Table 2-27. The crystal size was varied from 5 to 50 nm, and the dielectric constant changed from 50 for the original glass to near 550 for the transparent crystalline material, with an estimated volume percent of 70% $NaNbO_3$. The retardation in the glass-ceramic was found proportional to the square of the electric field *(E)*. If the refractive index difference *(Δn)* is written as:

$$\Delta n = B·\lambda·E^2$$

Table 2-27

Composition of SiO_2–Nb_2O_5–Na_2O–(BaO) Glass-Ceramics (wt%) and Properties (after Allen and Herczog, 1962)

	1	2	3	4
SiO_2	14.0	11.8	9.5	13.8
Nb_2O_5	70.0	68.8	62.3	68.6
Na_2O	16.0	9.7	2.5	14.4
BaO	–	9.7	25.7	–
CdO	–	–	–	1.0
K (dielectric constant)	340	290	220	517
Loss tangent	1.6	2.9	0.3	2.0

the Kerr constant B increased approximately proportional to the square of the dielectric constant as is observed in Fig. 2-53.

2.7.1.2 SiO_2–Al_2O_3–TiO_2–PbO

The controlled crystallization of the perovskite-type lead titanate, $PbTiO_3$, was first reported by Herczog and Stookey (1960), in the SiO_2–Al_2O_3–TiO_2–PbO system. High dielectric constants were achieved that allowed application as capacitor dielectric materials. The crystallization process of $PbTiO_3$ in glasses was studied by Russell and Bergeron (1965). They discovered that nucleation proceeded via glass-in-glass phase separation. Kokubo et al. (1969) studied glass formation and the subsequent controlled crystallization in the SiO_2–Al_2O_3–TiO_2–PbO system in the range of 15–45 mol% SiO_2, 0–26 mol% Al_2O_3, 16–33 mol% TiO_2, 27–52 mol% PbO. They determined that glass formation without secondary crystallization could occur during the cooling phase of glasses. Furthermore, if these glasses were subjected to heat treatment in a second step, crystallization took

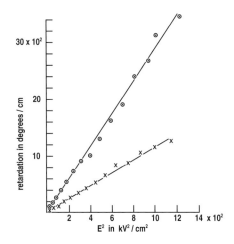

Figure 2-53 Retardation versus square of applied electric field. Measured at 6328 Å (Borrelli et al., 1965).
• samples with dielectric constant of 540
• samples with dielectric constant of 356

place between 620°C and 740°C. In the process, it was possible to precipitate $PbTiO_3$ of the perovskite-type, of the pyrochlore-type $PbTi_3O_7$, or both as the primary crystal. Kokubo et al. (1972) also conducted an in-depth study of the nucleation processes that led up to this primary phase. In the course of this study, they made an important discovery. At the time, it was very remarkable and required an even-more-detailed analysis and interpretation of the structure of these glasses. They discovered that glass-in-glass phase separation occurred in homogenous, Al_2O_3-free base glasses with the addition of Al_2O_3. A microheterogeneous glass structure with a glassy droplet phase and a glassy matrix was produced when increasing amounts of Al_2O_3 were added to the material. PbO and TiO_2 accumulated in the droplet phase and initiated a decisive nucleation process. With regard to phase separation produced by Al_2O_3, it should be noted that the addition of Al_2O_3 to alkali–silicate glasses reduces glass-in-glass phase separation. Exactly the contrary is the case for glasses with a high PbO content. Kokubo et al. (1972) illustrated this phenomenon with a type of "tricluster" model proposed by Lacy (1963) and with the field strength of the constituent ions. In this case, tricluster refers to the Al^{3+} in the alkali-free glass that utilizes the oxygen that already bridges two tetrahedra. Oxygen, therefore, is shared with three tetrahedral groups. The field strength of Al^{3+}, however, is as high as 1.77. Furthermore, the Al^{3+} tends to form its own phases and finds polarized oxygen ions in its vicinity by forming triclusters with Pb^{2+} or Ti^{4+} ions. This behavior may explain the reason for the increase in phase separation when Al_2O_3 is added to $PbO–TiO_2–SiO_2$ glasses.

Kokubo and Tashiro (1973) outlined the issue of controlling the dielectric properties of $PbTiO_3$ glass-ceramics for special requirements. In this publication, they presented a glass-ceramic containing large amounts of PbO as well as TiO_2, with which they achieved a high $PbTiO_3$ crystal content and a high dielectric constant. The glass-ceramic was composed of 40 mol% PbO, 40 mol% TiO_2, 10 mol% Al_2O_3, and 10 mol% SiO_2.

The highest dielectric constant of 400–500 of the glass-ceramic was achieved when the $PbTiO_3$ crystal grains measured 0.15 μm and 0.025 μm. This finding suggests that a transparent glass-ceramic with a considerable electro-optic effect can be produced. The maximum dielectric constant of the glass-ceramic containing crystals measuring 0.15 μm is produced by the high internal stress in the crystals. The high dielectric constant in combination with very small crystals measuring 250 Å may be attributed to an internal electric field produced by randomly oriented single domain ferroelectric crystals.

It must be noted at this point that transparent $PbTiO_3$ glass-ceramics are of interest for electro-optic and electro-luminescent applications (Kokubo and Tashiro, 1976). Non-transparent, opaque $PbTiO_3$ glass-ceramics, however, can be used for thick-film capacitors with a dielectric constant of 94 and tan δ of 0.0130 (at 10^6 c/s) on alumina substrates that have been heated to temperatures up to approximately 600°C. The dielectric constant of the capacitor as a function of temperature is almost linear from room temperature to 270°C, with a temperature coefficient of 0.00083 K^{-1}. Therefore, it was found that glass powders were suitable for preparing capacitors with a high dielectric constant (Kokubo and Tashiro, 1976).

Lynch and Shelby (1984) examined the reasons for the high dielectric properties of lead titanate glass-ceramics. They defined the matrix phase around the ferroelectric crystals. They also determined a special effect they called *clamping*. This phenomenon occurs when the glass matrix contracts and produces compressive stress within the ferroelectric particles, causing paraelectric-to-ferroelectric phase transition. As a result, the crystal size must be controlled for electro-optic applications. The crystals should preferably be smaller than 0.1 μm.

Saegusa (1996) developed a $PbTiO_3$ glass-ceramic derived from a borate system by a sol–gel process. The crystals precipitated in the matrix measured 1–2 μm. Saegusa (1997) also produced thin films of $PbTiO_3$ glass-ceramics derived from a silicate system. The dielectric constant was 219 and dielectric loss 0.04.

2.7.1.3 $SiO_2–Al_2O_3–K_2O–Ta_2O_5–Nb_2O_5$

Kokubo et al. (1973) were able to precipitate crystals of the perovskite-type in the $SiO_2–Al_2O_3–K_2O–Ta_2O_5–Nb_2O_5$ glass-ceramic system. Base glasses in the range of 0–17 mol% SiO_2, 0–20 mol% $AlO_{1.5}$, 25–50 mol% $KO_{0.5}$, 7–33 mol% $TaO_{2.5}$, 11–30 mol% $NbO_{2.5}$ were melted. Glasses with the highest K_2O ($KO_{0.5}$) content as well as the highest $TaO_{2.5}$ (Ta_2O_5) content were produced by undercooling. The subsequent heat treatment of melted glasses cooled to room temperature proceeded at temperatures between 650° and 1050°C. The K(Ta,Nb)O_3 main crystal phase of the perovskite-type was determined with X-ray diffraction in glasses with the highest K_2O ($KO_{0.5}$) content as well as the highest $TaO_{2.5}$ (Ta_2O_5) content at approximately 750°C. Prior to the formation of this main crystal phase, however, Kokubo et al. (1973) analyzed another primary crystal phase with the chemical composition of approximately $K_{1.5}(Ta_{0.65}Nb_{0.35})_2O_{5.75}$. The crystals correspond to those of the pyrochlore-type. The structure of the pyrochlore-type

($Cd_2Nb_2O_7$) crystals transform themselves into the main crystal phase of the perovskite-type when the temperature of the heat treatment of the glass-ceramic increases. Therefore, the main characteristics of the glass-ceramics at 1050°C are similar to those of the SiO_2–Al_2O_3–TiO_2–PbO system.

Reece et al. (1966) studied the microstructure of ferroelectric glass-ceramics containing (Pb, Sr, Ba, Nb_2O_6) crystals. They developed glass-ceramics by surface crystallization of glass frits. Samples that were hot-pressed at 800°C demonstrated a uniform microstructure containing crystals of approximately 0.1 µm. The dielectric properties of the glass-ceramics were related to the size, phase, and composition of the crystals. The glass-ceramics demonstrated wide Curie transitions.

2.7.2 Ilmenite-Type (SiO_2–Al_2O_3–Li_2O–Ta_2O_5) Glass-Ceramics

Beall (1971) and Ito et al. (1978) produced transparent glass-ceramics with this system. Glass formation and controlled crystallization for the precipitation of $LiTaO_3$ were investigated in the range of 10–77.5 mol% SiO_2, 2.2–55 mol% Al_2O_3, and 10–70 mol% $LiTaO_3$. Transparent glass-ceramics were produced by heat treating the base glasses between 800° and 1050°C. Primary crystals of $LiTaO_3$ were already formed at 800°C. This crystal phase remained stable until 1050°C was reached. The optical properties of the glasses were controlled by using different chemical compositions. Aluminum is required for transparency as Al_2O_3-free glass-ceramics are opaque. The $LiTaO_3$ crystal does not have the perovskite structure but a hexagonal structure similar to that of ilmenite: $FeTiO_3$. This is also true of $LiNbO_3$, an important optical crystal with nonlinear behavior that is not a good enough glass-former to produce glass-ceramics analogous to those of the tantalate system.

Hase et al. (1996) observed a preferential c-axis orientation of $LiTaO_3$ crystals in surface crystallized 35 mol% Li_2O, 30 mol% Ta_2O_5, and 35 mol% SiO_2 glass. They discovered a second harmonic generation at room temperature for 24h under a 0.7 kVdc voltage.

2.7.3 B_2O_3–$BaFe_{12}O_{19}$ (Barium Hexaferrite) or ($BaFe_{10}O_{15}$) Barium Ferrite

The interest in barium ferrite glass-ceramics is the preparation of high-quality magnetic powders. Shirk and Buessem (1970) prepared a glass-ceramic of the composition 40.5 mol% BaO, 33 mol% Fe_2O_3, and 26.5 mol% B_2O_3. The base glass was prepared by melting and pouring through two brass rollers. The glass was heat treated between 500° and 885°C to precipitate the crystalline magnetic phase of $BaFe_{12}O_{19}$. The intrinsic coercive force increased up to 5350 Oe after heat treatment of the glass. The crystal size of $BaFe_{12}O_{19}$ was determined to be approximately 0.5 μm. This chemical system allowed removal of the barium borate-rich matrix by etching.

Taubert et al. (1996) produced a base glass of the composition 40 mol% BaO, 27 mol% Fe_2O_3, and 33 mol% B_2O_3. The glass was melted at 1400°C and two kinds of samples were prepared. First, the glass was quenched between steel rollers with cooling rate of about 10^4–10^5 K/s. The results were flakes of 100 μm thick. These amorphous samples were heat treated at 820°C for 4–24 h. A nucleation and crystallization of $BaFe_{12}O_{19}$ in 20–500 nm scale took place. The glassy matrix was rich in BaB_2O_4.

Second, the glass flakes were ground and then heat treated as a powder compact. In contrast to the first sample, the crystallization was characterized by M-type of $BaFe_{12}O_{19}$ and Fe_3O_4. The magnetic hysteresis loop yielded a coercitivity Hc of 385 kA/m.

2.7.4 SiO_2–Al_2O_3–BaO–TiO_2 (Barium Titanate)

Barium titanate-sintered ceramics are developed with the objective of producing materials with special electrical properties, particularly a high dielectric constant. Barium titanate-based sintered ceramics are produced with reactive crystalline base materials.

In addition to $BaTiO_3$ sintered ceramics, glass-ceramics containing barium titanate crystals have also gained in importance. Herczog (1964) investigated the crystallization behavior of glass in the SiO_2–Al_2O_3–BaO–TiO_2–F system by adding only small amounts of F to the base glass. When these high melting glasses cooled, Herczog (1964) obtained crystal precipitates of the $BaTiO_3$-type.

The most important crystalline phases in the SiO_2–Al_2O_3–BaO–TiO_2 system are perovskite-type $BaTiO_3$, hexacelsian ($BaAl_2Si_2O_8$), and barium titanium silicate ($BaTiSiO_5$). Kokubo (1969) studied the glass formation tendency and crystallization behavior during reheating of cooled glass with the composition of 20–25 mol% SiO_2, 6–20 mol% Al_2O_3, 45 mol% $BaOTiO_2$. Nucleation and crystallization of the base glasses were initiated during heat treatment between 600 and 1100°C. Crystallization started at 850°C. X-ray diffraction analyses showed that a considerable amount of perovskite-type $BaTiO_3$ crystals were precipitated at 900°C. Crystallization was completed at 1100°C. The main crystal phases at 1100°C were the perovskite-type $BaTiO_3$, hexacelsian ($BaAl_2Si_2O_8$), barium titanium silicate ($BaTiSiO_5$), and some secondary phases. Kokubo (1969) determined the properties of the glass-ceramics by varying their composition and microstructure. He discovered that the dielectric constant decreased with an increasing SiO_2:(BaO·TiO_2)

molar ratio and a constant Al_2O_3 content. With a constant $BaO \cdot TiO_2$ content, it reached a maximum when the $Al_2O_3{:}SiO_2$ molar ratio was 35:65. In general, the dielectric constant increased when increasing amounts of $BaTiO_3$ crystals were precipitated in the glass-ceramic. The highest dielectric constant of 500 at a frequency of 10^6 c/s was demonstrated by a glass-ceramic with the following composition: 26 mol% SiO_2, 14 mol% Al_2O_3, 60 mol% $BaO \cdot TiO_2$.

Kokubo et al. (1969) studied the crystallization mechanism of the above glass-ceramic in detail. They found that a metastable benitoite-type crystal, $BaTiSi_3O_9$, was formed at 800°C, according to a mechanism of surface crystallization. Beginning at 950°C, the crystal was progressively transformed into the stable barium titanate and hexacelsian. The hexacelsian crystal grew anisotropically and parallel to the surface of the glass-ceramic. The preferred orientation of hexacelsian was attributed to the metastable benitoite-type crystals. Moreover, volume crystallization took place at the same time as surface crystallization. In volume crystallization, however, barium titanate and hexacelsian were formed as primary crystals.

The barium titanate glass-ceramic has been used to make thin films (Kokubo et al., 1968). The most favorable properties were demonstrated when the glass-ceramic measured between 50 and 100 μm. The dielectric constant was 500 and tan δ of 0.03 (at 10^6 c/s). The films were produced by pressing the melt between two rotating iron drums and subjecting the glass to heat treatment. The dielectric constant was reduced by half in a glass-ceramic of a thickness of 200 μm. This finding was attributed to the preferred orientation of hexacelsian precipitation near the surface of the specimens.

McCauley et al. (1998) synthesized series of $BaTiO_3$ glass-ceramics. The crystals were precipitated in a nanometer-size in the glassy matrix. It could be demonstrated by McCauley et al. (1998) that depolarization fields limited the dielectric polarizability of the crystals. A critical size of 17 nm for $BaTiO_3$ was determined.

Sol–gel processing became very important for $BaTiO_3$ thin film. Uhlmann et al. (1997) showed a great variety of sol–gel-derived coatings on glass. They demonstrated the formation of $BaTiO_3$ thin films in an economical manner by sol-gel processing. Sol–gel methods were also applied by Gust et al. (1997). They developed nanocrystalline $BaTiO_3$ in a 20–60 nm scale in thin films.

$BaTiO_3$ glass-ceramics were also developed with other additives. Aitken (1992) developed glass-ceramics consisting of a solid solution of $PbMg_{1/3}$, $Nb_{2/3}O_3$, and $BaTiO_3$. These perovskite-type phases were fine-grained at 0.2 μm. Additional phases of the glass-ceramic were pyrochlore and fresnoite,

$(Ba,Pb)_2TiSi_2O_8$. The glass-ceramic was characterized by a high dielectric constant of about 750 near 115°C.

2.7.5 Bi_2O_3–SrO–CaO–CuO

Developments in T_c-superconducting materials for applications in power engineering have also involved significant use of glass-ceramics. Abe et al. (1988) developed a superconducting glass-ceramic in the Bi_2O_3–SrO–CaO–CuO-system. Glasses of the composition of approximately $BiSrCaCu_2O_x$ were fabricated by melting. The crystal phase of $Bi_2Sr_2CuO_6$ was analyzed by X-ray diffraction (Abe et al., 1989) in a glass containing some Al_2O_3 following heat treatment at 845°C. Crystallization was carried out using a unidirectional crystallization method (see Section 1.5). Ca_2CuO_3 was determined as a secondary crystal phase. The superconducting glass-ceramic was characterized by a T_c of 62 K.

Applications of superconducting glass-ceramics require the material to be formed in the shape of pipes or hollow cylinders. Kasuga et al. (1996), therefore, developed a joining method for producing long pipes or rods. Flame joining enabled the fabrication of long products. The interface of the joining region of the glass-ceramics is characterized by the additional precipitation of a $Bi_2Sr_2CuO_x$ crystal phase. Furthermore, this method shows a great deal of potential for the development of new properties and applications of glass-ceramics.

CHAPTER 3

Microstructure Control

3.1 SOLID-STATE REACTIONS

As has already been observed (Section 2.2.2), metastable crystal phases are very common as the first crystals forming when glass is heat treated and devitrifies. In particular, metastable β-quartz solid solutions form in wide areas of the SiO_2–Al_2O_3–Li_2O, SiO_2–Al_2O_3–MgO, and SiO_2–Al_2O_3–ZnO systems. Metastable carnegieite, a stuffed derivative of cristobalite, forms first instead of stable nepheline, a stuffed derivative of tridymite, in regions near the $NaAlSiO_4$ composition in the SiO_2–Al_2O_3–Na_2O system. Similarly metastable lithium metasilicate precedes the development of lithium disilicate in composition areas where the latter phase is thermodynamically stable, and there are many other examples.

When glass-ceramics based on these metastable phases are heated to a high-enough temperature, solid-state reactions inevitably occur, and the result is the development of the stable crystalline phase or phase assemblage. It is interesting to examine some of the types of forms these reactions commonly take.

3.1.1 Isochemical Phase Transformation

A classical example of the approach from metastable to stable equilibrium occurs in the SiO_2–Al_2O_3–Li_2O system, in glass-ceramics on the SiO_2–$LiAlO_2$ join, regardless of the nucleating agent used. In the composition range Li_2O–Al_2O_3–$nSiO_2$ form n = 3.5–9, metastable quartz solid solutions will first form upon whatever nucleating agent crystal sites are used: e.g., TiO_2, ZrO_2, $ZrTiO_4$, Ta_2O_5, at temperatures generally in the range from 750° to 850°C. The β-quartz solid solutions are very persistent for long periods of time, unless they are heat treated above 900°C, or even 950°C in some compositions. In commercial compositions that contain other components such as MgO and ZnO, which can enter the β-quartz structure, even higher temperatures may be required to break down this phase.

Invariably, however, the hexagonal quartz crystals will transform to the stable tetragonal phase β-spodumene, or as it is sometimes referred to, stuffed keatite. This phase transformation is isochemical with compositions along the SiO_2–$LiAlO_2$ join, that is, the stable β-spodumene solid-solution crystals are identical or very similar to their parent metastable quartz solid-solution in chemical constituency. On the other hand, the morphology of the phases is different, especially in terms of grain size, which typically increases by a factor of 5–10. This increase in crystal size may be caused by a less-efficient nucleation of the transformation along the β-quartz grain boundaries than that of the original nucleation of β-quartz from the even smaller original nuclei.

3.1.2 Reaction between Phases

In most glass-ceramics, the multicomponent nature of the composition requires a more complex reaction between the metastable crystalline phases themselves, or in addition, a residual glass phase. An example of this way of approaching the stable crystalline assemblage occurs in some fluormica glass-ceramics, where norbergite, a metastable crystal, reacts with a potassium aluminosilicate residual glassy phase of composition close to the crystal leucite ($KAlSi_2O_6$) to yield the stable mica phase:

$$Mg_2SiO_4 \cdot MgF_2 + KAlSi_2O_6 \rightarrow KMg_3AlSi_3O_{10}F_2$$
$$\text{(norbergite)} \qquad \text{(glass)} \qquad \text{(phlogopite)}$$

Another example lies in the development of cordierite, a desirable and stable refractory, low-thermal-expansion crystal in the SiO_2–Al_2O_3–MgO system. Here a reaction occurs between metastable phases, spinel, α-quartz, and the nucleating titanate (pseudobrookite structure) to produce cordierite and rutile as follows:

$$MgAl_2O_4 + MgAl_2Ti_3O_{10} + 5SiO_2 \rightarrow Mg_2Al_4Si_5O_{18} + 3TiO_2$$
$$\text{(spinel)} \quad \text{(pseudobrookite-type)} \quad \text{(quartz)} \quad \text{(cordierite)} \quad \text{(rutile)}$$

In both these examples, the microstructure is completely changed by the solid-state reaction that produces the equilibrium crystal(s).

3.1.3 Exsolution

A rarer case of a solid-state reaction that allows attainment of equilibrium develops within a metastable crystal as an exsolution event. An example of this occurs in the SiO_2–Al_2O_3–ZnO–ZrO_2 system. In this case, the equilibrium crystals formed are of smaller grain size than the original β-quartz crystals formed on the primary nuclei of tetragonal zirconia. The metastable phase is a zinc-stuffed β-quartz of composition $ZnAl_2O_4 \cdot nSiO_2$, where n can vary from 2.5 to 4.0. Upon heat treatment, the β-quartz solid solution, of interest because of its low coefficient of thermal expansion combined with high electrical resistivity, breaks down the gahnite (zinc spinel) and α-quartz. The quartz then converts to cristobalite. The gahnite exsolves in tiny nanocrystals from the β-quartz solid solution leaving it highly siliceous and subject to inversion to α-quartz upon cooling. If the quartz–gahnite–zirconia assemblage is heated high enough, it yields the stable cristobalite–gahnite–zirconia assemblage as the α-quartz transforms to cristobalite. Thus in summary form:

$$ZnAl_2O_4 \cdot nSiO_2 \quad \rightarrow \quad ZnAl_2O_4 \quad + \quad nSiO_2$$

β-quartz solid solution   gahnite   cristobalite

3.1.4 Use of Phase Diagrams to Predict Glass-Ceramic Assemblages

Phase diagrams, by their very nature, predict equilibrium assemblages, and therefore can be used to forecast glass-ceramic crystalline assemblages, provided that the glass-ceramics have been heated to a high-enough temperature to erase any vestiges of metastable phases.

Because of the commercial importance of glass-ceramics in the quaternary system $SiO_2–Al_2O_3–Li_2O–MgO$, an isothermal section of a ternary slice of this system $SiO_2–LiAlO_2–MgAl_2O_4$ was obtained at 1230°C (Beall et al., 1967) about 60°C below the point of first melting (Beall 1994). This phase diagram is illustrated in Section 1.3 (Fig. 1-5). Two phases of low coefficient of thermal expansion, CTE, one positive (cordierite: CTE = $+13 \times 10^{-7}$ K^{-1}, 25°–300°C) and one negative (β-eucryptite: CTE = -5×10^{-7} K^{-1}, 25°–300°C) are both stable phases at this temperature. However, the phase diagram clearly shows that whereas each phase is stable, they do not coexist as a stable assemblage. If forced together, they would react to form β-spodumene plus spinel.

Therefore it would be sheer folly to attempt developing a glass-ceramic with a stable assemblage of cordierite ($Mg_2Al_4Si_5O_{18}$) and β-eucryptite ($LiAlSiO_4$), even though it would seem useful to have such a phase mixture to achieve a zero coefficient of thermal expansion in a high-temperature material.

In summary, phase diagrams can determine the feasibility of coexistence of crystal species in glass-ceramics. They cannot, however, yield any information regarding such issues as nucleation, grain growth, or other kinetic phenomena.

3.2 MICROSTRUCTURE DESIGN

In materials science, the design of a product involves a variety of factors such as its external contours, color or surface structure, and its microstructure. In this section, the design of the microstructure of glass-ceramic materials is discussed. The design of the microstructure represents one of the main priorities for materials engineers and forms the basis in the development of new materials with particular properties. The information in this section is intended to give those interested in materials science a better understanding of glass-ceramics as well as to make users of the materials aware of new applications. The further development and modification of glass-ceramics is often initiated by suggestions from users.

The microstructures of the glass-ceramics resemble well-known structures and phenomena in nature, science, technology, and daily life. In some cases, the microstructures that are discussed exhibit ultrafine crystal structures reminiscent of those of nanopowders. To help describe the appearance of certain microstructures, comparisons are made to the structure of a cell, cabbage head, or a house of cards. These examples are addressed in the following sections. The special properties produced by the specific designs will also be discussed in these sections.

3.2.1 Nanocrystalline Microstructures

Nanocrystals in glass-ceramics refer to crystallites below 100 nm. The development of glass-ceramics based on such small crystallites was one of the earliest examples of nanotechnology (Beall and Duke, 1969), a term now applied in the fabrication of carbon nanotubes, sintered ceramics, sol–gel glasses, and ormocer composites made of, or from, similar-sized particles. The microstructure of nanophase glass-ceramics may show a wide range of crystallinity, from a few volume percent to more than 90 vol%. In most cases, the nanocrystals are surrounded by glass. To achieve such a fine microstructure, the nucleation rate must be very high, but the secondary crystal growth must be suppressed.

Spinel and β-quartz solid solution glass-ceramics represent typical nanostructured glass-ceramics. The crystal structure of spinel is shown in Appendix Fig. 17 and the crystal structure of β-quartz is shown in Appendix Fig. 2. The microstructure of spinel glass-ceramics is shown in Fig. 3-1 as ultrafine scaled crystals embedded in a glass matrix. In the SiO_2–Al_2O_3–ZnO–MgO glass-ceramic system, spinel crystal phases are developed through effective precipitation on fine oxide particles. The spinel types range from gahnite, $ZnAl_2O_4$, to the classical spinel, $MgAl_2O_4$. The spinel crystals of the glass-ceramic (see

Section 2.2.7) possess the structural formula $(Zn,Mg)Al_2O_4$ although some TiO_2 may be incorporated as $(Mg,Zn)_2TiO_4$ and Ti^{3+} as $(Mg,Zn)(Al,Ti)_2O_4$. These crystals measure approximately 10 nm. Thus they represent the characteristic nano-structure. Figure 2-14 of Section 2.2.7 shows

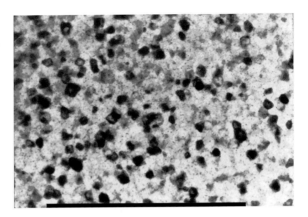

Figure 3-1 Ultrafine grained microstructure of ß-quartz solid-solution glass-ceramic. Black bar = 1µm.

that the crystals are precipitated in an isolated fashion in the glass matrix. This reaction was produced by controlling the nucleation processes with ultrafine glass-in-glass phase separation. The nucleation in the phase-separation process was initiated by the heterogeneous nucleating agent TiO_2. As a result, the spinel crystals did not grow beyond the ten-nanometer range. Nevertheless, a crystal content of 30–40 vol% was achieved.

Spinel glass-ceramics demonstrate special properties by virtue of the extremely fine crystals and the nature of the glass phase. A high degree of visible transparency is achieved because the dimensions of the crystals are far below those of the wavelengths of visible light. This reduces optical scattering to very low levels, even though the refractive index difference between spinel and the siliceous host glass is substantial. The glass matrix for its part provides the glass-ceramic with excellent chemical durability. Furthermore, the high SiO_2 content of the glass matrix ensures that the material is stable to temperatures in excess of 900°C. As a result, this glass-ceramic is suitable for flat-panel displays in the electronic industry (Beall and Pinckney, 1999; Pinckney 1999).

Because of the fine crystals in β-quartz glass-ceramics near 50 nm, these materials also fall in the nanostructured category. The characteristic ultrafine crystallite microstructure is produced with the nucleation process, as shown for spinel glass-ceramics. Following heat treatment of 50 K above the transformation range of the SiO_2–Al_2O_3–Li_2O base glass with the nucleating oxides (TiO_2)–(ZrO_2), the primary crystal phase $ZrTiO_4$ is formed. A total content of approximately 4 mol% of TiO_2 and ZrO_2 is sufficient for primary crystal formation. Because of the nucleation process and the fine distribution of these crystals throughout the entire volume of the glass matrix, these crystals measure less than 100 nm. Subsequent heat treatment at 850°C causes the precipitation of a β-quartz solid solution main crystal phase by heterogeneous growth on zirconium titanate crystals. Figure 3-1 shows the nanoscaled microstructure of this type of glass-ceramic. The individual crystals have grown in an isolated fashion in the glass matrix. Interlocking or twinning of the crystals is not discernible, but impingement of crystals is not observed. Beall (1992) established the crystal number density of the quartz solid solution phases in the silicate matrix to be approximately 5×10^{-21} m^{-3}.

Excellent properties have been achieved with this ultrafine grained microstructure, which is produced with only one major crystal phase and a high percentage of crystallinity. The thermal properties are especially outstanding. For example, the coefficient of thermal expansion is 7×10^{-7} K^{-1} in the range of 0°–500°C. Because of the small and uniform dimensions of the crystals, which measure about 500 nm in the glass-ceramic product, and because both the birefringence in β-quartz solid-solution and the index difference between crystals and residual glass are small, practically no scattering of visible light occurs. Thus, the glass-ceramics are highly transparent and suitable for a variety of technical and household applications.

3.2.2 Cellular Membrane Microstructures

Another type of microstructure in glass-ceramics can best be described by comparing it to the structure of an organic cell. A cell is composed of a very thin membrane that separates and protects the contents of the cell from the neighboring cells. Similarly, in a cellular membrane microstructure, a very thin membrane surrounds the contents of a cell-like entity.

This type of cellular membrane microstructure can be developed in glass-ceramics that exhibit a very low coefficient of thermal expansion because of the precipitation of β-quartz solid-solution or β-spodumene crystals (Beall 1992). The glass matrix, which is very thin and surrounds the crystals, plays the part of the cellular membrane. As a result, the crystals are separated from each other. The membrane accounts for approximately 10 vol% of the microstructure and can, for example, act as a diffusion barrier between the crystals.

Figure 3-2 shows the cellular membrane microstructure of a β-quartz solid solution glass-ceramic. The individual phases are presented in connection with their effect on the different solid-state reactions to provide a better understanding of this microstructure. As described in Section 2.2.2, the nucleation of β-quartz solid-solution crystals is heterogeneously initiated by $ZrTiO_4$ nucleating agents. These $ZrTiO_4$ crystals represent the nucleus of the cell, on which the entire contents of the cell grow heterogeneously. In the case of glass-ceramics, β-quartz solid solution grows. In the high-resolution TEM image by Maier and Müller (1989) in Fig. 3-2, the $ZrTiO_4$ nucleation phase appears as a small dot measuring a few nanometers in the middle of various β-quartz solid solution crystals. Further excess $ZrTiO_4$ crystals that do not initiate β-quartz crystals are also discernible as black dots in the glass matrix cellular membrane.

Beall (1992) attributes a great deal of importance to the glass matrix representing the cellular membrane in the solid-state reactions of β-quartz solid solution and β-spodumene solid solution glass-ceramics, the tenacious residual

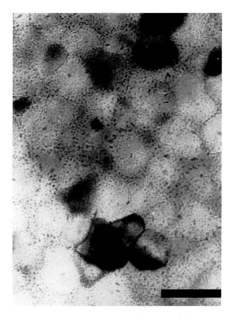

Figure 3-2 Cellular membrane microstructure of a β-quartz solid-solution glass-ceramic (bar 200 nm). TEM (Maier and Müller, 1989). Residual glass with $ZrTiO_4$ precipitates is observed with cellular structure.

glass acting as a barrier to the diffusion of Al ions. Chyung (1969) determined that the diffusion of Al ions influences the secondary crystal growth of β-spodumene solid solution. In other words, because of the further diffusion of Al^{3+} in the cell, β-spodumene solid solution crystals form as a result of the rapid growth accompanying the transformation of the metastable crystals to the stable β-spodumene phase. In this growth process, proceeding from primary β-quartz solid solution with crystallites measuring between 100 and 200 nm in size (see Fig. 3-2), it is important that β-spodumene crystals do not exceed a crystallite size of 5 μm. If these dimensions are exceeded, the anisotropic properties of the crystals may initiate microcracks in the glass-ceramic. These microcracks can produce uncontrolled creep or mechanical weakening. As a result, repeated temperature cycling may have a negative effect on glass-ceramics with complex geometrical shapes. In the worst case, defects such as fractures may form in the glass-ceramic. Controlled secondary crystal growth, however, prevents this type of reaction by controlling the solid-state reactions with the glass phase. Furthermore, minor stresses are absorbed at the crystal-glass matrix interface (Raj and

Chyung, 1981). The microstructure of β-spodumene glass-ceramics is shown in Sections 1.5.1 and 2.2.2.

By controlling the solid-state reactions with the glass phase, it has become possible to partially dissolve and re-precipitate crystals under bending strength at temperatures well below the initiation of melting of the β-spodumene solid solution. Thus, vacuum forming of highly crystalline glass-ceramic sheet has permitted products of complex shapes, such as laboratory bench tops with seamless basins, to be produced.

3.2.3 Coast-and-Island Microstructure

Since the mechanism of surface crystallization can be controlled, additional microstructures demonstrating favorable characteristics have been developed. One of the microstructures produced in this way goes by the apt name *coast-and-island microstructure*. This particular crystallization process, which has already been addressed in Section 2.2.9, deals with several different microstructures and their specific characteristics.

Beall (1992) reported on typical coast-and-island microstructures in cordierite (Section 2.2.5) and pollucite (Section 2.2.4) glass-ceramics. The cordierite crystallized from grain boundaries and became the predominant crystalline phase. In pollucite glass-ceramics, the mullite-type remains of the glass matrix became enveloped by the pollucite matrix.

In the technical manufacturing of wollastonite glass-ceramics, which are produced as architectural materials on a large scale, the coast-and-island microstructure represents an important intermediate step. Following primary crystallization of β-wollastonite, a dense aggregation of crystals grows along the inner boundary of the former glass grains. The glass phase within the former glass grains remains intact. The crystals, therefore, form a coast, while the glass phases enclosed in the crystals appear like small islands. As the crystallization process continues, however, the phase boundaries of the former glass grains disappear and the coasts can no longer be distinguished from the islands (see Section 2.2.6) (Wada et al., 1995; Kawamura et al., 1974). A similar intermediate stage featuring the coast-and-island microstructure is achieved by the controlled surface crystallization of apatite–wollastonite glass-ceramics (see Section 2.4.2) (Kokubo 1993).

Glass-ceramics with a leucite main crystal phase (Section 2.2.9) are also produced according to the mechanism of controlled surface crystallization of the opal base glass since volume nucleation cannot be controlled. In this case, it is important for the crystallites to achieve a high nuclei density and to uniformly precipitate into the glassy matrix. The coast-and-island microstructure has been specifically developed as the transitional stage (Höland et al., 1996b).

In a subsequent processing stage of the glass-ceramic, however, the glass and crystal phases are mixed as a result of the viscous flow phenomenon. Consequently, the coast-and-island microstructure disappears.

The coast-and-island microstructure has also been used in the development of opal glass-ceramics. It has been used to combine specific properties of glass-ceramics such as opalescence with a high coefficient of thermal expansion (see Section 2.2.9). Figure 3-3 shows a typical coast-and-island microstructure in an opal glass-ceramic, according to Höland et al. (1996b). The glass-ceramic specimen was etched first to make the features in Fig. 3-3 more clearly visible. During the 10-sec etching procedure with 2.5% hydrofluoric acid, the leucite crystals were almost entirely dissolved. As a result, the etching patterns measuring approximately 1 µm, which are visible in Figs. 3-3 and 3-5, represent leucite-type crystals. Therefore, the coast in the coast-and-island microstructure in Fig. 3-3 is represented by leucite crystals and the islands are represented by the glassy phase. The individual areas were examined by SEM at high magnification.

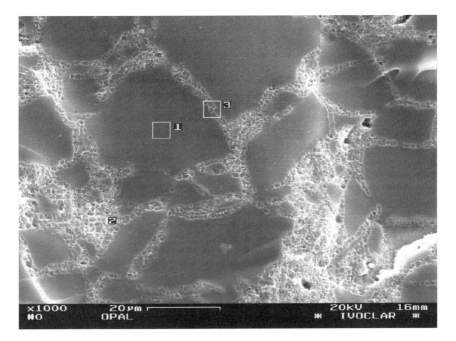

Figure 3-3 SEM overview of the coast-and-island microstructure in opal leucite glass-ceramics.
1) island (glass)
2) coast (crystal)
3) coast-and-island interface

Figures 3-4, 3-5, and 3-6 show the individual phases and the interface magnified 20,000, 30,000, and 50,000 times. The glass phase (Fig. 3-4) exhibits phase-separation processes in the form of droplet phases less than 200 nm in size. This phase separation creates the opal effect of the glass-ceramic. Although the crystals of the leucite type (Fig. 3-5) in the coastal areas (marked 2 in Fig. 3-3) measure only approximately 1 μm, they produce a highly translucent effect in the glass-ceramic. The crystals provide the material with a very high coefficient of thermal expansion. The crystal-glass interface is shown in Fig. 3-6. Clearly, crystal growth was interrupted at a specific stage of growth once a crystal front of some micrometer thickness had formed.

3.2.4. Dendritic Microstructures

The formation of a dendritic microstructure by controlled crystallization in glasses with specific properties may seem rather surprising at first, since dendritic crystal growth is generally considered to be a defect in the fabrication of conventional glasses. Vogel (1992), for example, observed this type of phenomenon when he examined glass defects in high-performance optical glasses. Dendritic crystal growth, however, also produces defects in glasses for technical applications (Jebsen-Marwedel and Brückner, 1980). The following crystals typically grow in a dendritic manner causing defects in glass:

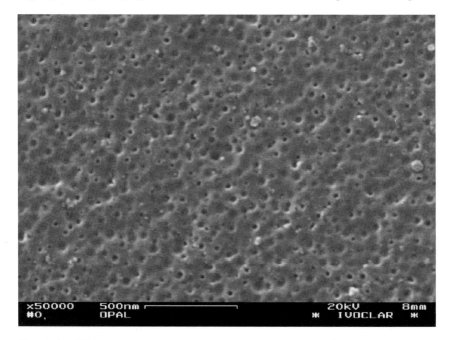

Figure 3-4 SEM image of the glassy phase (marked 1 in Fig. 3-3).

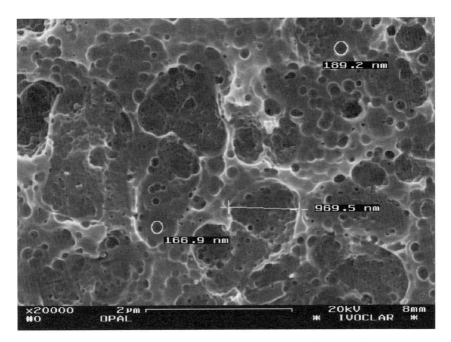

Figure 3-5 SEM image of the crystal phase area (marked 2 in Fig. 3-3).

cristobalite (SiO_2), tridymite (SiO_2), baddeleyite (ZrO_2), and carnegieite ($NaAlSiO_4$).

The development of the Fotoform® glass-ceramic involves a particularly pertinent example of controlled dendritic crystal growth. This glass-ceramic was developed from the SiO_2–Li_2O system (Beall 1992). It was one of the first glass-ceramics developed by Stookey (1953). As a result, it is all the more spectacular that the precisely controlled crystallization proceeded dendritically. Figure 3-7 shows the dendritic growth of lithium metasilicate in the Fotoform® glass-ceramic. The characteristic feature of dendritic growth is that crystals grow preferentially in one axial direction and occasionally split off short parallel branches.

Figure 3-7 also shows that the crystals grow according to the normal crystal morphology. The characteristic feature of this microstructure, however, is that crystal growth proceeds in specific lattice directions or along lattice planes. The resultant microstructure demonstrates a high glass-phase content interspersed with dendritic crystals. Growth proceeds in a typical skeletal fashion, with the crystals joined in the area where growth nucleated, and barely touching or not touching at all in the areas where growth terminated.

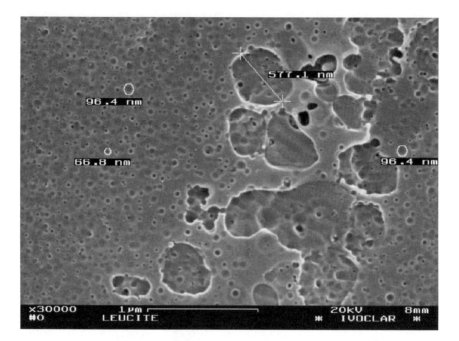

Figure 3-6 SEM image of the coast-and-island interface (marked 3 in Fig. 3-3).

It is also interesting to note that the sections between the crystals exhibit a typical hexagonal morphology.

Since the dendritic crystals in the Fotoform® glass-ceramic demonstrate typical skeletal crystal growth (Fig. 3-7), a path of crystals is established through the glass-ceramic. This path allows a particular property of the glass-ceramic to be used in the production of glass-ceramic preforms. The course of the crystallization path can be controlled by the use of special masks and photosensitive nucleation methods. Thus, the lithium metasilicate crystals are particularly easy to etch, while the surrounding aluminosilicate glassy matrix is considerably more resistant to acid attack. In a further fabrication step, these paths are removed by etching, resulting in high-precision components. This controlled process is superior to many other production procedures, such as mechanical or laser machining because of its accuracy and minimal heating of the glass product. The specific application of the dendritic microstructure is presented in Section 4.1.2 using the Fotoform® example.

Apart from the dendritic growth of crystals according to the mechanism of controlled crystallization, dendritic growth has also been observed for the mechanism of surface crystallization. When dendritic growth occurred, however, it produced an undesirable, disordered and defective structure, rather than

Figure 3-7 SEM image showing dendritic crystallization of lithium metasilicate in Fotoform®
glass-ceramic. Size of dendrites = approx. 2 microns.

a controlled dendritic one. This typical example of dendritic crystal growth in a leucite glass-ceramic was observed in the SiO_2–Al_2O_3–K_2O system and in a leucite–apatite glass-ceramic in the SiO_2–Al_2O_3–K_2O–CaO–P_2O_5–F system. The crystal structure of leucite is shown in Appendix Fig. 14. Following the nucleation of flat, two-dimensional microcrystals, the leucite crystals grow from the surface of the glass specimen toward the interior at a high rate (Höland et al., 1995a; Pinckney 1998). The microstructure produced in this way is shown in Fig. 3-8 (Höland et al., 2000c). This microstructure is not desired in the development of leucite or leucite–apatite glass-ceramics, since the surface of the glass-ceramic is weakened by the crystals formed by the high reaction rate. This example merely serves to illustrate dendritic growth by clearly showing the crystal morphology. The growth of small, densely packed crystals, however, is ideal for the controlled surface crystallization of leucite. In various instances, twinning of crystals has also been observed under these growth conditions, but to date the potential of using this type of growth process to control surface crystallization in glass-ceramics has not yet been explored.

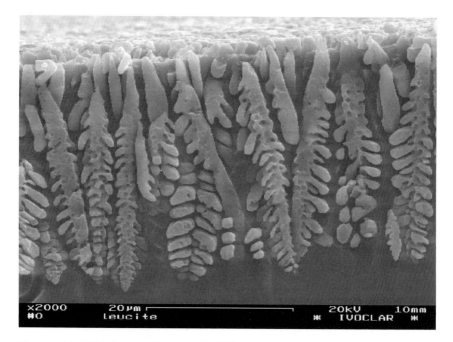

Figure 3-8 SEM image showing dendritic surface crystallization of leucite in leucite–apatite glass-ceramics at 900°C, heat-treated for 1 h, etched for 10 sec with 3% HF (Höland et al., 2000c).

3.2.5 Relict Microstructures

The name of this special microstructure refers to the fact that the phase separation in the base glass is preserved to a large extent in the glass-ceramic. In other words, one amorphous phase crystallizes without change in geometry. Therefore, the glass-ceramic microstructure is considered to be a relict of the glass microstructure. As a result, specific properties of the base glass, such as high optical transmission, may be maintained in the glass-ceramic. A mullite glass-ceramic demonstrates a typical relict microstructure. The crystal structure of mullite is shown in Appendix Fig. 16.

The base glass used in the development of mullite glass-ceramics (see Section 2.2.1), for example, contains small droplets of a glass phase measuring less than 100 nm in diameter and a glass matrix, both produced by microphase-separation processes. During controlled crystallization in the temperature range of 900°C, very small mullite crystals form within the glass droplets. These crystals measure less than 100 nm. This microstructure of the glass-ceramic is shown in Fig. 3-9 (Beall 1993; MacDowell and Beall, 1969; Beall and Rittler, 1982).

Special properties such as translucency and luminescence, the latter following the incorporation of Cr^{3+} ions, were achieved with this particular microstructure (see Section 4.3.3).

3.2.6 House-of-Cards Microstructures

Mica crystals of phlogopite-type (see the crystal structure in Appendix Fig. 13) precipitate in a house-of-cards microstructure in glass-ceramics. The precision machinability of mica glass-ceramics such as commercial MACOR® with conventional metal-working tools is the direct result of the house-of-cards microstructure, as illustrated in Fig. 3-10 (Beall 1992). Randomly oriented and flexible flakes tend to either arrest fractures or cause deflection or branching of cracks; therefore, only local damage results as tiny polyhedra of glass are dislodged. Another important property dependent on this microstructure is high dielectric strength, typically around 40 kV/nm. Such insulating qualities are possible because of the continuous interlocking form of the mica sheets. Very low helium permeation rates, important in high-vacuum applications, are also related to this microstructure. Helium can, of course, permeate most glasses, but it is severely slowed even at high temperature by mica crystals, whose basal plane is

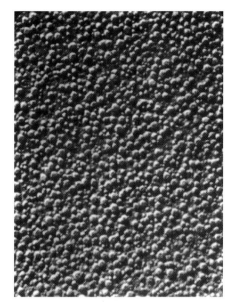

Figure 3-9 TEM image showing microstructure of a relict mullite glass-ceramic. Base of figure = approx. 1 micron.

Figure 3-10 House-of-cards structure in machinable fluormica glass-ceramic. Note phase-separated residual borosilicate glass and affinity of siliceous droplets for mica flakes. Black bar = 1 µm.

composed of an anion network of oxygens arranged in a virtual hexagonal close-packed state. The house-of-cards microstructure also contributes to the relatively high fracture toughness values in mica glass-ceramic (2 MPa·m$^{0.5}$), which is also the result of crack blunting, branching, and deflection.

The analysis of the phase-formation processes of glasses derived from the SiO_2–Al_2O_3–MgO–Na_2O–K_2O–F system has shown that the phase-separation processes of the base glasses have a remarkable influence on the nucleation processes in the development of mica-type glass-ceramics. The results in Section 1.5 clearly demonstrate that the process of mica precipitation was diffusion controlled. The activation energies of both crystal growth and viscosity were similar. The mica glass-ceramic reported upon in Section 1.5 demonstrates a house-of-cards microstructure. The process of microstructure control was characterized by the following solid-state reactions.

3.2.6.1 Nucleation Reactions

The base glass consisted of the chemical composition 50.4 wt% SiO_2, 28.1 wt% Al_2O_3, 12.0 wt% MgO, 3.3 wt% Na_2O, 3.1 wt% K_2O, and 3.0 wt% F.

Monolithic glass samples have been melted and cooled in the transformation range at a rate of 3–4 K/min. The glass microstructure is characterized by phase separation. Figure 3-11 shows that the base glass is phase separated into a glassy matrix and a droplet phase, in contrast to the flat MoO_3 test area. The droplet phase is rich in network-forming elements and structural units of network-forming structural groups.

3.2.6.2 Primary Crystal Formation and Mica Precipitation

Heat treatment of the glass at 750°C/5 h resulted in the nucleation of norbergite, as reported in Section 1.5, and the beginning of mica precipitation (Fig. 3-12). The norbergite crystals supplied the ions, magnesium and flourine in particular, to the mica crystal. Thus, norbergite acted as the intermediate crystal phase in the solid-state reaction of mica formation in glass-ceramics. When the base glass of the mentioned composition was heat treated at 980°C/1 h rather than at 750°C/5 h, a mica glass-ceramic with a house-of-cards structure was formed (Vogel 1978; Höland et al., 1982a). These results are shown in Fig. 3-13.

To sum up, the formation of the house-of-cards microstructure in mica glass-ceramics is characterized by the following solid-state reactions, which take place as consecutive reactions:

- Phase-separated glass ⇒ primary crystal formation of the norbergite-type as the intermediate phase ⇒ mica crystal formation.

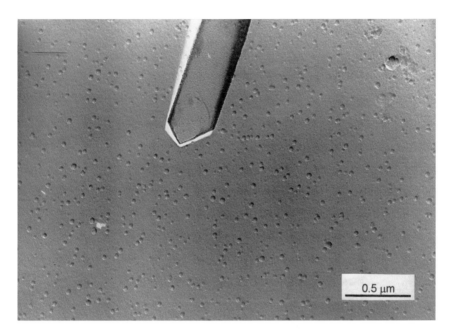

Figure 3-11 TEM replica of the microstructure of the base glass, showing a phase-separated microstructure and an almost-flat MoO_3-crystal test area.

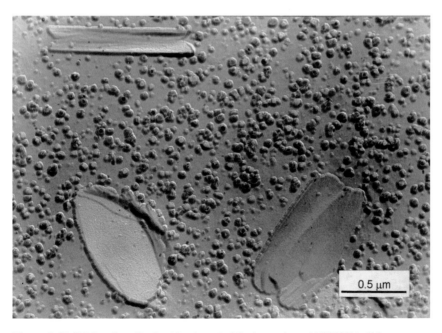

Figure 3-12 TEM replica after heat treatment of the base glass at 750°C/5 h. Primary crystal formation takes place within the droplet phase and at the beginning of mica growth.

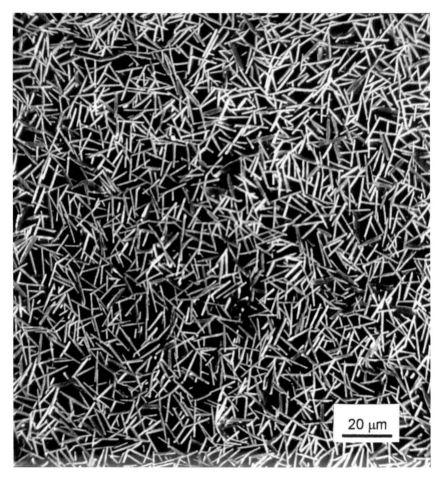

Figure 3-13 House-of-cards microstructure following heat treatment of the base glass at 980°C/1 h.

- The precipitated micas show a typical flat, sheet-like habit similar to micas of the phlogopite-type that are common in nature.

3.2.7 Cabbage-Head Microstructures

In the course of the development of mica glass-ceramics of the phlogopite-type, zweier layer silicates (mica-like layer) with a new type of appearance were unexpectedly found (Höland et al., 1981). The new mica crystals demonstrate a curved shape. During nucleation and crystallization, the curved crystals group together like the leaves of a cabbage head. Hence, this type of microstructure is also given the name of *cabbage head*.

Many investigations have been conducted to define the composition range of the glasses and the resulting glass-ceramics that are responsible for the formation of this new type of mica. This composition range is mentioned in connection with the synthesis of BIOVERIT® II in Section 2.4.4.

The formation and control of the microstructure is explained here. Moreover, with regard to the chemical composition of the base glass, it was discovered that the composition was very similar to that required for the formation of the flat mica crystals of the house-of-cards microstructure. Hence, overlaps of the main components SiO_2, Al_2O_3, and MgO are evident with regard to the basic composition. In different examples, however, it was discovered that a reduction in the SiO_2 content of the glasses and an increase in their Al_2O_3 and MgO content led to the preferred formation of curved micas. At the same time, however, the fluorine content and the Na_2O, K_2O content influenced the formation of both straight and curved micas. It was possible to form curved micas without Na_2O in some special compositions. But preferred compositions of curved micas contained both Na_2O and K_2O.

Table 3-1 shows the composition of the glass that was used for studying the microstructure formation of cabbage-head phlogopite. Once the difference in the chemical composition of straight and cabbage-head glass-ceramics had been determined as being small, it was all the more interesting to note that the microstructures of the base glasses were quite different. Compared with the glass for flat micas, the base glass for curved micas demonstrated almost no phase separation. Certainly, phase separation was substantially reduced (Höland et al., 1983b). After heat treating the glass at a temperature range below 1000°C, the primary crystal phase was phlogopite in the form of a curved crystal. No other primary crystal phase was visible.

Figure 3-14a shows that after having heat treated the glass at 790°C/3 h, the phlogopite formed a cabbage-head microstructure. It must be noted that the mica crystals formed spherical aggregates as early as in the initial stage of crystallization. There were no isolated crystals present that would become

Table 3-1

Base Glass Composition (Wt%)	
SiO_2	48.9
Al_2O_3	27.3
MgO	11.7
Na_2O	3.2
K_2O	5.2
F	3.7

curved at a later stage as a result of a high crystal density. Curved micas demonstrating defects were formed only at low temperature ranges when the ion diffusion reactions were kinetically low. This type of defective curved mica was formed by heat treating the glass near 750°C.

At temperatures in the range of 790°– 1000°C, curved micas demonstrating a cabbage-head microstructure were formed. A typical example of this microstructure is shown in Fig. 3-14b. The growth of a second crystal phase in the form of a secondary phase was observed in the final stage of the formation of curved micas in different glass-ceramics of this type. This phase was located between the mica crystals (Fig. 3-15a). X-ray diffraction investigations have shown the phase to be composed of cordierite crystals. The formation of

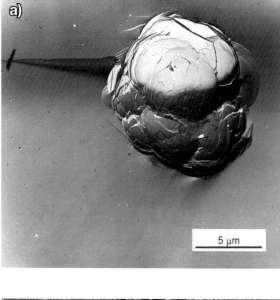

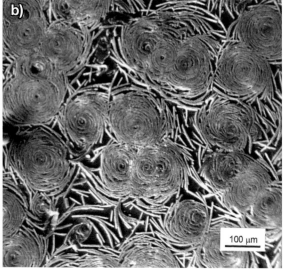

Figure 3-14 SEM image of the formation of a cabbage head microstructure. Polished sample etched for 2.5 sec in 2% HF.

a) TEM replica of the primary crystal formation of curved micas. Heat treatment of 790°C/3 h.

b) SEM cross-section through globular arrangements showing advanced stage of the precipitation of curved phlogopite crystals of cabbage-head microstructure. Heat treatment of 960°C/5 h. HF-etched samples.

cordierite became even more obvious when the content of Al_2O_3 and MgO was increased and that of SiO_2 reduced within the chemical composition of the base glass. Figure 3-15b shows a special case in which a large cordierite has grown around the curved micas. This is an interesting phenomenon because TiO_2 is usually needed to nucleate fine cordierite. Kasten et al. (1997) investigated the behavior of polyvalent ions on the crystallization of mica glass-ceramics. The occurrence of V^{5+} showed a preferred crystallization of indialite (high-temperature modification of cordierite) and a V^{3+}/V^{4+} content resulted in smaller micas.

It is interesting to note that the glass with the chemical composition of Table 3-1 can be converted into mica glass-ceramic with a house-of-cards microstructure when heat treated at temperatures above 1000°C. During the heat treatment above 1000°C, certain transitional stages were noted for the shapes of the mica precipitation. Hence, curved, slightly curved, and straight micas were evident at the same time. In any case, it was possible to produce a house-of-cards microstructure demonstrating only straight micas by heat treating the glass at 1040°C/1 h for example, with rapid heating rate through 750°C region.

The fact that curved mica crystals were formed at low temperatures and straight ones at temperatures above

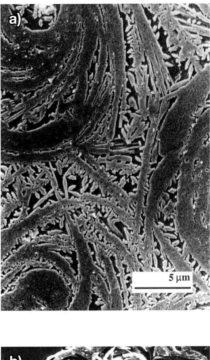

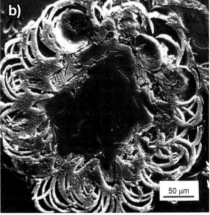

Figure 3-15 SEM image showing precipitation of curved micas and additional crystallization of cordierite. Polished sample etched for 2.5 sec in 2% HF.
a) Mica–cordierite glass-ceramic (BIOVERIT®
   II). The precipitation of cordierite takes place between the curved mica sheets.
b) Precipitation of a relatively large cordierite crystal and crystallization of curved phlogopite.

1000°C in glasses with an identical composition indicated that the ion diffusion processes were significant in the formation of crystals.

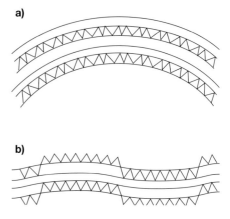

Figure 3-16 Reduction of strain between octahedral layers (white bars) and tetrahedral layers (triangles) in hydrous phyllosilicates with kaolinite-like arrangements.
a) chrysotile
b) antigorite (Liebau 1985).

From a crystallochemical point of view, the question of why these curved micas are formed naturally arises. Curved multilayer silicates are only known for materials such as chrysotile, $Mg_3[Si_2O_5](OH)_4$, and antigorite, $Mg_{48}[Si_4O_{10}]_{8.5}$ $(OH)_{62}$, (Liebau 1985), both of which are zweier sheet minerals. Liebau (1985) presented a clear structural explanation for the curving of the zweier layer minerals in comparison with kaolinite, $Al_2[Si_2O_5](OH)_4$. Liebau (1985) reported that the tetrahedra of the $[Si_2O_5]$ layer of cation-rich phyllosilicates with b_{oct} were only slightly smaller than b_{tetr}. Therefore, the nonbridging basal O atoms of the tetrahedra-type deviated from a flat plane. However, since only one side of the octahedral sheet was linked to a tetrahedral layer, the resulting strain caused by the misfit between the two parts of the composite layer was reduced by bending the tetrahedral–octahedral pair of sheets (Fig. 3-16).

Curved layers are also evident in vermiculites, $Mg_{2.36}Fe^{III}_{0.48}Al_{0.16}[(OH)_2/Al_{1.28}Si_{2.72}O_{10}]^{0.64-}Mg_{0.32}(H_2O)_4$ (Rösler 1991). These sheets, however, had been formed by an evolution of water during heat treatment. Electroneutrality was achieved with the subsequent bending of the layers.

Now the cause for the curving of the phlogopite-type mica crystals in the form of dreier layer silicates must be addressed. It is unclear as to what extent the structural change continues into the next layer or the layer after that. These issues cannot be fully answered in crystallochemical terms at this time. Important findings, however, have been made. They are presented here.

The crystal structure and chemical composition of curved micas have been studied in comparison with flat micas in glass-ceramics, both of which had the same composition as shown in Table 3-1. The flat crystals were formed at 1040°C and the curved ones at 960°C. The chemical composition of the crystals has been determined with electron microprobe analyses (Höland et al., 1983).

Flat crystals: $(Na_{0.21}K_{0.81})(Mg_{2.52}Al_{0.44})(Si_{2.80}Al_{1.20})(O_{10.18}F_{1.82})$

Curved crystals: $(Na_{0.18}K_{0.82})(Mg_{2.24}Al_{0.61})(Si_{2.78}Al_{1.22})(O_{10.10}F_{1.90})$

These results demonstrate that the key difference is the substitution of aluminum for magnesium in the octahedral sites, which reduces the size of the octahedral layer.

The crystal structure of the flat fluorophlogopite has been defined using single crystal X-ray data. The results for flat fluorophlogopite in glass-ceramics were compared to those in technical literature. Flat phlogopite crystals formed in glass-ceramics at 1040°C exhibit the following crystal data: space group $C2/m$, a = 5.281 (2) Å, b = 9.140 (2) Å, c = 10.085 (2) Å, β = 100.17 (2), V = 479.2 (7) Å^3, Z = 2 (Elsen et al., 1989).

The lateral dimensions of a unit of a tetrahedral layer of micas were usually larger than those of an octahedral layer. The rotation of the $(\alpha Si, Al)O_4$-tetrahedra about the c-axis reduced the size of the tetrahedral sheet and the misfit between the layers. The extent of this reduction was limited by a blocking interlayer cation. The phlogopite formed at 1040°C had a tetrahedral rotation of 10.6° as the α value. This value was higher than that found in other fluorophlogopites. The total calculated strength (S) of the bond between the interlayer cation and the hypothetical structures with different α values is given in Fig. 3-17, together with the experimental values from the structural determination of other micas. Micas with a high aluminum content (Y) tended to have greater S and α values than those in which the octahedral cation was mainly Mg^{2+}. Figure 3-17 shows that the phlogopite formed at 1040°C lies near the upper stability limit. Therefore, it corresponds to an upper limit of aluminum substitution in fluorophlogopite.

The curved fluorophlogopite formed at 960°C had an even higher octahedral aluminum content. Therefore, it would be located high on the curve shown in Fig. 3-17. It would also have an impossibly high α value. The crystal structure, which should result with a very high α value of 20°, is shown in Fig. 3-18 (Höland et al., 1991a). The net result is, that because of the high α value of the cabbage-head mica, the misfit can only be relieved by bending the crystal (Elsen et al., 1989).

3.2.8 Acicular Interlocking Microstructures

Glass-ceramics with an acicular interlocking microstructure were developed with the objective of producing high-strength, high-toughness products. To achieve this objective, a microstructure that resists crack propagation in the glass-ceramic must be developed.

Bond strength, S

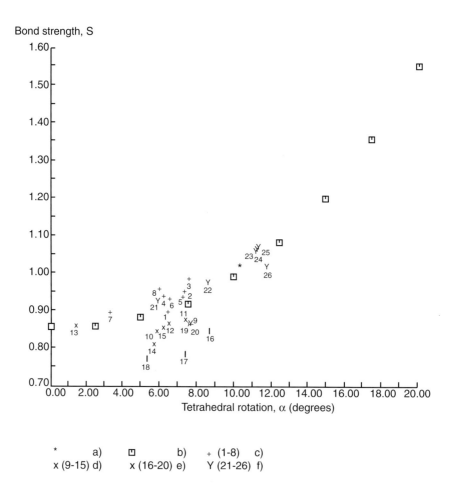

Figure 3-17 Bond strength, S, versus tetrahedral rotation, α, of micas. (a) curved phlogopite, (b) calculated structures, (c) lepidolites, (d) fluorophlogopites, (e) phlogopite–biotite series, (f) dioctahedral micas.

Since interesting developments have been conducted in the field of sintered ceramics and inorganic–inorganic or inorganic–organic composites, a comparison between these materials and glass-ceramics is appropriate at this stage. For example, whisker-reinforced composites or composites reinforced with long fibers have been produced. These materials simultaneously exhibit high strength and toughness. The principle behind these properties is crack deflection within the material, which causes a large amount of energy to be consumed during the fracture process before the material actually fails. The strength and toughness were found to be directly proportional to the aspect ratio of the filler. The highest increase in these properties was achieved with a high content of components demonstrating a high aspect ratio, such as long fibers. The fibers in the material, however, were best arranged in different directions. Drawing on the experience of the textile industry, they had to be woven together in multiple layers. The mechanical properties of the composite structure are therefore highly anisotropic.

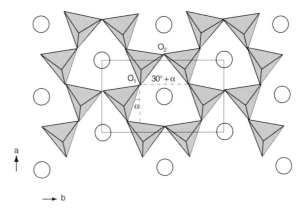

Figure 3-18 c-axis projection of a calculated phlogopite structure with tetrahedral rotation of $\alpha = 20°$.

In recent developments of long-fiber-reinforced composites described by Prewo (1989), the toughness was increased to more than 8 MPa·m$^{0.5}$. In this case, however, the fiber pull-out properties from the matrix contributed to the increase in the toughness of the material.

After the initial developments in the 1950s and 1960s, attempts were made to increase the strength of glass-ceramics by surface compression. The materials, however, were difficult to realize practically. In response to these problems, Beall (1991) developed a new generation of tough glass-ceramics, the microstructure of which is discussed in this section. This type of microstructure had not been previously formed in glass-ceramics and it is composed of a uniform, dense network of interlocking crystals throughout the volume of the material. Chain-silicate glass-ceramics such as canasite (Section 2.3.3) and richterite glass-ceramics (Section 2.3.4 and crystal structure in Appendix Fig. 10) provide excellent examples of this microstructure, which is shown in Fig. 3-19 for canasite glass-ceramic

Figure 3-19 SEM image of the microstructure of canasite glass-ceramics. The crystals measure approximately 10–20 µm. Etched sample. Scale bar = 1 µm.

(see Section 2.3.4 for richterite glass-ceramic). A close examination of these figures reveals that a large number of fairly long crystals are contained in the glass matrix. These crystals, however, are not added to the matrix as in the case of the mentioned long-fiber-reinforced composites developed by Prewo (1989). Instead, the crystals grow in the matrix in a process of controlled crystallization. The crystals not only demonstrate an acicular microstructure, but also an interlocking one (Fig. 3-19).

A microstructure containing acicular interlocking canasite or richterite crystals has a considerable effect on the properties of the material. First, however, the physical nature of the crystals must be examined in detail. Given their structure, crystals grow preferentially along one crystallographic axis. Thus, practically needlelike crystals are produced. This structure and the high aspect ratio of 10:1 render the crystals anisotropic in their thermal properties. For example, canasite crystals demonstrate different coefficients of thermal expansion along the three crystallographic axes (a, b, or c):

(α) $a = 159 \times 10^{-7}$ K^{-1},
(α) $b = 82 \times 10^{-7}$ K^{-1},
(α) $c = 248 \times 10^{-7}$ K^{-1} (0–700°C).

Therefore, considerable stress develops around the crystals during crystallization and subsequent cooling. This stress, in turn, is exerted on the glass matrix and small microcracks occur. However, since the crystals demonstrate a dense interlocking microstructure, the microcracks do not propagate. Thus, the entire material is rendered considerably tougher than the base glass. In this way, fracture toughness of approximately 5.0 MPa·m$^{0.5}$ was achieved. This degree of toughness is the highest in glass-ceramics to date and comes close to that of sintered ZrO_2. In addition to this high toughness, canasite glass-ceramics also exhibit a high bending strength of 300 MPa. The temperature resistance of canasite glass-ceramics, however, is very low. At approximately 600°C, the glass matrix becomes plastically deformable and the strengthening mechanism disappears.

3.2.9 Lamellar Twinned Microstructures

The term *lamellar twinned* refers to the twinning of crystals in the microstructure of a glass-ceramic by controlled crystallization. This process of polysynthetic twinning can be initiated in silicate crystals during the cooling of the glass-ceramic, in other words, following the heat treatment that permits the controlled precipitation of crystals. Enstatite phases, $MgSiO_3$, are an example of this type of crystal. The crystal structure of enstatite is shown in Appendix Fig. 7. Protoenstatite, the high-temperature orthorhombic form, is subject to martensitic transformation during the cooling phase of the glass-ceramic process. Martensitic transformation refers to a spontaneous and nonquenchable transformation, in this case protoenstatite to clinoenstatite. In coarse crystalline materials such as plutonic rocks or meteorites, this transformation cannot be quenched, but in finer-grained materials such as glass-ceramics, the transformation may only be partially completed. The partial proto-to-clino transformation exhibits a volume shrinkage that is manifested in polysynthetic or lamellar twinning. The type of microstructure shown in Fig. 3-20 results. This image shows a fracture surface of a glass-ceramic sample composed of enstatite and zircon

← 1 μm →

Figure 3-20 SEM image showing microstructure of enstatite–zircon showing a lamellar twinning microstructure.

($ZrSiO_4$). The microstructure exhibits fine polysynthetic twinning as well as cleavage steps, which are orthogonal to the crystallographic (100) twin plane. Fracture propagation in this enstatite glass-ceramic is clearly influenced by two processes: twinning of the crystals and unusual stepwise crack propagation, both of which indicate that significant energy is necessary to fracture the crystals. These processes cause the glass-ceramic to splinter when it fractures, as shown in Fig. 3-20. This energy-absorbing mechanism is responsible for the high fracture toughness, approximately 4.6 MPa·m$^{0.5}$, found in this enstatite–zircon glass-ceramic (Beall 1987; Echeverría and Beall, 1991) (see also Section 2.2.5).

To avoid confusion, it should be noted that the high fracture toughness of partially stabilized zirconium oxide ceramics is caused by a totally different phenomenon associated with martensitic transformation than in the case of enstatite. Upon cooling zirconia ceramics, a martensitic transformation from tetragonal to monoclinic ZrO_2 involves an expansion, as opposed to the contraction in enstatite. Thus if the tetragonal phase is partially stabilized to room temperature by alloying additions of CaO, MgO, or Y_2O_3, the phase transformation to the monoclinic structure can be initiated through the stress field in the crack-tip zone. The resulting compressive stresses reduce and eventually stop the propagation of the crack. This toughening mechanism, of course, does not require lamellar twinning, and is totally distinct from the situation of enstatite, where the transformation involves contraction, not expansion.

3.2.10 Preferred Crystal Orientation

The oriented precipitation of crystals in base glasses that are used to produce glass-ceramics with new properties has been pursued since the beginnings of glass-ceramic research. A significant development in this field is the controlled surface crystallization of β-quartz solid solution in a thin skin, measuring a maximum of 100 μm in width, around the glass-ceramic. The crystals are oriented in such a way to increase the strength of the glass-ceramic (Beall 1971) (see Section 2.3.3).

The basis of unidirectional crystallization and the development of high-strength glass-ceramics was given by Abe et al. (1984) (Section 2.6.1). Halliyal et al. (1989) conducted further investigations into the oriented precipitation of crystals during the development of a glass-ceramic for electronic applications. Halliyal et al. (1989) showed in detail that polar glass-ceramics could be developed in the following materials systems in particular: Li_2O–SiO_2, Na_2O–SiO_2, Li_2O–GeO_2, Li_2O–B_2O_3, SrO–TiO_2–SiO_2, Li_2O–P_2O_5, BaO–TiO_2–SiO_2, BaO–TiO_2–GeO_2, SrO–PbO–B_2O_3, and ZnO–P_2O_5.

Grain-oriented polar glass-ceramics are clearly presented in these systems. Kokubo et al. (1979) and Halliyal et al. (1989) showed a particularly preferred orientation in virtually a single crystallographic axial direction (c-axis) in the Li_2O–SiO_2 system. The orientation of the crystals in the main crystal phase of lithium disilicate, $Li_2Si_2O_5$, proceeded according to the unidirectional crystallization mechanism. Glass-ceramics produced in this way demonstrated dielectric properties.

Halliyal et al. (1989) developed pyroelectric properties in glass-ceramics by the precipitation of the $Ba_2TiGe_2O_8$, $Ba_2TiSi_2O_8$ (fresnoite) and $Li_2B_4O_7$ main crystal phases in glasses. In this connection, glass-ceramics from the fresnoite family are considered to be excellent candidate materials for hydrophones. The properties, which have been determined for fresnoite glass-ceramics are comparable to those of polyvinylidene fluoride. The magnitudes of the hydrostatic coefficient, for example, are approximately 100×10^{-3} VmN^{-1} and the dielectric constant is approximately 10. The electromechanical coupling coefficient k_p and k_t ranges between 15 and 20%. Furthermore, fresnoite glass-ceramics have been shown to demonstrate acoustic-wave properties.

The orientation of crystals in an electric field during the crystallization process in the SiO_2–TiO_2–BaO system is described by Keding and Rüssel (1997) and Rüssel (1997). The controlled crystallization of fresnoite,

$Ba_2TiSi_2O_8$, has been achieved in a glass with the following composition: $2.75SiO_2 \cdot 1TiO_2 \cdot 2BaO$.

In a study on the effects of oxidation on crystallization, Keding and Rüssel (1997) found that reduction produced the Ti^{3+} ion, which acted as a nucleating agent. Rüssel (1997) managed to control the orientation of the main crystal phases in a specific axial direction in various glass-ceramic systems. The fresnoite system as well as apatite glass-ceramics and lithium disilicate glass-ceramics were particularly suitable.

3.2.11 Crystal Network Microstructures

Various methods have been used to develop very strong and tough glass-ceramics, as shown by the development of glass-ceramics with chain-silicate structures in Sections 2.3.3 and 2.3.4. This section describes a very special microstructure that enables the fabrication of glass-ceramics that are both very strong and very tough. This microstructure is composed of a dense network of tightly interlocked crystals. In contrast to glass-ceramics containing chain silicates, this glass-ceramic consists of layered silicates of the lithium disilicate type. The processes of nucleation and crystallization of this type of glass-ceramic are described in Section 2.1.1. A typical microstructure of the final glass-ceramic product is shown in Fig. 3-21. To produce the SEM image shown in Fig. 3-21, a fracture surface of the glass-ceramic was prepared and subsequently etched with diluted hydrofluoric acid. In contrast to ground sections, this technique clearly shows that the crystals are tightly interlocked. A large number of microstructures of this glass-ceramic were evaluated using a special quantitative image analysis method. The volume content of the lithium disilicate crystals was established as 70 ± 5 vol% (Höland et al., 2000b). Thus, it is clearly shown that high levels of crystal interlocking permit high strength and toughness to be obtained in glass-ceramics of high crystalline content. Unlike in oriented glass-ceramics, the mechanical properties of these materials are isotropic. Because the index of refraction of the crystal phase is very similar to that of the glass phase ($n_D = 1.5323$), these glass-ceramics are also translucent. The way in which these findings can be used to develop a biomaterial for restorative dental applications (IPS Empress® 2) is described in Section 4.4.2.

3.2.12 Nature as an Example

As shown in Section 2.4.6, natural hydroxy/carbonate apatite in human bones and teeth structures exhibits a needleshaped habit. It has been a long-standing objective in the development of biomaterials to reproduce these

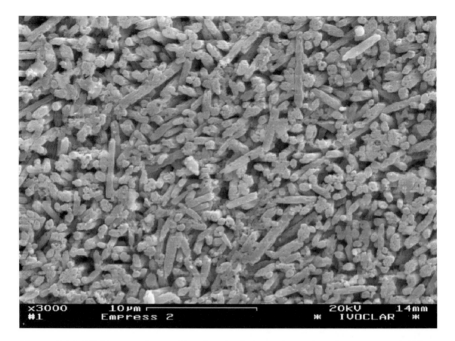

x3000 10µm 20kV 14mm
#1 Empress 2 ✳ IVOCLAR ✳

Figure 3-21 SEM image after etching (2.5% HF, 10 sec) showing high content of lithium disilicate crystals of 70 vol% crystallinity in a network microstructure.

needlelike crystals. Sections 2.4.6 and 2.4.7 show the phase reactions in the glass-ceramic, which are responsible for the formation of needleshaped crystals. Clearly, a very complex reaction mechanism is involved. Figure 3-22 shows a glass-ceramic containing needlelike fluoroapatite crystals in an apatite–leucite glass-ceramic. An additional glass-ceramic containing needlelike fluoroapatite as the main crystal phase is presented in Section 2.4.7. The microstructures of the two glass-ceramics show the substantial progress that has been made in attempting to reproduce apatite crystals in needleshaped habit as they are found in nature. This type of apatite is special in that its length in the glass-ceramic can be controlled in the nanometer-to-micrometer range. The method of controlled volume crystallization is particularly useful in this respect.

Another significant difference between needlelike apatite in glass-ceramics and other apatites is found in the fluoride content. The crystal structure of fluoroapatite compared with other apatites is shown in Section 1.3.2. The crystal structure of fluoroapatite is shown in Appendix Fig. 19.

Ways in which the findings related to the formation of needlelike apatite have been used in the development of restorative dental biomaterials are presented on the basis of two materials, IPS Empress® 2, IPS ERis for E2, and IPS d.SIGN®, in Section 4.4.2.

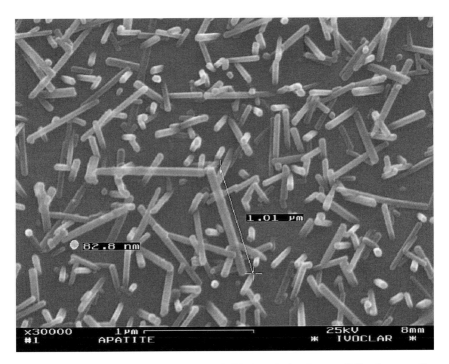

Figure 3-22 SEM image after etching (2.5% HF, 10 sec) showing needlelike fluoroapatite crystals in a glass-ceramic.

3.3 CONTROL OF KEY PROPERTIES

A significant advantage of glass-ceramics is the fact that even though they are multicomponent materials, their properties can be controlled. The aim in controlling these properties is to develop tailor-made products for use in technical fields such as microelectronics and engineering or in medicine and dentistry. The complex nature of glass-ceramics, that is, their multiphase character and wide spectrum of microstructures, chemical compositions, and processing technologies, give researchers and technicians a variety of options for controlling these key properties. Two examples will illustrate the variety of possibilities that are available for controlling properties and will show some preferred methods and techniques.

First, the chemical system must be selected. This system should provide a basis for the properties that are of interest for the desired application. For example, if the aim is to develop a glass-ceramic demonstrating a very low coefficient of thermal expansion, it would be necessary to select the SiO_2–Al_2O_3–Li_2O system. This system is selected because it contains glass-forming crystalline phases characterized by low—or even negative—coefficients of thermal expansion, and techniques of efficient nucleation of these phases in glass are well known. This first step is relatively simple, and detailed research really begins after this choice is made. Next, questions must be asked regarding the desired chemical durability and thermal and dimensional stability of the product envisioned. Furthermore, the possibilities for processing the material economically must be examined. This would involve a hot-glass-forming process that requires certain viscosity–temperature relationships above the top temperature of crystallization (liquidus). A set of experiments is necessary to find the best solution for combining all these properties. Usually, a wide variety of chemical compositions must be tried. For example, additive components may have to be selected to control nucleation, help stabilize the glass for forming purposes, develop colors or tints, etc. All the major properties must then be determined for each composition. Often, several hundred chemical compositions have to be tested and several hundred thermal treatment cycles have to be conducted to control nucleation and crystallization.

A second example involving developing a new dental material illustrates a different situation with other prerequisites for controlling the properties of a material. The new material must demonstrate high mechanical strength, good chemical durability, and the aesthetic character of natural teeth. Furthermore, the aim is to develop a material that can be shaped using the viscous flow method, which is currently desirable in the market. As a result, research must focus on combining high strength with translucent optical properties and viscous flow processing. Because viscous flow is characteristic of glass-ceramics, these materials are leading candidates. As before, it is necessary to control nucleation and crystallization. Two different means are available. A very slow rate of nucleation and crystallization generally enables very good viscous flow, but requires additional heat treatment. Rapid nucleation and crystallization, however, does not allow the glass-ceramic to flow sufficiently in the shaping process. Several hundred chemical compositions and heat treatment cycles in combination with special processing methods must be tested in this case as well.

These two simplified examples illustrate some of the complexities of controlling properties in the development of glass-ceramic materials. The following methods are used to control and measure individual properties in the first stages of applied research.

3.4 METHODS AND MEASUREMENTS

3.4.1 Chemical System and Crystalline Phases

The selection of the chemical system and appropriate crystalline phases form the basic potential of the material. The properties of the constituent crystalline phases in combination with those of the glassy matrix determine the key properties, but microstructure may also play a critical role. As shown in the previous examples, other accessory properties must also be considered at this stage.

3.4.2 Determination of Crystal Phases

The preferred method used to determine the crystal phases and monitor their development and stability is X-ray diffraction. A glass is typically heated to a certain temperature for a specific period of time, and the crystal phases are analyzed at room temperature after heat treatment. New methods involving heat treatment and simultaneous X-ray diffraction allow an *in situ* investigation of the crystallization process (high-temperature X-ray diffraction, HT-XRD). Figure 3-23 shows the results of this technique applied to the crystallization of a lithium disilicate glass-ceramic (Cramer von Clausbruch et al., 2000). In this case, the composition of the glass-ceramic was 63.2 mol% SiO_2, 29.1 mol% Li_2O, 2.9 mol% K_2O, 3.3 mol% ZnO, and 1.5 mol% P_2O_5 (see Section 2.1.1). At 500°C, lithium metasilicate, Li_2SiO_3, and lithium disilicate, $Li_2Si_2O_5$, started to precipitate. The intensity of these phases slightly increased with increasing temperature up to 650°C. In the temperature range between 610° and 650°C, SiO_2, cristobalite, was present as an intermediate phase along with the two prior phases. At 650°C, lithium metasilicate and cristobalite decomposed while the intensity of the lithium disilicate peaks (130, 040, 111) strongly increased until this phase finally dissolved at about 960°C. The crystallization of minor lithium phosphate began after the transformation of lithium metasilicate and cristobalite at approximately 650°C, showing a very low intensity. Therefore, the faster growth of lithium disilicate is a result of a solid-state reaction between lithium metasilicate and cristobalite (Section 2.1.1).

Identification of crystal phases in glass-ceramics can also be made through solid-state nuclear magnetic resonance (NMR) (Brow et al., 1995;

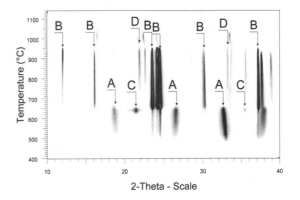

Figure 3-23 High-temperature X-ray diffraction, HT-XRD, of a multicomponent glass-ceramic of lithium disilicate type. A = Li_2SiO_3, B = $Li_2Si_2O_5$, C = SiO_2 (cristobalite), and D = Li_3PO_4.

Gee and Eckert, 1996). This technique also yields information concerning short-range structure of the residual glassy matrix. Schmedt auf der Günne et al. (2000) used NMR investigations to analyze phase-formation processes in the multicomponent system SiO_2–Al_2O_3–K_2O–Na_2O–CaO–P_2O_5–F (Section 2.4.6). In addition to the crystal phase characterization by XRD, differential scanning calorimetry (DSC), and SEM investigations, it was also possible in this case to distinguish and compare crystal phases and the glassy matrix by NMR. First results of this investigation showed sharp peaks characteristic of crystal phases and flat peaks of ^{19}F MAS–NMR spectra characteristic of the glassy phase. These preliminary results demonstrate that a very small number of small primary crystal phases was formed in the base glass, even when the melt was rapidly quenched. An ongoing research program is focused on NMR of different nuclei and using double-resonance NMR experiments, e.g. [19]F–[31]P.

3.4.3 Kinetic Process of Crystal Formation

The phase sequence and solid-state reactions involving the formation of different phases within a glass-ceramic can be determined by thermal analysis. Differential thermal analysis (DTA) and DSC are particularly useful methods for studying these processes. Sestak (1996) determined phenomenological kinetics based on DTA and Donald (1998) analyzed the crystallization process of iron-containing glasses using the same technique.

Ray and Day (1990) used DTA investigations of the crystallization peak as a rapid method to determine the temperature for maximum nucleation of glasses to develop glass-ceramics. The height of the exothermic signal of the crystallization of lithium disilicate, $Li_2Si_2O_5$, was measured at different temperatures for a constant time of 3 h. The results clearly showed a maximum of the crystallization peak with a sample heat treated at 453° ± 3°C for nucleation of the base glass. Therefore, the temperature of 453°C was determined as the temperature for maximum nucleation.

The DTA method is also very useful to distinguish between surface and volume crystallization. Ray et al. (1996) demonstrated in different types of glasses that the height of the exothermic peak decreased with increasing particle size when surface crystallization is the dominant mechanism. But on the other hand, the peak intensity will increase with increasing particle size when volume crystallization of glasses is dominant.

Furthermore, Ray and Day (1997) determined the influence of heterogeneous nucleation on the crystallization rate of a $Na_2O \cdot 2CaO \cdot 3SiO_2$ glass in comparison to an undoped glass. Platinum (0.1 wt%) was used as a heterogeneous dopant. In comparison to the undoped base glass, the Pt-doped glass showed a shifting of the DTA crystallization peak to lower temperatures (Fig. 3-24) than the undoped glass. This result is a very good indication that heterogeneous nucleation (in this case Pt) increases the crystal growth rate

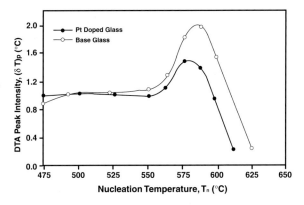

Figure 3-24 DTA peak intensity as a function of nucleation temperature of an undoped and a doped (0.1 wt% Pt) $Na_2O \cdot CaO \cdot 3SiO_2$ glass (Ray and Day, 1997).

of a glass. Ray and Day (1997) also calculated DTA curves as a function on the overlap of nucleation and crystallization rate (see Fig. 1-26 in Section 1.4). The calculated results were compared to experimental investigations. The authors also investigated the influence of P_2O_5 and Ag_2O on the overlap of the nucleation and crystallization curves using doped and undoped glasses. The overlap of the curves increased at all temperatures for $Na_2O \cdot 2CaO \cdot 3SiO_2$ glasses with both dopants, but the influence of P_2O_5 was greater.

Methods of processing powder compact glass-ceramics are also very complex by nature. A large variety of gas reactions are involved in these processes. Gases are formed in the bulk of the sample as well as from the surface of the grains. Gaber et al. (1995, 1998) investigated the process of volatile species emission during thermal processing of glass-ceramic powder compacts by EGA (evolved gas analysis). Figure 3-25 shows the degassing behavior (EGA) of an alkali–silicate glass powder (1 mm in thickness) involving H_2O, CO_2, CO, and C up to 1200°C (heating rate $40K \cdot min^{-1}$) in comparison to evidence from DTA (Gaber et al. 1998). The reaction processes (a–f) were determined as follows:

a) release of water (up to 450°C), accompanied by CO_2 and CO degassing
b) degassing and sintering reaction (560°–620°C)
c) degassing and first crystallization (650°C)
d) degassing and second crystallization (820°–840°C)
e) degassing and melting of a crystalline phase (950°C)
f) bursting of gas bubbles (1000°–1200°C)

3.4.4 Design of Microstructure

Scanning electron microscopy and transmission electron microscopy are the preferred methods for determining the microstructure of glass-ceramics. Special results regarding the determination of microstructures were shown in Section 3.2. At this stage, it should be emphasized that the technique used to prepare the samples is very critical to obtaining the best results. Often, etching is necessary to prepare samples. Etching patterns, however, require a detailed interpretation. Furthermore, a large number of additional investigations and additional etching procedures have to be conducted. For TEM and other surface analysis methods, the ion release during sputtering is very critical. Völksch et al. (1998) showed a preferred form of TEM investigations for leucite–apatite glass-ceramics. It is very important to consider the sample preparation methods. Thus, Brow (1989) demonstrated ion-beam effects on the composition and structure of glass surfaces. These results are also very important for glass-ceramic investigations.

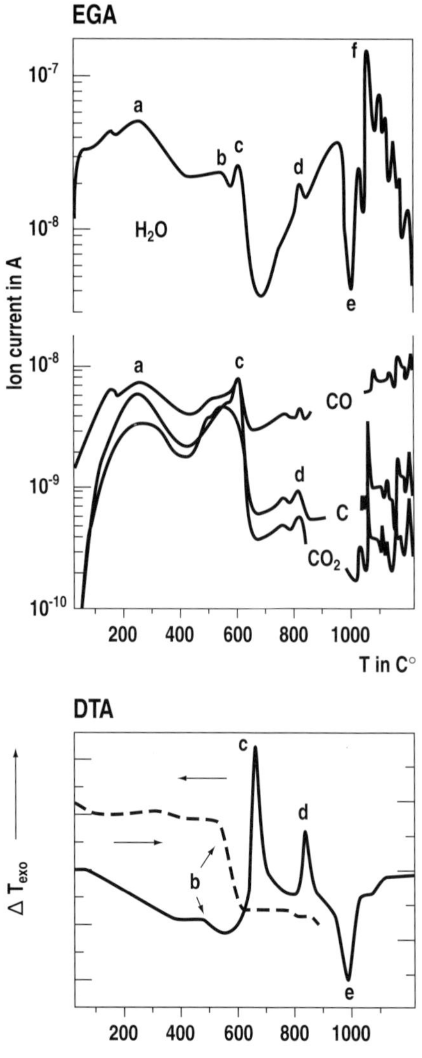

Atomic force microscopy (AFM) is also a very sensitive method of investigating the microstructure of glass-ceramics. This method can even be used for the analysis of the nucleation process (Rädlein and Frischat, 1997; Pinckney 1998).

Figure 3-25 EGA of a lithium disilicate glass-ceramic powder compact in comparison to DTA.

3.4.5 Mechanical, Optical, Electrical, Chemical, and Biological Properties

The different complex characteristics such as the mechanical, optical, electrical, chemical, and biological properties are standardized. That is, the properties are determined according to ISO, United States, or EN (European standards) standards. It is essential to prepare all the samples using the procedure reports in these documents or to follow the special instructions. The sample preparation of glass-ceramics for mechanical testing, e.g., 3-point flexural strength, has to be conducted very carefully. Edge chipping of samples (which would lead to lower strength values) must be avoided.

The determination of properties should provide information and some scientific conclusions helpful in further designing and improving glass-ceramics. Based on the results of even preliminary properties, new insight generally suggests changes in the chemical composition or directions in modification of the microstructure.

CHAPTER 4

Applications of Glass-Ceramics

4.1 TECHNICAL APPLICATIONS

4.1.1 Radomes

The first commercially viable glass-ceramics were developed in the aerospace industry in the late 1950s. The glass-ceramics were used to manufacture radomes to allow the use of and to protect radar equipment (internal antennae) in the noses of aircraft and rockets. According to McMillan (1979), materials used for this type of application must exhibit a very homogeneous and low dielectric constant and a low coefficient of thermal expansion, low dielectric loss, high strength, and high abrasion resistance. These properties are also demonstrated by sintered Al_2O_3. This material, however, is difficult to process and quality control is problematic.

Stookey (1959) managed to fabricate an aircraft nose with a complex compound curvature using Corning 9606 cordierite glass-ceramic. The properties and the microstructure of this glass-ceramic are covered in Section 2.2.5. Stookey (1959) reported that the nose was manufactured by centrifugal casting from a base glass using processes from glass spinning technology. Because of the transparency of the glass, inhomogeneous areas were easily detected, which facilitated quality control. Crystallization of the material followed to generate the glass-ceramic. Care had to be taken not to deform the preform during this step. The outstanding properties of this glass-ceramic are described by Grossman (1982). The high bending strength of 240 MPa combined with a coefficient of linear thermal expansion of 57×10^{-7} K^{-1} at 25°C–800°C and high thermal shock resistance must be mentioned in particular. These are critical in situations of rain erosion and atmosphere re-entry. Because of these properties, this glass-ceramic is now used to manfucture noses for high-performance aircraft.

4.1.2 Photosensitive and Etched Patterned Materials

Microtechnology refers to a field of development concerning tiny components for a variety of applications. The developments are concentrated on reducing the products to the smallest possible size. Special procedures are also being developed to integrate high-precision holes, channels, or patterns into these small components. Highly accurate materials of miniature dimensions are produced for use in the fields of electronics, chemistry, acoustics, optics, mechanics, and biology.

4.1.2.1 Fotoform® and Fotoceram®

A) Processing and Properties

The first glass-ceramic in the history of materials science was developed by Stookey (1953, 1954). The glass-ceramic in question was photosensitive and it could be etched. The main crystal phase of this glass-ceramic was lithium disilicate $Li_2Si_2O_5$. The basic steps of the development, the fundamental principles, and the reaction mechanisms of the precipitation of lithium disilicate crystals are described in Section 2.1.1. Fotoform® is a glass and Fotoceram® is a lithium disilicate glass-ceramic. Processing of Fotoform® and Fotoceram® is characterized by the following reactions and intermediate products:

- melting of a base glass and casting to monolithic products or thin films
- exposure to UV light (using masks)
- heat treatment of the glass and precipitation of lithium metasilicate crystals, Li_2SiO_3
- etching of the lithium metasilicate glass and producing of Fotoceram® materials as glass products
- additional heat treatment and precipitation of lithium disilicate crystals to form Fotoceram®

The base glasses for these glass-ceramics are derived from the SiO_2–Li_2O system. The base glasses contains Ce^{3+}- and Ag^+-ions. The compositions are shown in Table 4-1. During exposure to UV light, photoelectrons cause the oxidation of Ce^{3+} to Ce^{4+}. As a result, Ag^+ is reduced to Ag(0).

$$Ce^{3+} + h \cdot v \ (312nm) \rightarrow Ce^{4+} + e^-$$
$$Ag^+ + e^- \rightarrow Ag^0$$

This metallic silver is the nucleating agent for the lithium metasilicate Li_2SiO_3 phase. As a result, this crystal phase can be precipitated by controlled crystallization at 600°C (Beall 1993). The lithium metasilicate crystals are easily etched with dilute hydrofluoric acid (HF). Thus, defined structures can be etched into the finished products. These specially structured glassy bodies

Table 4-1

Composition of Fotoform®/Fotoceram® (Corning 8603) (wt%)	
SiO_2	79.6
Al_2O_3	4.0
Li_2O	9.3
K_2O	4.1
Na_2O	1.6
Ag	0.11
Au	0.001
CeO_2	0.014
SnO_2	0.003
Sb_2O_3	0.4

were produced and distributed by Corning Inc., USA, under the Fotoform® trademark.

If UV exposure and heat treatment are conducted a second time, approximately 40 wt% of the main crystal phase lithium disilicate is produced along with α-quartz with a total crystal content of approximately 60%. This product called Fotoceram® was also manufactured by Corning Glass Works, USA. The chemical composition of Fotoform®/Fotoceram® is shown in Table 4-1. Selected properties are listed in Table 4-2 (Beall 1992).

B) Applications

The exposure to UV light of the base glass can be precisely controlled with specially tailored patterns that permit high-precision components to be fabricated (Fig. 4-1). The accuracy (Table 4-3) of these parts is higher than that which could have been achieved using mechanical or even laser procedures, although less accurate than photolithographic techniques used to pattern silicon and certain metals.

This technique has been used to produce glass-ceramic components that are particularly useful in equipment manufacturing, micro-mechanics, and in the electrical industry. The material can be used for the following applications: gas discharge panels, ink-jet printer plates, fluid devices, and magnetic recording head pads. Examples of these applications are shown in Fig. 4-2.

Table 4-2

Properties of Fotoform®/Fotoceram®

Properties	Fotoform® Glass	Fotoform® Opal Glass-Ceramic	Fotoceram® Glass-Ceramic
Mechanical			
Density (g/cm^3)	2.365	2.380	2.407
Young's modulus ($\times$ 10^6 psi)	11.15	12.00	12.62
(GPa)	77	83	87
Modulus of rupture, abraded			
(psi)	8690	12100	21500
(MPa)	60	83	148
Poisson's ratio	0.22	0.21	0.19
Hardness (Knoop), KHN$_{100}$ Wilson Turkon	450	500	500
Thermal			
Coefficient of thermal expansion (25°–300°C) (K^{-1})	8.4·10^{-6}	8.9·10^{-6}	10.3·10^{-6} up to 16·10^{-6}
Thermal conductivity, (W·m^{-1}·K^{-1})			
25°C	0.75	1.5	2.6
200°C	1.1	1.5	2.0
Maximum safe processing temperature (°C)	450	550	750
Specific heat, (J·g^{-1}·K^{-1})			
25°C	0.88	0.88	0.88
200°C	1.2	1.2	1.2

(continued)

4.1.2.2 Foturan®

A) Processing and Properties

Foturan® is a photostructurable glass that is produced according to the above photosensitive techniques used for manufacturing glass-ceramics. The glass-ceramic of lithium metasilicate, Li_2SiO_3, however, represents an intermediate product. The final product is a glass with the properties shown in Table 4-4. The mechanical, electrical, and chemical properties of the glass, as well as the processing parameters for the production of very small components are noteworthy. The parameters for capabilities and tolerances are shown in Table 4-5. This list of parameters emphasizes the high accuracy of the production of the intermediate glass-ceramic product.

Table 4-2 (continued)

Properties of Fotoform®/Fotoceram®

Properties	Fotoform® Glass	Fotoform® Opal Glass Ceramic	Fotoceram® Glass Ceramic
Electrical			
$\log_{10}$ volume resistivity ($\Omega \cdot cm$)			
250°C	6.27	8.81	8.76
350°C	4.90	7.23	7.07
Dissipation factor, at 100 kHz			
21°C	0.008	0.004	0.003
150°C	0.050	0.014	0.008
Dielectric constant, at 100 kHz			
21°C	7.62	5.73	5.63
150°C	9.06	6.14	6.00
Loss factor, at 100 kHz			
21°C	0.061	0.023	0.017
Dielectric strength, (volts/mil) to 10 mil sample thickness, DC under oil			
25°C	4500	4000	3800

In the process, a monolithic or thin-layer product with the following composition is produced: 75–85 wt% SiO_2, 7–11 wt% Li_2O, 3–6 wt% K_2O, 3–6 wt% Al_2O_3, 1–2 wt% Na_2O, 0–2 wt% ZnO, 0.2–0.4 wt% Sb_2O_3, 0.05–0.15 wt% Ag_2O, 0.01–0.04 wt% CeO_2 (Speit 1993, Dietrich et al., 1996). The glass-melting process is carried out in reduced conditions to produce Ce^{3+} in the glass $2Ce^{4+} + Sb^{3+} \rightarrow 2Ce^{3+} + Sb^{5+}$. Once the glass has been fabricated, a glass-ceramic intermediate product is formed in a subsequent step. Using masks, areas of the glass in which a particular pattern, engravings, cavities, or holes must be produced are exposed to UV light. This process is the same as was reported for Fotoceram®/Fotoform®.

After the exposure to UV light, the glass is heat treated. Crystals are produced only in the areas that had previously been exposed to light. The Ag(0) functions as a nucleating agent and lithium metasilicate, Li_2SiO_3, as the main crystal phase. The glass-ceramic thus produced can be etched with a hydrofluoric aqueous solution. Consequently, the patterns such as holes, engravings, channels, etc., are produced in microscopic dimensions. The microstructured glass is the end product.

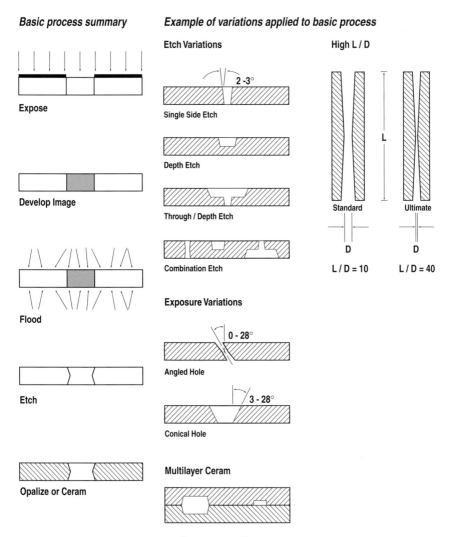

Figure 4-1 Processing of Fotoform®/Fotoceram®.

B) Applications

Foturan® is produced by the Schott Glaswerke and mtg mikroglas technik AG, both in Mainz, Germany. The material is used in automotive and precision engineering. Furthermore, Foturan® is particularly suitable for micro-optic and integrated optic systems. Microchannels in optical fibers, ink-jet printer heads, substrates for pressure sensors, and acoustic systems in headphones are additional applications.

Table 4-3

Precision Capability for Fotoform®/Fotoceram®

Standard Process Capability	Fotoform® Glass		Fotoform® Opal and Fotoceram®	
Hole and slot size tolerances	± 0.025 mm	± 0.001"	± 0.025 mm	± 0.001"
Centerline tolerance	± 0.025 mm	± 0.001"/"	± 0.038 mm	± 0.0015"/"
Edge to vertical angle of etch		2°–3°		2°–3°
Minimum feature size	0.063 mm	0.0025"	0.0762 mm	0.003"
Maximum ratio of thickness to feature size		8:1		8:1
Thickness control	± 0.05 mm	± 0.002"	± 0.05 mm	± 0.002"
Flatness	0.025 mm	0.001"/"	0.025 mm	0.001"/"
Minimum thickness	0.508 mm	0.020"	0.508 mm	0.020"
Ultimate Precision Capability				
Hole and slot size tolerances	± 0.0076 mm	± 0.0003"	± 0.01016 mm	± 0.0004"
Centerline tolerance	± 0.0025 mm	± 0.0001"/"	± 0.005 mm	± 0.0002"/"
Edge to vertical angle of etch		1°–2°		1°–2°
Minimum feature size	0.025 mm	0.001"	0.038 mm	0.0015"
Maximum ratio of thickness to feature size		40:1		40:1
Thickness control	± 0.025 mm	± 0.001"	± 0.0127 mm	± 0.0005"
Flatness	0.0025 mm	0.0001"/"	0.0025 mm	0.0001"/"
Minimum thickness (dependent on size)	0.1016 mm	0.004"	0.076 mm	0.003"

4.1.2.3 Additional Products

Additional glass-ceramics based on lithium disilicate are produced and distributed by General Electric, USA, under the name RE-X®; and Pfaudler under the name Nucerite® (special application for enamel of steel) (Pincus 1971)

Figure 4-2 Application of Fotoform®/Fotoceram® in the electronic industry, micro-mechanics, and in apparatus manufacturing. Fotoceram® glass-ceramic head pads.

Table 4-4

Properties of Foturan® (Schott, Mikroglas, 1999)

Properties	Glass	Glass-Ceramic (brown)
Mechanical		
Young's modulus (GPa)	78	88
Poisson's ratio	0.22	0.19
Hardness, Knoop (MPa)	4600	5200
Modulus of rupture (MPa)	60	150
Density (g/cm^3)	2.37	2.41
Thermal		
Thermal expansion 20°–300°C (10^{-6} K^{-1})	8.6	10.5
Thermal conductivity, 20°C (W·m^{-1}·K^{-1})	1.35	2.73
Specific heat 25°C (J·g^{-1}·K^{-1})	0.88	0.92
Transformation temperature (°C)	465	–
Maximum safe processing temperature (°C)	450	750
Electrical		
Electrical conductivity		
at 25°C (Ω·cm)	8.1·10^{12}	5.6·10^{16}
at 200°C (Ω·cm)	1.3·10^{7}	4.3·10^{7}
Dielectric constant at 1 MHz, 20°C	6.5	5.7
Loss factor tan δ at 1 MHz, 20°C [× 10^{-6}]	65	25
Chemical		
Water durability DIN/ISO 719 ((μg)Na$_2$O/g)	468	1300
Acid durability DIN 12116 (mg/dm^2)	0.4	0.9
Alkali durability DIN/ISO 695 (mg/dm^2)	96	250

It should be mentioned that the dc resistivity of $Li_2Si_2O_5$ is five orders of magnitude greater than that of its parent glass—hence the insulating value of RE-X® despite high Li^+ content.

4.1.3 Machinable Glass-Ceramics

4.1.3.1 MACOR® and DICOR®

Machinable glass-ceramics are based on internally nucleated fluoromica crystals in glass (Beall 1971a). One commercial material has been marketed for 20 years under the trademark MACOR® and has found wide application in such diverse and speciality areas as precision electrical insulators, vacuum feedthroughs, windows for microwave-type parts, samples holders for field-ion microscopes, seismograph bobbins, gamma-ray telescope frames, and

Table 4-5

Processing Capabilities of Foturan® (Schott, Mikroglas, 1999)

Process Capabilities

Minimum hole and slot sizes (mm)	Minimum hole and slot depth (mm)	Minimum hole and slot distances* (mm)
0.025 ± 0.005	0.2 ± 0.01	0.035 ± 0.01
0.05 ± 0.015	0.4 ± 0.02	0.06 ± 0.015
0.12 ± 0.02	1.0 ± 0.03	0.14 ± 0.02
0.18 ± 0.03	1.5 ± 0.04	0.21 ± 0.03

Tolerances

Roughness of etched structures	1–3 µm
Maximum hole density	10,000 holes/cm², depth
Tolerance of hole distance per 100 mm	<0.2% (100 mm; ± 20)
Maximum unfinished holes	<0.3% (3 per 1000)

*Hole spacing (center to center)

boundary retainers on the space shuttle. The precision machinability of the MACOR® material with conventional metal-working tools, combined with high dielectric strength (~40 kV/nm) and very low helium permeation rates, are particularly important in high-vacuum applications.

Although the MACOR® glass-ceramic is based on the fluorine–phlogopite phase ($KMg_3AlSi_3O_{10}F_2$), this stoichiometry does not form a glass. The bulk composition had to be altered largely through additions of B_2O_3 and SiO_2 to form a stable although opalized glass (Table 4-6). The parent glass is composed of a dispersion of aluminosilicate droplets in a magnesium-rich matrix (Chyung et al., 1974). The crystallization begins near 650°C when a metastable phase chondrodite, $2Mg_2SiO_4 \cdot MgF_2$, forms in the magnesium-rich matrix at the interfaces of the aluminosilicate droplets. The chondrodite subsequently transforms to norbergite, $Mg_2SiO_4 \cdot MgF_2$, which finally reacts with the components in the residual glass to produce fluorphlogopite mica and minor mullite.

$$Mg_2SiO_4 \cdot MgF_2 + KAlSi_2O_6 \rightarrow KMg_3AlSi_3O_{10}F_2$$
$$\text{(glass)}$$

$KAlSi_2O_6$ represents the glassy droplet phase, having near leucite composition.

Table 4-6

Commercial Fluoromica Glass-Ceramic Compositions

	MACOR® (Corning) (wt%)	DICOR® (Corning/Dentsply) (wt%)
SiO_2	47.2	56–64
B_2O_3	8.5	
Al_2O_3	16.7	0–2
MgO	14.5	15–20
K_2O	9.5	12–18
F	6.3	4–9
ZrO_2		0–5
CeO_2		0.05
Mica type		
MACOR	$K_{1-x}Mg_3Al_{1-x}Si_{3+x}O_{10}F_2$	
DICOR	$K_{1-x}Mg_{2.5+x/2}Si_4O_{10}F_2$	

Note: x <0.2

The mica grows in a preferred lateral direction because the residual glass is fluidized by the B_2O_3 flux and is also designed to be deficient in the crosslinking species potassium.

The thermal, electrical, mechanical, and chemical properties of the glass-ceramic are shown in Fig. 4-3 and Table 4-7. MACOR® glass-ceramics are produced by Corning Inc., USA, and distributed in Europe by Corning Europe, Wiesbaden, Germany, and in Switzerland by FiberOPtics P.+P. AG, Spreitenbach. Figure 4-3 shows the individual properties as a function of temperature. These properties are particularly important for applications in the manufacture of equipment and installations, as well as in the very demanding aerospace and aeronautical industries. The following industrial applications of MACOR® glass-ceramics in high-performance fields must be mentioned in particular (Fig. 4-4):

- Aerospace industry

More than 200 special parts of the U.S. space shuttle orbiter are made of this glass-ceramic. These parts include rings at all hinge points, windows, and doors.

- Medical equipment

The accurate machinability of the material as well as its inert character are particularly important in the production of specialized medical equipment.

• Ultrahigh applications

MACOR® glass-ceramics make excellent insulators. They are widely used to manufacture equipment for vacuum technology. Compared with sintered ceramics, glass-ceramics are pore-free.

• Welding

MACOR® is used in welding equipment, as the material exhibits excellent nonwetting properties with regard to oxyacetylene.

• Nuclear-related experiments

MACOR® is not dimensionally affected by irradiation. As a result, applications in this field are possible.

This wide spectrum of application as a high-performance material demonstrates the importance of MACOR® glass-ceramics in technology and medicine. Further potential applications must be considered.

More recently, another commercial material has been developed for use in DICOR® dental restorations (Malament and Grossman, 1987). This glass-ceramic, with improved chemical durability and translucency over the MACOR® material, is based on the tetrasilicic mica, $KMg_{2.5}Si_4O_{10}F_2$. Good strength (~150 MPa) is associated with the development of anisotropic flakes

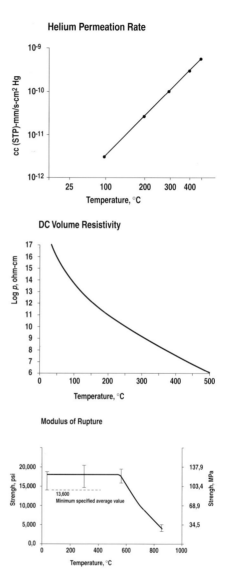

Figure 4-3 Selected properties of MACOR® glass-ceramic (MACOR® 1992).

at relatively high temperatures (>1000°C). Translucency is achieved by roughly matching crystal and glass indices and maintaining a fine-grained (~1 μm) crystal size. Ceria is added to simulate the fluorescent character of natural teeth.

Table 4-7

Properties of MACOR® Glass-Ceramic (MACOR® 1992)

Properties	MACOR™ Machinable Glass-Ceramic
Mechanical	
Density	2.52 g/cm^3
Porosity	0%
Hardness (Knoop)	250 NA
Compressive strength	50,000 psi 350 MPa
Flexural strength	15,000 psi 104 MPa
Thermal	
Coefficient of thermal expansion	5.2×10^{-6} K^{-1} (T in °F) 9.4×10^{-6} K^{-1} (T in °C)
Maximum use temperature (no load)	1832°F / 1000°C
Electrical	
Dielectric strength (a.c.)	1,000 volts / (10^{-3} inch)
Volume resistivity	> 10^{14} Ω·cm

The unique feature of DICOR® dental restorations include the close match to natural teeth in both hardness and appearance. The glass-ceramic may be accurately cast using a lost-wax technique and conventional dental laboratory investment molds. The high strength and low thermal conductivity of the material provide advantages over conventional metal-ceramic systems. The application of DICOR® glass-ceramics as a biomaterial in dentistry is discussed in Section 4.4.2.2.

Figure 4-4 Examples of the uses of MACOR® glass-ceramics.

4.1.3.2 Vitronit™

Vitron Spezialwerkstoffe GmbH, Jena, Germany, manufactures a machinable glass-ceramic called Vitronit™. It is used for technical applications.

This glass-ceramic (Glaskeramik 1995), like Bioverit® (see Section 4.4.1), exhibits a cabbage-head microstructure of mica crystals. The special properties of this glass-ceramic are:

- high temperature resistance
- electrical insulation capabilities
- stable for 40 kV/mm
- vacuum density
- high corrosion resistance
- hydrolytic class 2 (DIN 12111)
- acidic class 3 (DIN 12116)
- basic class 1 (DIN 52322)

These properties make this glass-ceramic suitable for joining with other materials, for example, metals.

4.1.3.3 Photoveel™

A mica glass-ceramic of the fluorogold mica type is produced in Japan. Apart from the crystallization of mica crystals containing gold, the crystallization of zirconia microcrystals has also been achieved in the glass matrix (Photoveel 1998). This glass-ceramic is produced by Sumikin Photon Ceramics Co., Ltd., Japan. It is distributed in two versions under the trademark of Photoveel™. The properties of the two products (Photoveel™ and Photoveel™ L) are shown in Table 4-8. The two versions mainly differ in their coefficient of thermal expansion: that of Photoveel™ L is $5.5 \times 10^{-6} K^{-1}$, while that of Photoveel™ is $8.5 \times 10^{-6} K^{-1}$. The sum of all these properties makes these glass-ceramics suited for applications in the electrical industry and in the production of equipment. Their isolation and vacuum properties are particularly significant for these applications. The materials are used to manufacture products such as electrically insulating components for semiconductors, heat insulating, vacuum sealing parts, microelectronic substrates, and electrical insulators.

4.1.4 Magnetic Memory Disk Substrates

Four types of glass-ceramics are used as magnetic memory disk substances (spinel–enstatite, spinel, lithium disilicate, and canasite). The development of spinel glass-ceramics from the SiO_2–Al_2O_3–ZnO–MgO–TiO_2 system represents a significant contribution to the fabrication of magnetic memory disk substrates (Beall and Pinckney, 1995, 1998). The material contains a very special nanostructure. The crystallites of the gahnite type, $ZnAl_2O_4$, or spinel, $MgAl_2O_4$, or the solid solution of both types of crystals are smaller

Table 4-8

Properties of Mica Glass-Ceramic Photoveel™

Properties	Photoveel™	Photoveel™ L (low expansion)
Mechanical		
Density (g/cm³)	2.59	2.90
Flexural strength (MPa)	150	90
Compressive strength (MPa)	500	
Young's modulus (GPa)	67	62
Thermal		
Maximum service temperature	1000	750
Coefficient of thermal expansion ($\times 10^{-6}$ K^{-1})	8.5	5.5
Thermal conductivity (W·m^{-1}·K^{-1})	1.6	2.5
Thermal shock resistance (K)	150	250
Electrical		
Volume resistivity (Ω·cm)	1.8×10^{15}	8.0×10^{13}

than 0.1 μm. Enstatite is an important accessory phase, increasing fracture toughness above 1 MPa·m$^{0.5}$ Furthermore, the crystals grow in an isolated manner in a glass matrix (Sections 2.2.7 and 3.2.1.).

As a result, the spinel–enstatite material is suitable as a substrate for magnetic memory disks. It demonstrates a favorable low surface roughness compared with other materials. Following the polishing process, the local roughness of the glass-ceramic is 5 Å. This is important because the electromagnetic recording head can fly well within 20 nm of the spinning disk surface.

Spinel glass-ceramics also demonstrate properties that meet with the requirements of the desired applications. For example, the Young's modulus of 100–165 MPa exceeds that of most other glass-ceramics. The glass-ceramic article exhibits a fracture toughness in excess of 1.0 MPa·m$^{0.5}$. The coefficient of linear thermal expansion measures 6–7 $\times$ 10^{-6} K^{-1} (0°C–300°C). The glass-ceramic can be used as a substrate in a magnetic memory storage device composed of a head-pad and a rigid information plate. On the surface of the plate is a sputtered magnetic material, while the base is a spinel–enstatite glass-ceramic (Fig. 4-5).

The glass-ceramic also demonstrates a high strain point (temperature resistance) of more than 1000°C. This property is particularly important in the fabrication of various types of substrates, such as that of the magnetic memory disk described previously, and for a variety of other applications. Spinel-enstatite glass-ceramics are translucent to opaque. High modulus is important, e.g., $E = 2.1 \times 10^6$ psi (14.5 GPa), because stiffness prevents flutter of the magnetic memory disk at high rotational speeds, e.g., 5000–10,000 rpm.

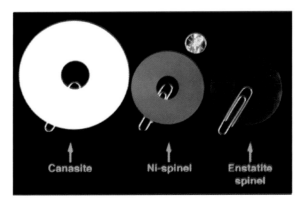

Figure 4-5 Spinel–enstatite glass-ceramic as a magnetic memory disk substrate, in comparison to a canasite disk and a nickel-spinel glass-ceramic.

Pure spinel glass-ceramics show other properties distinct from enstatite–spinel glass-ceramic. These include spinel glass-ceramics and are characterized by CTE of 3.4×10^{-7} K^{-1} (0°C–300°C) and transparency in the visible spectrum. They are also suitable for the production of highly polishable disk substrates (Fig. 4-5) (Pinckney 1999).

Beall (1998) developed another family of glass-ceramics for disk substrates requiring high mechanical strength. These materials are derived from the chemical system SiO_2–Li_2O–K_2O–P_2O_5–Al_2O_3. The crystal phases lithium disilicate and tridymite were precipitated in the glassy matrix.

In Section 2.3.3, canasite glass-ceramics were introduced. These materials are characterized by their outstanding mechanical resistance and their high fracture toughness of approximately 5 MPa·m$^{0.5}$ in particular (Beall 1991). Furthermore, rolling presents an economical glass forming process. In the rolling process, a glass strip is produced by feeding a liquid glass melt through two rollers. The glass strip is cooled, and disks are cut out of the viscous glass and heat treated. During the subsequent heat treatment, controlled crystallization takes place and the glass is transformed from a glass into a strong glass-ceramic. This process is particularly suitable for producing magnetic memory disk substrates.

The canasite glass-ceramic demonstrates high fracture resistance, which is favorable for the disk substrate application, in addition to a number of other favorable properties compared with standard nickel phosphide-coated aluminum ceramic substrate materials. These properties are:

- higher rigidity and shape maintenance
- higher data densities
- higher processing temperatures
- smoother surface texture
- lower cost for preparation of the disk

Canasite glass-ceramics are covered with a lubricant layer when they are used as memory disk substrates. Onyiriuka (1993) established the characteristics of this layered canasite glass-ceramic. The lubricant layer is an important part of the disk, since it prevents the disk from being damaged. Perfluoropolyester (PFPE) was used to develop the lubricant layer. The results of Onyiriuka's (1993) study showed that this type of thin layer with a controlled thickness of 2–5 nm could be applied to uncoated as well as carbon-coated canasite glass-ceramics.

Summarizing the results on canasite glass-ceramics, it can be mentioned that canasite glass-ceramic is less smooth than spinel–enstatite. Also Na^+ migration up through the columnar Cr-crystals below the magnetic layer can cause Na_2CO_3 to form on the disk surface.

All four types of glass-ceramics (spinel–enstatite, spinel, lithium disilicate, and canasite) have been produced by Corning Inc., USA.

In Japan, glass-ceramics and special glasses have been developed with the objective of increasing the storage capacity of computer hard disk drives (HDD). Potential substrates were compared with both metallic aluminum and Al_2O_3 materials. It was determined that glass or glass-ceramic substrates are more suited for these applications than ceramics, since they are flatter and smoother. Moreover, it is easier to produce sheets with glass according to the thin-layer method than with ceramics. Glass-derived materials have a major advantage over aluminum metal substrates in terms of shape maintenance under impact or bending stresses.

Goto (1995), for example, developed a lithium disilicate glass-ceramic. The glass-ceramic was derived from the SiO_2–Li_2O–P_2O_5–ZnO system. It was produced from a base glass of this materials system. This glass was pressed to form substrates immediately after it was melted. Subsequently, controlled crystallization was initiated in the pressed base glass. The microstructure of the glass-ceramic achieved in this fashion contained α-$Li_2Si_2O_5$ (α-lithium disilicate) and α-quartz main crystal phases in addition to a large amount of glass matrix. Lithium disilicate was precipitated as monospherical grains measuring approximately 0.1 μm. Although the quartz was the same size as the lithium disilicate crystals, it agglomerated and crystallized to formations measuring approximately 1 μm.

Goto (1995) achieved this specific microstructure by producing an Na_2O-free glass-ceramic from a lithium disilicate glass-ceramic that he had developed from the SiO_2–Li_2O–P_2O_5–Na_2O–MgO system. To produce magnetic disk substrates, the surface of this glass-ceramic is subjected to lapping and polishing, and it is cleaned. This specific glass-ceramic is manufactured by Ohara Inc., Japan, under the product name TS-10™. It demonstrates particularly low surface roughness R_a of 10–25 Å. Measurements of the surface roughness were conducted with an atomic force microscope. The roughness of the glass-ceramic is less than 5 μm. Additional properties of the TS-10™ glass-ceramic are shown in Table 4-9.

Another glass-ceramic containing a lithium disilicate main crystal phase is manufactured by Nippon Electric Glass Co., Ltd., Japan, for magnetic memory disk substrates. This glass-ceramic is produced under the product name ML-05™ (Electronic Glass Materials 1996). The material demonstrates a high coefficient of thermal expansion, and it is heat resistant. Furthermore, its surface is very smooth. The most important properties of the glass-ceramic are shown in Table 4-10. This opaque white glass-ceramic can also be used for magnetic substrates. A clear glass-ceramic with the product name ML-08 has also been developed by modifying the original material.

4.1.5 Liquid Crystal Displays

Transparent glass-ceramics of the SiO_2–Al_2O_3–Li_2O system are used to produce color filter substrates for polysilicon thin-film-transistor liquid crystal displays for laptop color PCs. Nippon Electric Glass Co., Ltd., Japan,

Table 4-9

Physical Properties of Lithium Disilicate Glass-Ceramic (TS-10™) Ohara Inc., Japan	
Properties	**TS-10**
Mechanical	
Young's modulus (MPa)	92
Flexural strength (MPa)	200–240
Vickers hardness (MPa)	700
Thermal	
Coefficient of thermal expansion ($\times 10^{-6}$ K^{-1})	8–9.2 $\times$ 10^{-6} K^{-1}
Electrical	
Volume resistivity	1.8–2.7 $\times$ 10^{15} Ω·cm

Table 4-10

Properties of a White Glass-Ceramic for Magnetic Disk Substrates (ML-05™)

Properties	ML-05™
Mechanical	
Density (g/cm³)	2.39
Hardness (Vickers)	650 HV(0.2)
Flexural strength (MPa)	300
Young's modulus (GPa, at 25°C)	95
Thermal	
Coefficient of thermal expansion 30°–500°C ($\times 10^{-6}$ K^{-1})	11.2
Chemical	
Acid durability 5% HCl, 90°C, 24 h (mg/cm²)	0.01
Alkali durability 5% Na_2CO_3, 90°C, 24 h (mg/cm²)	1.06

manufactures glass-ceramic disks for this application under the product name Neoceram™ N-0. The properties of this type of glass-ceramic are reported in Section 4.2.2. The microstructure of this glass-ceramic exhibits a main crystal phase of the β-quartz solid solution type with a crystallite size of approximately 0.1 μm (Neoceram 1992, 1995). Since the crystallite size is smaller than the wavelength of the visible light, the glass-ceramic is transparent. In addition, crystal possesses almost the same refractive index as the glass matrix.

The glass-ceramic is fabricated in two steps. First, the base glass is melted and formed into small plates by rolling, drawing, or pressing the material. Once the glass plates have cooled to room temperature, they are heat treated again. In this step, a β-quartz solid solution is formed by controlled crystallization. Finally, the material is left to cool to room temperature again. The glass-ceramic end product (Electronic Glass Materials 1996) measures 320 ± 0.2 mm (in length), 350 ± 0.2 mm (in width), and 1.1 ± 0.05 mm (in thickness). The following specifications must also be observed to ensure the quality:

- parallelism: max. 0.02 μm tolerance
- squareness: max. 0.17 mm/100 mm
- surface roughness, R_a: 100 Å.

In addition to the application mentioned, the Neoceram™ N-0 glass-ceramic may also be used to produce cylinders, tubes, and a variety of other objects.

In addition to Neoceram™ N-0, Nippon Electric Glass Co., Ltd., also produces another glass-ceramic that demonstrates zero expansion. Neoceram™ N-11 is a white, nontransparent glass-ceramic that contains β-spodumene solid solution primary crystals measuring approximately 1 μm. This glass-ceramic is used for induction cooker top plates, Kitchenware, (Section 4.2.1) or for optical components, such as the coupler housings of optical fibers (Neoceram 1992, 1995).

4.2 CONSUMER APPLICATIONS

4.2.1 β-Spodumene Solid Solution Glass-Ceramic

Stookey (1959) developed one of the first glass-ceramic materials marketed worldwide for use as household crockery. The glass-ceramic in question was called Pyroceram® 9608 (it is also known as Corning Ware® 9608). The material contains the main crystal phase β-spodumene solid solution, with minor rutile. This particular glass-ceramic is white and exhibits a low coefficient of thermal expansion of 7×10^{-7} K^{-1} (Fig. 4-6). The economical manufacturing techniques of Pyroceram® 9608 in the Corning Glass Works, USA, allowed the material to be used for low-cost kitchen applications and thermal-shock-resistant cooking dishes that could withstand high temperature fluctuations. These products represented a new generation of consumer household goods (Fig. 4-7).

Other β-spodumene solid solution glass-ceramics were developed under the trademarks Cer-Vit™ (Owens-Illinois) and Hercuvit™ (PPG) in the United States and in the United Kingdom. The latter was used to produce cooking surfaces (Pincus 1971).

In Japan in 1962, β-spodumene solid-solution glass-ceramics were also produced on the basis of developments achieved by Tashiro and Wada (1963). The products

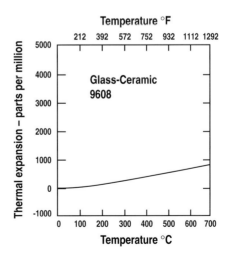

Figure 4-6 Thermal expansion of Corning Ware® 9608.

are manufactured by Nippon Electric Glass, Japan, under the trademark Neoceram™ N-11. The development of this glass-ceramic is based on the discovery of the nucleation reaction of ZrO_2 by Tashiro and Wada (1963) (see Section 2.2.2). The properties of

Figure 4-7 Pyroceram Corning Ware® 9608 dishes.

Neoceram™ N-11 glass-ceramic are shown in Table 4-11. Because of these properties, this white glass-ceramic can be used in a number of consumer-oriented applications, including flat, smooth products such as stove tops, hot plates in microwave ovens, or the inner cladding of microwave ovens. At the same time, however, it is possible to produce glass-ceramics of curved or almost any desired shape (Wada 1998).

4.2.2 ß-Quartz Solid Solution Glass-Ceramic

Following the development and production of white glass-ceramics for household applications on the basis of β-spodumene solid-solution, translucent and even highly transparent products were demanded by the market. It must be noted that more convenient products of better surface quality can be produced with transparent glass-ceramics. These demands followed on the success of domestic products such as bowls, tea pots, glasses, and cups made of a transparent borosilicate glass called Pyrex®. Considerable improvements in thermal shock resistance were made with borosilicate glasses compared with conventional silicate glasses. The coefficient of linear thermal expansion of these glasses of approximately $33 \times 10^{-7} K^{-1}$, however, still had not reached the ideal value of zero expansion. To increase the domestic applications of glass-ceramics, therefore, new materials were developed with the objective of achieving close to zero expansion in a wide temperature range.

According to Beall and Pinckney (1999), three important criteria must be observed in the development of glass-ceramics with zero expansion. The first two requirements involve low scattering, while the third requirement involves low ionic or atomic absorption. First, the index of refraction of the glass and the crystal phase must be almost identical. Second, the crystallite

Table 4-11

Properties of ß-Spodumene Solid-Solution Glass-Ceramic Neoceram™N-11 and the ß-Quartz Solid-Solution Glass-Ceramics Neoceram™N-0, Ceran®, and Robax®

Properties	Neoceram™ N-11	Neoceram™ N-0	Ceran®	Robax®
Mechanical				
Flexural strength (MPa)	~170	~140	110 ± 25	~75
Hardness (Knoop)	600 (HK 0.2)	500 (HK 0.2)	~600 (HK 0.1/20)	
Young's modulus (GPa)	90	90	≤ 95	~92
Thermal				
CTE α ($\times 10^{-6}$ K^{-1}) 20°–700°C	approx. + 1.3	approx. − 0.3	0 ± 0.15	0 ± 0.3
specific thermal capacity, c_p ($J \cdot g^{-1} \cdot K^{-1}$)	0.7	0.7	~0.8	
thermal conductivity, λ ($W \cdot m^{-1} \cdot K^{-1}$)	0.9	0.9	1.6 (100°C)	
Max. service T, long-term (°C)	850	700	700	680
Thermal shock resistance, no stress fracture up to T_0 (°C)	600	800	700	~700
Chemical				
Hydrolytic durability	0.15 (ASTM-stand.)	0.19 µl (ASTM-stand.)	class 1 (DIN/ISO 719)	class 1 (DIN 12111)
Alkali durability	0.9 (90°C, 24 h; 5% Na_2CO_3)	0.3 mg/cm^2 (90°C, 24 h; 5% Na_2CO_3)	class 2 (DIN 52322)	class 2 (DIN 52322)
Acid durability	0.6 (90°C, 24 h; 5% HCl)	0.1 mg/cm^2 (90°C, 24 h; 5% HCl)	class 3 (DIN 12116)	class 2 (DIN 12116)
Electrical				
Dielectric constant, ε at: (1 MHz, 25°C)	6.4	7.6	~7.8	
Volume resistivity (log) (ρ in $\Omega \cdot cm$) at 250°C	6.7	6.4	≥ 6.7	
loss tangent, tan δ at (1 MHz, 25°C)	4.1	22×10^{-3}	$\sim20 \times 10^{-3}$	

size must be considerably smaller than the shortest wave length of the visible spectrum. Based on the theoretical equation (Eq. 4-1) of Rayleigh-Gans (Kerker 1969)

$$\sigma_p \cong 2/3 \ NVk^4a^3(n\Delta n)^2 \qquad \text{(Eq. 4-1)}$$

(where σ_p is the total turbidity, N is the particle number density, V is the particle volume, a is the particle radius, k is $2\pi/\lambda$ (λ being the wavelength), n is the refractive index of the crystal, and Δn is the index difference between the crystal and the host), the crystallites must be smaller than 15 nm in size. However, according to the theoretical equation (Eq. 4-2) of Hopper (1985):

$$\sigma_c \cong ((2/3 \cdot 10^{-3})k^4\theta^3)(\ n\Delta n)^2 \qquad \text{(Eq. 4-2)}$$

where σ_c is turbidity, θ is the mean phase width (a + $W/2$) and W in the interparticle spacing), the particles must measure less than 30 nm in size, provided the spacing is not more than 6 times the crystal size.

The third requirement concerns the low absorption of light by the ions and atoms in the glass-ceramics.

In the development of glass-ceramics of the β-quartz solid solution type, all three requirements were met: similarity of the index of refraction between the glass matrix and the crystals, very small particle size, and low absorption. The solid-state reactions that are involved in the formation of β-quartz solid solution glass-ceramic are presented in Section 2.2.2. The chemical composition of different products are also discussed in this section. The special nanostructure of this glass-ceramic is described in Section 3.2.1.

The most important and economically significant β-quartz solid solution glass-ceramics for domestic applications are called Vision® (Corning USA), Keraglas® (Eurokera: Corning/St.Gobain, USA/France), Ceran®, and Robax® (Schott, Germany), as well as Neoceram™ N-0 (Nippon Electric Glass, Japan).

The optical and electrical properties of Keraglas® glass-ceramics are shown in Fig. 4-8a and 4-8b. The small coefficient of linear thermal expansion of $0 \pm 1.5 \times 10^{-7}$ K^{-1} and the associated high temperature resistance of 700 K are particularly important for the specific applications of the material. With regard to the optical properties, the glass-ceramic can be successfully heated by radiative heating. At 1100 nm (1.1 μm), for example, the heat transfer efficiency is 61% and at 2400 nm (2.4 μm) it is 79% (Fig. 4-8a). In addition to Fig. 4-8b, the electrical properties are characterized by a dielectric constant (1 MHz, 25°C) of 7.3 and a loss tangent δ (1 MHz, 25°C) of

9.8 × 10⁻³. Apart from passing standardized chemical durability tests, the material also had to fulfill special tests regarding the resistance to popular foods and to stains. The material met these standards and was deemed particularly suitable for producing consumer goods. Eurokera also produces a high visible transmission version of Keraglas®, called Eclair®, used for fire doors and woodstove windows.

Vision® glass-ceramic has been successfully used to make household goods such as bowls, pots, and pans for cooking and frying (Beall and Pinckney, 1999). Fig-

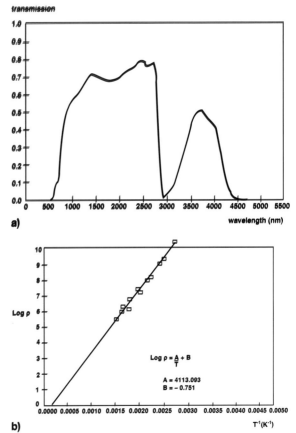

Figure 4-8 Keraglas® glass-ceramic.
a) optical transmission as a function of wavelength (4-mm thickness of the sample).
b) electrical properties (log volume resistivity).

ure 4-9 shows several examples. Keraglas® glass-ceramics are particularly suitable for producing cooking surfaces for electric and gas stoves. This application is shown in Fig. 4-10. Apart from the mentioned properties, this application also requires the glass-ceramic to demonstrate a high-quality surface finish. That is, the surface must be scratchproof. The stovetops are supplied in various decorative finishes with knobs on the reverse side. The pigmentation of the glass-ceramic makes the stovetop look dark, almost black, in incident light. In transmitted light, that is, when the heating coil of the electric range becomes incandescent, the glass-ceramic appears red. This effect is produced by the high transparency of the glass-ceramic material in the visible spectrum (Eurokera 1995).

The most important properties of Ceran® glass-ceramic are shown in Table 4-11 (Ceran® 1996). In the 1990s, Schott, Germany, produced two types of this glass-ceramic, characterized by different optical transmission (Pannhorst 1992 and Nass et al., 1995). The products are called Ceran Color® and Ceran Hightrans®. Ceran® glass-ceramic is used for stovetops. The fundamental developments for this technically interesting and improved product for domestic applications are described by Sack (1974). The practical development of the material, however, was closely linked to the

Figure 4-9 Vision® glass-ceramic in the form of domestic products.

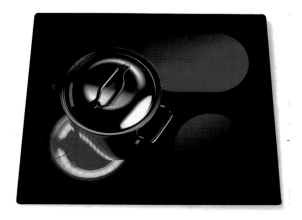

Figure 4-10 EuroKera® cooktop panels.

development of a precision-processing method for the fabrication of stovetops. It must be emphasized that the processing of the base glass by drawing as well as the precisely controlled heat treatment of the glass to achieve a uniform coefficient of thermal expansion in the glass-ceramic plate posed very high technological requirements. The precisely controlled manufacturing process of this type of glass-ceramic surface, measuring approximately 4–5 mm in thickness, by drawing and rolling liquid glass from a glass melt and the subsequent processing by crystallization of the glass plate to the glass-ceramic and its decoration is described in detail by Nass et al. (1995).

Robax® glass-ceramic is the nonpigmented version of Ceran® glass-ceramic. Its properties are summarized in Table 4-11 (Robax® 1998).

This glass-ceramic is characterized by high thermal shock resistance. It is predominantly used to produce oven doors, through which burners can be monitored and problems detected visually. In addition, the glass-ceramic can be used as an alternative in every application using borosilicate or silicate glass.

Japanese glass-ceramics of the Neoceram® N-0 type are manufactured by Nippon Electric Glass. The properties of this glass-ceramic are shown in Table 4-11. Compared with Neoceram® N-11, this glass-ceramic exhibits a very high translucency. In the visible wavelength range, the glass-ceramic demonstrates a slight yellowish tinge. The glass-ceramic is used as a viewing window in various household appliances, e.g., in gas burners for boilers or portable stoves. It is also used to clad fireplaces and coal stoves. As a result, the fireplace in a domestic setting is both safe and cozy, since the wood or coal fire is visible behind the glass pane.

4.3 OPTICAL APPLICATIONS

4.3.1 Telescope Mirrors

4.3.1.1 Requirements for Their Development

Glass-ceramics with very low coefficients of thermal expansion are highly suitable for precision optical equipment. In observatories, for example, large reflective telescopes are required to explore the far reaches of the universe. High precision of the telescope's optical components, the mirror in particular, must be ensured to achieve this goal. Furthermore, the optical equipment must be resistant to temperature fluctuations because observatories are often located in areas that experience temperature fluctuations of almost 100 K. Clearly, glass-ceramics demonstrating close to zero expansion are highly suitable for manufacturing telescope mirrors.

4.3.1.2 Zerodur® Glass-Ceramics

A) Properties

The main crystal phase of the Zerodur® glass-ceramic is composed of β-quartz solid solution This phase is produced by volume crystallization. Special thermal properties develop as a result of the formation of very small crystallites and the simultaneous high crystallite count per unit of glass volume.

Figure 4-11 shows the course of the linear coefficient of thermal expansion (α) as a function of temperature. According to the curve in Fig. 4-11, minimum thermal expansion, that is, almost zero expansion, is achieved in the temperature range of 0°–100°C in particular. Depending on the requirements, glass-ceramics demonstrating three different types of expansion behavior can be developed. In the zero expansion category, a value of $0 \pm 0.02 \times 10^{-6}$ K^{-1} is achieved.

Another important thermal property of glass-ceramics is a thermal conductivity of 1.46 W·m^{-1}·K^{-1} at 20°C.

The mechanical properties of Zerodur®

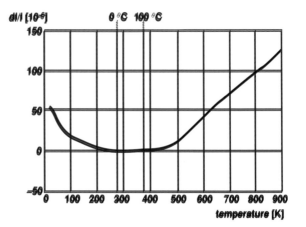

Figure 4-11 Linear thermal expansion as a function of temperature of Zerodur® glass-ceramic (Zerodur® information 1991).

(Zerodur® information 1991) are characterized by a Young's modulus of 90.3 GPa (20°C), bending strength of approximately 110 MPa, and Knoop hardness HK of 0.1/20 (ISO 9385) of 620.

A high rate of light transmission is also a favorable optical property of glass-ceramics. Figure 4-12 shows the light transmission of the Zerodur® glass-ceramic in layers of 5 and 25 mm. Furthermore, the glass-ceramic also demonstrates very good chemical durability and helium permeability.

B) Processing

The Zerodur® glass-ceramic is produced with a base glass of 57.2 wt% SiO$_2$, 25.3 wt% Al$_2$O$_3$, 6.5 wt% P$_2$O$_5$, 3.4 wt% Li$_2$O, 1.0 wt% MgO, 1.4 wt% ZnO, 0.2 wt% Na$_2$O, 0.4 wt% K$_2$O, 0.5 wt% As$_2$O$_3$, 2.3 wt% TiO$_2$, and 1.8 wt% ZrO$_2$, using controlled crystallization (Petzoldt and

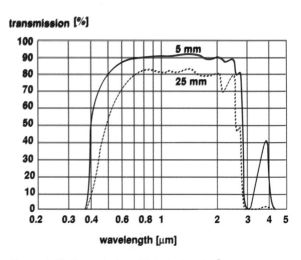

Figure 4-12 Transmission of light in Zerodur® in thicknesses of 5 mm and 25 mm.

Pannhorst, 1991). The nucleation and crystallization processes up to the formation of the β-quartz solid solution main crystal phase are addressed in Section 2.2.2. The processes are described in detail by Müller (1995) and Pannhorst (1995).

The Schott Glaswerke developed a special technique for producing large mirrors for reflective telescopes. The procedure up to the production of the Zerodur® mirror measuring 8.2 m in diameter is described by Höness et al. (1995). This procedure involves pouring the melted material from a 70-ton tank and spincasting it. As a result, the raw product becomes slightly curved. Hence, complicated mechanical processing after cooling of the glass is eliminated. At this stage, however, the mirror still requires several months of work (controlled crystallization and mechanical processing).

In addition to the largest monolithic glass-ceramic telescope mirrors in the world today, measuring 8.2 m in diameter, light-weight structures made up of individual segments are also produced (Schott product information 1988).

C) Application

Mirrors measuring 8.2 m in diameter (Fig. 4-13) were produced by the Schott Glaswerke for the Very Large Telescopes (VLT) of the European Southern Observatory (ESO). They are being installed in mirror telescopes at the observatory in La Sila, Chile.

The mirrors, measuring 7.6 m in diameter, for the Keck I and Keck II telescopes at the Mauna Kea Observatory on the Hawaiian islands of Manoa, are composed of 36 individual glass-ceramic segments.

4.3.2 Integrated Lens Arrays

As a result of the increased miniaturization of components in the electrical and electronics industries, optical

Figure 4-13 Processing of 8.2 m Zerodur® glass-ceramic telescope mirror.

elements are also required in microscopic dimensions. For example, optical lenses that measure several centimeters in diameter in traditional microscopes are being reduced in size. Therefore, lenses with micrometer-scale diameters and a thickness of approximately 20 µm are in demand.

To meet this demand, lithium metasilicate/lithium disilicate glass-ceramics of the Fotoform®/Fotoceram® type from Corning Glass Works were used. The traditional manufacturing process for Fotoform®/Fotoceram® products was further developed by Borrelli and Morse (Beall 1993) and specially adapted.

The development of the lithium disilicate glass-ceramic, its mechanisms of controlled crystallization, as well as the microstructure formation, and its properties are described in Section 2.1.1. The application of the glass-ceramics in high-precision equipment components and in the electrical industry is addressed in Section 4.1.2.

It must be noted that the initiation of crystallization is achieved by metal colloids, which are generated by UV irradiation and heat treatment. To apply this glass-ceramic for the production of integrated lens arrays, its manufacturing procedure had to be refined, and Borrelli, et al. (1985) achieved this objective (Beall 1992, 1993).

This new procedure featured a unique step. Following irradiation with UV light, the formation of the microstructure was minutely examined during the heat treatment required for the crystallization of Li_2SiO_3. The examination revealed that additional shaping of the material is possible. Therefore, it is possible to produce curvature in cylindrical bodies of microscopic dimensions during the manufacturing procedure. This curvature corresponds to a lens-shaped geometry. The material curves because the crystals are denser than the glass. A detailed description of the process follows.

The base glass is produced in the form of a very thin glass film of a specific thickness (e.g., 20 µm). Specific areas of the material are irradiated through a pattern. This pattern ensures that microscopic cylindrical micro areas are not irradiated. Figure 4-14 shows a schematic diagram of the process. The irradiation and the subsequent heat treatment produce lithium metasilicate, which places compressive strain on the non-irradiated cylindrical glass areas. In the process, the cylindrical areas protrude on both sides of the glass film. As a result, an optical lens is created in the microscopic glass cylinders. This shaping process is used in the production of microscopic lenses for three different applications (Beall 1993).

1) One-to-one arrays for reproducing images in the same ratio. These lenses are used to reproduce printed pages. In this case, the compactness of the optical system is important. Lenses of this type are used in photocopiers and fax machines.

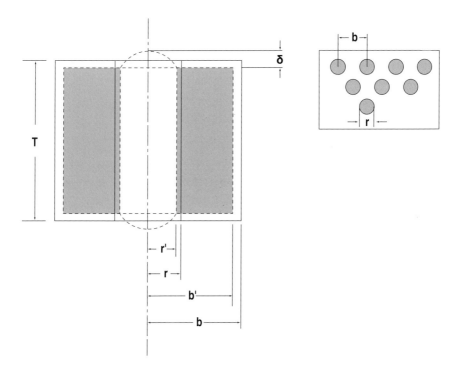

Figure 4-14 Process for producing integrated lens arrays made of Fotoform®.

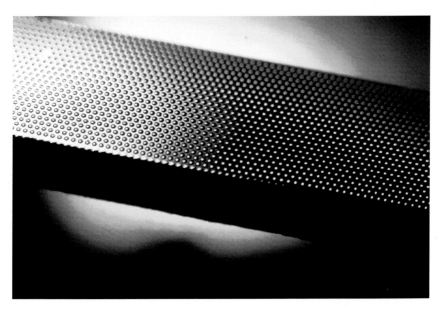

Figure 4-15 Electron micrograph showing spherical protrusion in perspective. The optical micro lenses produced according to the procedure in Fig. 4-14 measure 400 μm in diameter and approximately 20 μm in height.

2) Lenses for interfaces in optical fiber systems. The microscopic lens forms the interface for light entering and exiting optical fiber and between arrays.

3) Special applications of the lenses in light-transmission systems in the event that the division of a single light beam into a number of beams is necessary on the way to a detector. This application is necessary in the auto-focus of a photographic camera, for example. This type of special device contains a micro lens.

To demonstrate the diminutive size of the microscopic lens and to show the high precision of its shape, electron microscopic investigations of the glass foil from which the lens had been produced were conducted. The results are shown in Fig. 4-15. The holes in the glass film, from which the high-precision lenses measuring 400 µm in diameter and 20 µm in height are produced, are also visible.

4.3.3 Applications for Luminescent Glass-Ceramics

4.3.3.1 Cr-Doped Mullite for Solar Concentrators

In solar engineering, sunlight is transformed into other useful forms of energy, such as electricity. For this purpose transparent materials may be desired. At the same time, these materials must enable the transfer of energy. Glass-ceramics can exhibit both high transmission and luminescent optical properties.

For applications of this kind, mullite, $3Al_2O_3 \cdot 2SiO_2$, glass-ceramics with chromium ions as luminescent centers were developed (Beall 1992, 1993). The composition of this glass-ceramic is described in Section 2.2.1 and the formation of its microstructure in Section 3.2.5. The special features of this material were developed on the basis of the applications described in this section.

Apart from using this mullite glass-ceramic for luminescent solar concentrators, it was also intended for laser applications, such as gain media of high-power lasers. Both applications of the glass-ceramic require the development of similar properties. These properties are:

- Strong absorption band, ranging from visible to near-infrared light. Efficient pumping by flash-lamps, in the case of laser applications, or solar radiation, in the case of solar-energy applications, must be possible.
- Incident solar rays must be effectively transformed into longer wave luminescence (for laser and solar energy applications).
- A high-quality optical medium is necessary to avoid energy losses due to processes such as scattering during absorption and the transformation of energy.

- To use the glass-ceramics in technological processes, it is necessary to produce large parts. Therefore, the manu-facturing procedures involv-ing the melting and process-ing of the glasses to form glass-ceramics must be eco-nomically feasible.

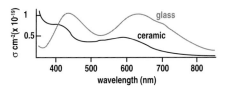

Figure **4-16** Comparison of the visible absorption spectra of mullite glass-ceramics with the base glass used to fabricate the glass-ceramic.

- Silicon photovoltaic strips must be effectively bonded to the edges of the glass-ceramic sheet (window) to efficiently convert the near-infrared lumi-nescence to electric energy.

These requirements have been met in the development of glass-ceramics. An effective production process in technical dimensions has been developed. Furthermore, most of the required properties have also been achieved.

As a result, it was possible to incorporate chromium ions into

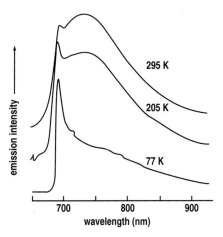

Figure **4-17** Spectrum of the mullite glass-ceramic in the near-IR range.

the crystals of the mullite phase during heat treatment of the base glass for the development of the glass-ceramic. The mullite crystals containing Cr^{3+} exhibited a crystallite size that was clearly below the smallest wavelength of light in the visible spectrum. The mullite crystals doped with chromium ions were at most 100 nm in size. Moreover, these very small crystals were isolated in the glass matrix. Thus, the glass-ceramic was highly transparent.

Andrews et al. (1986) were able to compare the optical absorption in the visible range of the optical spectrum of the base glass and of the glass-ceramic. Their findings are shown in Fig. 4-16. The spectra in Fig. 4-16 substantiate what is visible to the human eye. The base glass appears green; the glass-ceramic, gray. The absorption of the glass-ceramic permits parallelism in the absorption of the sun's rays in solar energy applications.

The optical properties of the glass-ceramic in the near-infrared range are shown in Fig. 4-17. These spectra exhibit two characteristics: a wide absorp-tion band near 730 nm and a sharp feature at 695 nm:

- The band at 730 nm characterizes the $^4T_2 \rightarrow {}^4A_2$ fluorescence.
- At 695 nm, the $^2E \rightarrow {}^4A_2$, R-line of phosphorescence is determined.

Further investigations regarding the optical effect of Cr^{3+} in mullite and β-quartz glass-ceramics were conducted by Reisfeld et al. (1984) and Kiselev et al. (1984).

In sum, numerous glasses have been melted and glass-ceramics synthesized (approximately 700 in total) to optimize the following properties:

- processing parameters to melt the glass
- setting of the reduction–oxidation ratio
- low impurity absorption
- IR luminescence
- minimum overlap between absorption and luminescence

As a result, the material is now suitable for consideration as solar energy and laser applications. Minimum scattering has also contributed to this result. Scattering is less than 0.04 measure, therefore, five times smaller than for other mullite glass-ceramics. The absorption of the glass-ceramic resulting from the composition of the microstructure can be tolerated for the indicated applications.

Nevertheless, quantum efficiency, measured by the integrated sphere technique, was 32%, not yet good enough for the luminescent collection application, which requires near 60% conversion of absorbed to luminescent energy. Also optical scattering should be lower.

Perhaps these shortcomings can be overcome by purer materials, in particular less impurity FeO, and also finer crystals.

4.3.3.2 Rare-Earth-Doped Oxyfluorides for Upconversion and Amplification

Transparent oxyfluoride glass-ceramics were first reported by Wang and Ohwaki, (1993) in the basic system SiO_2–Al_2O_3–CdF_2–PbF_2, to which they added rare-earth fluorides. A typical composition designed for upconversion is given in Table 4-12. The parent glass was heat treated at 470°C, and 20-nm crystallites of $(Pb,Cd)F_2$ with the fluorite structure were precipitated. No observable decrease in transparency from the parent glass was observed. With heavy doping of YbF_3 (10 mol% in the bulk composition) in combination with ErF_3 (1 mol%), excitation at 970 nm produced strong upconversion to 545 nm and 660 nm from energy transfer from Yb to Er in the glass-ceramic, roughly two orders of magnitude more than that observed in the original glass (see Fig. 4-18). This upconversion was also roughly double that of similar upconversion in the best aluminofluoride

Table 4-12

Composition and Properties of Some Transparent Oxyfluoride Glass-Ceramics (Tick et al., 1995)

m/o	Wang-Ohwaki (Upconversion)	Tick et al. (1995) (1.3 μm amplifier)	
SiO_2	30	30	30
Al_2O_3	15	15	15
CdF_2	20	29	29
PbF_2	24	17	22
ZnF_2	–	5	–
YF_3	–	4	4
YbF_3	10	–	–
ErF_3	1	–	–
PrF_3	–	0.01	0.01
Refractive index	–	1.75	1.75
CTE ($\times 10^{-7}$ K^{-1})	–	110	–
T_g (°C)	400	395	408
Cubic lattice edge (Å) (Pb, Cd, Re)F_{2-3}	5.72	5.75	5.75

glasses and similar to that observed in the best Er–Yb co-doped single crystal: BaY_2F_8. Clearly, strong partitioning of the rare earths into the crystal phase was responsible.

Tick et al. (1995) modified this system by adding ZnF_2 and YF_3 and doping solely with the rare-earth ion Pr^{3+} (Table 4-12) at the low level of 0.1 mol%. The luminescent quantum efficiency of Pr^{3+} emission at the 1300 nm telecommunication band was measured at 7%, an improvement over similar Pr emission from the well-known fluoride glass, ZBLAN®, measured at 4%. These glass-ceramics

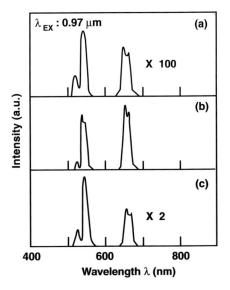

Figure 4-18 Fluorescence spectra of a) Er–Yb oxyfluoride glass, b) Er–Yb oxyfluoride glass-ceramic, c) aluminofluoride glass.

showed slightly smaller crystallites than those of Wang and Ohwaki, nearer to 15 rather than 20 nm in diameter. The optical scattering was estimated using polished 1-cm^3 samples and a 647-nm laser beam. The Rayleigh ratio and depolarization was measured at right angles to the transmitted beam. It is remarkable that this ratio, a general measure of scattering, was only 10% above that of the parent glass, and only 33% above that of a commercial optical glass. In fact, fiber losses of less than 1 dB/m have recently been measured in these glass-ceramics, albeit at the longer wavelengths of tele-communication interest (Tick et al., 2000). The differential loss between glass and glass-ceramic fibers was estimated to be in the range of tens of dB/km.

Transparent LaF$_3$ glass-ceramics, the latest of the oxyfluoride materials, were developed in the SiO$_2$–Al$_2$O$_3$–Na$_2$O–LaF$_3$ system (Dejneka 1998a,b). LaF$_3$ is a preferred host for rare-earth ions in a glass-ceramic because exten-sive solid solution with all rare-earth ions is allowed. This suggests strong par-titioning of these ions into the crystal. Moreover, LaF$_3$ is known to have a lower phonon energy (350/cm) than ZBLAN® fluoride glass (580/cm), which helps reduce or prevent multiphonon decay. The glass-forming region in the SiO$_2$–Al$_2$O$_3$–Na$_2$O–LaF$_3$ system is extensive, but the LaF$_3$ solubility limit is given by the relation: La$_2$F$_6$ = (Al$_2$O − Na$_3$O)/2. Levels far below this limit produce stable glasses that do not form glass-ceramics, whereas levels above this limit produce translucent or opal glass-ceramics. Levels near this limit produce excellent transparent glass-ceramics with chemical dura-bility in water 10,000 times better than fluoride glasses. The crystal sizes can be controlled by heat treatment in the range of 750°C–850°C for 4 h (see Fig. 4-19). They range from 7.2 nm at 750°C to 33.3 nm at 825°C. Curiously, the extinction coefficient of the glass-ceramics crystal-lized at or below 775°C was below that of the parent glass, which though appearing perfectly transparent, may have conceivably been already phase-separated.

Applications envisioned for these LaF$_3$ glass-ceramics are in the area of amplification of telecommunication signals, both at 1300 nm with Pr-doping and at 1550 nm with Er-doping. The ionic size difference between Er^{3+} and the larger La^{3+} induces a greater distortion of the crystal field, which pro-duces a broad emission at 1550 nm. This gives an improved gain spectrum for a LaF$_3$ glass-ceramic fiber as compared to a conventional commercial alumina-doped silica glass fiber.

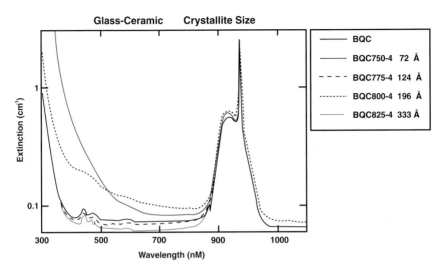

Figure 4-19 Extinction of LaF$_3$ glass and glass-ceramics (BQC) base glass (BQC 750-4) to (BQC 525-4) base glass heat treated at various temperatures (750°–825°C) for 4 h and average crystallite size (in Å). Absorption bands at 460, 590, and 950 nm are due to Pr^{3+} and Yb^{3+} doping.

4.3.4 Optical Components

4.3.4.1 Glass-Ceramics for Fiber Bragg Grating Athermalization

Ceramic materials with a strong negative coefficient of thermal expansion were discovered by Hummel (1951) in the SiO$_2$–Al$_2$O$_3$–Li$_2$O system. These were based on the mineral phase β-eucryptite originally described by Winkler (1948). β-Eucryptite is a 1:1:2 Li$_2$O–Al$_2$O$_3$–SiO$_2$ composition of simple formula LiAlSiO$_4$. This phase is hexagonal with a structure related to β-quartz but with extremely anisotropic thermal expansion characteristics. Gillery and Bush (1959) originally measured the crystallographic thermal expansion by X-ray diffraction and found the coefficients of thermal expansion to be -176×10^{-7} K^{-1} parallel to the c-axis and $+82.1 \times 10^{-7}$ K^{-1} perpendicular to the c-axis. Both Murthy (1962) and later Petzoldt (1967) attempted to sinter powdered glass of the eucryptite composition and arrived at bulk measured thermal expansion values near -90×10^{-7} K^{-1}. Internally nucleated glass-ceramics based on β-eucryptite using a nucleating agent such as titania were never made in bulk because of the great tendency of these glasses to crystallize in an uncontrolled fashion upon cooling. In fact, negative CTE glass-ceramics in general were never pursued because there were no obvious applications.

Recently Beall and Weidman (1995) described a unique application incorporating a negative expansion β-eucryptite substrate for athermalizing an optical fiber reflective grating device. Index of refraction changes induced by ultraviolet light are useful in producing complex narrow-band optical components such as filters and channel add/drop devices. These devices can be important parts of multiple wavelength telecommunications systems. The prototypical photosensitive device is a reflective grating (or Bragg grating) that reflects light over a narrow band. Typically, these devices have channel spacings measured in nanometers. One practical difficulty in the use of these gratings is their variation with temperature. Since the frequency of light reflected by the fiber grating varies with the temperature of the grating region, this basic filter cannot be used where the reflected light frequency is to be independent of temperature. One method of athermalization is to create a negative expansion substrate for the fiber grating, which compensates for the dn/dT. In general, a Bragg optical fiber, for example, a germania silica fiber, is affixed to the negative expansion substrate at two points along the length of the fiber. The increase in the index of refraction of the fiber caused by thermal drift is then compensated by a negative mechanical expansion.

The first effort to study bulk glass-ceramics based on β-eucryptite and related solid solution toward silica and $AlPO_4$ were made by Murthy of the Ontario Research Foundation under contract to then Corning Glass Works in 1962. Glasses containing no nucleation agent were cast over a wide composition range in the system eucryptite–silica–berlinite ($LiAlSiO_4$–SiO_2–$AlPO_4$). He confronted crystallization at the eucryptite stoichiometry and along the binary eucryptite–silica join up to about 58.1 wt% SiO_2. He found that additions of aluminum phosphate greatly helped glass stability but reduced measured thermal expansion coefficients on crushed, sintered, and crystallized materials.

Petzoldt (1967) was more successful in quenching glasses along the $LiAlSiO_4$–SiO_2 join and measured high negative thermal expansion coefficients below -50×10^{-7} K^{-1} at silica levels below 48 wt%. In both cases, these materials, lacking a nucleation agent, were relatively coarse-grained with pits, voids, and microcracks present.

Rittler (1980) subsequently described a method for strengthening optical fibers by passing them through a molten glass composed of stoichiometric 1:1:2 β-eucryptite, plus sufficient titania to cause internal nucleation. Since the glass was basically a thin film coating the fiber, it was rapidly quenched and could be recrystallized to a glass-ceramic. When he measured the coefficient of thermal expansion of this glass-ceramic, however, the negative value was low, about -10×10^{-7} K^{-1}. He did not, however, heat these materials at high temperatures like 1300°C, within 100°C or 150°C of the melting point of β-eucryptite.

A) Structure of ß-Eucryptite and Related Solid Solutions

It has long been known that β-eucryptite is a stuffed derivative of β-quartz (Winkler 1948; Roy and Osborn, 1949; and Buerger 1954). This means that in order to derive β-eucryptite from β-quartz, one-half of the Si^{4+} ions of quartz must be replaced by Al^{3+} ions with the interstitial vacancies filled by Li^+ ions (Appendix Fig. 5). Although the unit cell is doubled along the c-axis, the space group remains $P6_222$ or $P6_422$, antiomorphic pairs. In true β-eucryptite, the aluminum tetrahedra are all joined at the corners to silica tetrahedra, thus preserving Lowenstein's rule whereby tetrahedral Al^{3+} cannot be joined to another tetrahedral Al^{3+} through a single oxygen. According to the phase equilibria data of Roy and Osborn, β-eucryptite has a compositional range of stability from the 1:1:2 stoichiometry upward in silica to 1:1:3, or roughly 58 wt%. The same phase diagram gives a thermal stability range of roughly 900°–1400°C.

α-Eucryptite, with a similar structure of phenakite or silicon nitride, is believed stable below 900°C. This phase, however, has never been observed in heating ceramic materials made from glass below 900°C, or at any other temperature for that matter. Apparently β-eucryptite is persistently metastable in the range from 900°C to room temperature. Other solid-solution components may occur in significant abundance in β-eucryptite. These include $AlPO_4$ for $2SiO_2$ replacement and Mg^{2+} for $2Li^+$ cation replacement as well as up to 40% B^{3+} for Al^{3+} replacement (Mazza et al., 1994). Excess alumina may also be possible. In the first two cases, the strong negative thermal expansion coefficient along the c direction is adversely affected. Palmer (1994) has attempted to explain the expansion of β-eucryptite in the (001) plane with increasing temperature with a corresponding contraction along c so that the overall unit cell volume decreases. This anomalous behavior is explained by the edge sharing of Li–(Al,Si)-containing tetrahedra. Under ambient conditions, the Li–(Al,Si) distance is very small (2.63–2.65 Å). The four atoms associated with the edge sharing (i.e., Li, Al/Si, and two oxygen atoms) are all co-planar. The repulsive force between the cations Li and Al/Si can be reduced by thermal expansion in the xy plane, but to maintain the Li–O and the (Al,Si) bond distances, the shared tetrahedral edge length (O–O) must decrease. This is also seen manifested in the torsional contraction of double helices of Al/Si-tetrahedra parallel to the c-axis. This causes a net decrease in the c-cell edge as shown in Section 1.3.1.

The classical structural study of β-eucryptite by Pillars and Peacor (1973) shows a distinct ordering of alumina and silica tetrahedra in accordance with Lowenstein's rule. However, it is not known for certain that the first β-quartz structure crystallizing from an eucryptite composition glass will be so ordered. One must consider the possibility that some disordering of aluminum and silicon tetrahedra may be permitted. Indeed, a new structural study of β-eucryptite by Xu et al. (1999) compared the structure and thermal expansion behavior of ordered β-eucryptite formed by a long sintering process at 1300°C with a highly disordered β-eucryptite formed by reheating quenched glass at 800°C for 1 h. They found that the anisotropic thermal expansion was far more pronounced in the ordered form (CTE from 25° to 600°C: $+72.6 \times 10^{-7}$ K^{-1} along the a-axes and -163.5×10^{-7} K^{-1} along the c-axis) than in the disordered form (+59.8 along the a-axes and −38.2 along the c-axis). They attributed the anomalous thermal expansion behavior of β-eucryptite to several interconnected processes, including tetrahedral tilting and flattening, and Si(Al)–O bond shortening. The difference in CTE behavior between the ordered and disordered forms was attributed to the fact that rigid rotational tilting of framework tetrahedra is only experienced by the ordered form.

The understanding of the position of the lithium ions in β-eucryptite has evolved over the years. Pillars and Peacor determined that there are four six-fold channels in the unit cell, each containing Li ions. In only one channel do the Li atoms reside within the layers of Al atoms, the so-called Li(1) site. The remaining three channels, which are symmetrically equivalent, contain Li(2) and Li(3) sites, which reside within layers of SiO_4 tetrahedra (see Section 1.3.1). While at low temperatures the Li ions are observed to reside in these three distinct sites, at temperatures above 482°C, the lithium ions become disordered over the two sites Li(1) and Li(2). This order–disorder transformation has not been observed to have any effect on the thermal expansion behavior.

Xu et al. (1999) also discovered that the thermal contraction of the c-axis of β-eucryptite is inhibited as the crystal is cooled below room temperature. By contrast, c continues to increase until it is saturated near 20°C. This yields average CTE values, from 20 to 298 K for ordered crystals, of $+18.1 \times 10^{-7}$ K^{-1} along the a-axes and $+109.5 \times 10^{-7}$ K^{-1} along the c-axis. Curiously, in the disordered crystals over the same temperature range, both axes are negative, -10.4×10^{-7} K^{-1} for a and -85.3×10^{-7} K^{-1} for c. Xu et al. (1999) attribute the inhibition of a-axes contraction with cooling below room temperature to strengthening localization of Li ions.

B) ß-Eucryptite Glass-Ceramics

In order to achieve consistent properties in a glass-ceramic it is necessary to start with a glass composition of at least reasonable resistance to devitrification. While this is not a problem in most systems, it is a major issue with glasses close to the eucryptite (LiAlSiO$_4$) (Li$_2$O:Al$_2$O$_3$:SiO$_2$ = 1:1:2) stoichiometry. For this reason, a solid solution with some excess SiO$_2$, namely Li$_2$O:Al$_2$O$_3$:SiO$_2$ = 1:1:2.5, was chosen. This composition still has sufficient CTE anisotropy to achieve a negative CTE with controlled microcracking (see Section 4.3.4.1C Microcracking and its Biasing Effect on CTE), and it forms a stable glass providing additions of TiO$_2$ and excess Al$_2$O$_3$ are made in roughly equimolar proportions at the proper level to produce internal nucleation. Typically, the weight percent of TiO$_2$ is near 4%, sufficient to cause internal nucleation, and the minor phase tielite (Al$_2$TiO$_5$) develops with the ß-eucryptite major phase on crystallization.

Typical crystallization cycles involve a nucleation hold in the range of 720°–770°C followed by a crystal growth hold at 1300°C or above in order to ensure large enough crystals to yield a strongly negative CTE. Figure 4-20 shows the dramatic difference between holding 0.5 h at 1300°C versus 4 h at

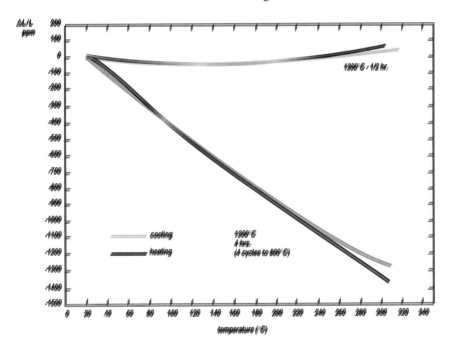

Figure 4-20 Thermal expansion/contraction behavior of a ß-eucryptite solid solution glass-ceramic heated at 1300°C for different lengths of time and differing resulting grain size (4 h in a microcrack on cooling).

this temperature, a CTE of roughly 0 vs. -50×10^{-7} K^{-1}. Clearly the near-zero thermal expansion is closer to the X-ray diffraction calculated average CTE for β-eucryptite (hexagonal) of $(2a + c)/3$, or -4×10^{-7} K^{-1}. Moreover, examination of the two samples showed that the one heated for 4 h was coarser and heavily microcracked, whereas the 0.5 h sample was finer with barely a hint of microcracking. Crystallization of this glass at lower temperatures (800°C–1200°C) even for up to 24 h always gave the near-zero CTE.

C) Microcracking and Its Biasing Effect on CTE

The phenomenon of microcrack-induced bias in the CTE of ceramic materials was first described by Bussem et al. in 1952 (Bussem et al., 1952) for Al_2TiO_5 ceramics and subsequently by Gillery and Bush (1959) for β-eucryptite ceramics. As the crystal size of β-eucryptite microcrystals increases by secondary grain growth at high temperatures like 1300°C, less than 100°C below the initiation of melting, a critical size is reached. This creates negative expansion bias toward crystals oriented with their c-axis in the direction of the CTE measurement via the following logic. If the crystal size is large enough, say between 5 and 10 μm, and particularly above 10 μm, strong stresses and resulting strains develop along crystal boundaries where the a- and c-axes of adjacent grains are nearly parallel. With large enough crystals in a randomly oriented arrangement, the anisotropic strain mismatch $(\alpha_a - \alpha_b)$ Td, where α is the CTE and d is the crystal size, can cause fracture initiation at appropriately oriented interfaces due to the resulting stored elastic strain energy. The effect of the normal positive CTE along the a-axes is to cause shrinkage on cooling, and this can be accommodated by opening of the microcracks. The expansion in the c-direction, on the other hand, cannot be accommodated and therefore becomes the dominant contributor to the bulk C.T.E., allowing a strong bias toward the negative c-axis component. This can be shown in the simple two-dimensional illustration of Fig. 4-21, where square crystals separated by microcracks are illustrated during a cooling cycle for both separate a-axis conditions of thermal expansion.

CTE hysteresis is usually observed in such microcracked materials when subjected to thermal cycles (Gillery and Bush, 1959), and this is also with β-eucryptite glass-ceramics. On initial cooling, the random network of microcracks develop, and those nearly perpendicular to the a-axes of adjacent grains of similar orientation open significantly, accommodating contraction, while those perpendicular to the c-axes of adjacent grains with similar orientations remain closed and bulk expansion results. A variety of random microcracks, some open to various degrees and some closed can be seen in Fig. 4-22,

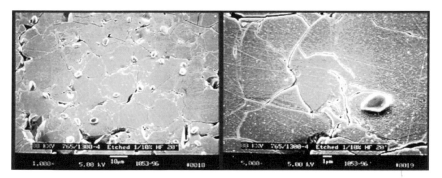

Figure 4-21 Microcracking in ß-eucryptite solid solution glass-ceramic.

an SEM photo of a typical β-eucryptite glass-ceramic with a strong negative CTE of about -70×10^{-7} K^{-1}. On reheating the glass-ceramic, the material contracts as the *c*-axis component shrinks and the cracks close and eventually heal. While considerable hysteresis occurs in the high-temperature cycle from 0°C to 800°C because crack initiation and healing occur at different

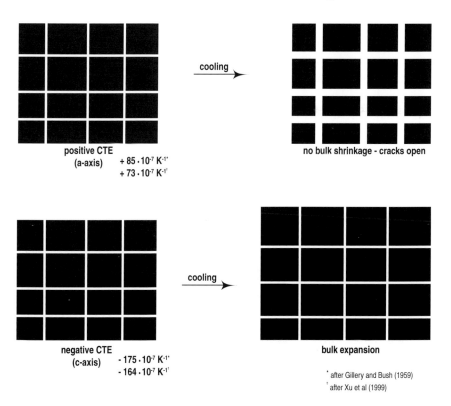

positive CTE
(a-axis) $+ 85 \cdot 10^{-7}$ K^{-1*}
            $+ 73 \cdot 10^{-7}$ $K^{-1\dagger}$

cooling

no bulk shrinkage - cracks open

negative CTE
(c-axis) $- 175 \cdot 10^{-7}$ K^{-1*}
            $- 164 \cdot 10^{-7}$ $K^{-1\dagger}$

cooling

bulk expansion

[*] after Gillery and Bush (1959)
[†] after Xu et al (1999)

Figure 4-22 Microcrack bias toward negative thermal expansion.

temperatures (Gillery and Bush, 1959), in practice, the low-temperature hysteresis (–50°C to +150°C) can be virtually eliminated by subjecting the material to three or more high-temperature cycles.

The microcrack network can be studied indirectly by sonic modulus measurements (Beall et al., 1998). Microcracking decreases the Young's modulus, and the hys-

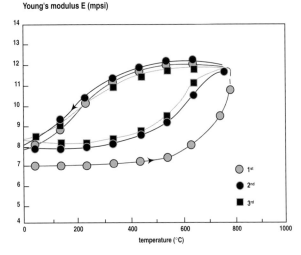

Figure 4-23 Sonic modulus hysteresis loops of ß-eucryptite (1:1:2.5) glass-ceramic during the first three cycles to 800°C. 7.98 mpsi = 55 GPa.

teresis effects of thermal cycles can be observed. Figure 4-23 illustrates the change and ultimate stabilization of elastic modulus at low temperatures for a typical composition. The flexural strength and Young's modulus of this microcracked glass-ceramic at ambient-use temperatures are 40 MPa and 55 GPa (7.98 mps:), respectively, adequate for service in the intended application.

D) Fiber Bragg Grating Athermalization

Fiber Bragg gratings are optical devices that can reflect a specific and narrow wavelength segment while transmitting all other wavelengths undisturbed. In practice, a waveguide fiber is treated with ultraviolet light from interfering beams in such a way as to produce a sinusoidal interference pattern with a characteristic and regular interval. The light acts to change the refractive index of the fiber core (GeO_2-doped SiO_2) producing a permanent grating. Germania-doped silica is particularly susceptible to this photorefractive effect, the refractive index typically being altered by the order of 1×10^{-4} in practice. The Bragg wavelength is related to the grating spacing Λ as follows:

$$\lambda = 2 \, N_{eff} \, \Lambda \qquad \text{(Eq. 4-3)}$$

where N_{eff} is the effective refractive index of the core and Λ is the grating spacing. A major issue is the relationship of the Bragg wavelength to temperature. As the temperature increases, the refractive index of the glass fiber increases significantly, and over an interval of 10K, this increase may be as large as the variation in the grating originally produced by the photorefractive effect. Therefore a technique for athermalization, preferably passive, must be introduced. The change of the Bragg wavelength with temperature is given by:

$$d\lambda_B/dT = 2\Lambda[dn/dT + n\alpha] \tag{Eq. 4-4}$$

where n is the average refractive index of the core, about 1.461 for Ge:SiO$_2$, λ_B is Bragg wavelength, α is the average CTE, roughly 5.2×10^{-7} K^{-1}, and dn/dT, the change of index with temperature, is about 1.1×10^{-5}. In order to completely athermalize the device, it can be seen from the above expression that the effective CTE must be made equal to dn/ndT, or about -75×10^{-7} K^{-1}. This is close to what is observed in β-eucryptite glass-ceramics as illustrated in Fig. 4-22.

In practice the fiber Bragg grating must be rigidly attached to a β-eucryptite substrate in order to achieve this athermalization. An inorganic glass frit, typically a tin–zinc pyrophosphate can be used as a bonding agent (Morena and Francis, 1998). This frit is filled with a suitable low-temperature phase-transforming mineral such as Co–Mg pyrophosphate to create a compatible thermal expansion behavior in the critical temperature range. An illustration of this packaging technique is given in Fig. 4-24. The resulting athermalized grating performance is shown in Fig. 4-25.

Applications for glass-ceramic athermalized fiber Bragg gratings include wavelength channel dropping filters (with 0.4 nm bandwidth at 1550 nm already demonstrated), laser stabilization, receiver filters, gain flattening filters for optical amplifiers, dispersion compensation, and multiplexing/demultiplexing components.

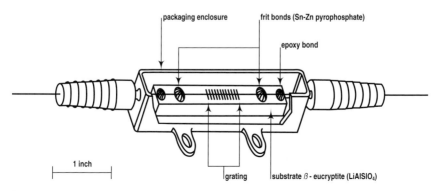

packaging enclosure   frit bonds (Sn-Zn pyrophosphate)

epoxy bond

1 inch

grating   substrate β - eucryptite (LiAlSIO₄)

Figure 4-24 Fiber Bragg grating package.

4.3.4.2 Glass-Ceramic Ferrule for Optical Connectors

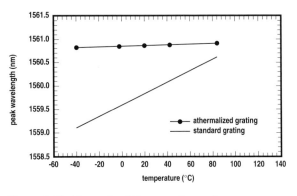

Figure 4-25 Athermalized Bragg grating performance.

Optical connectors are necessary for branching optical fibers to transfer data to different locations. The materials used to produce these connectors must fulfill several important requirements. For example, they must demonstrate a high dimensional accuracy in the micrometer range. Optical losses must be kept to a minimum. Furthermore, the connector should demonstrate excellent scratch resistance. These requirements were generally fulfilled by ZrO_2 ceramics. The development of a β-spodumene glass-ceramic by Sakamato and Wada (1998), however, produced optimized properties. Furthermore, the associated processing technology was also more favorable than that required for the fabrication of ZrO_2 ceramics.

To develop this material, Sakamato and Wada (1998) drew a crystallized cylindrical β-spodumene glass-ceramic blank with a hole drilled through the middle to form thin capillaries at 1150°C. These capillaries demonstrated the dimensions of 2.5 or 1.25 mm (outer diameter) and 0.125 mm (inner diameter).

This method prevented undesired crystallization processes from taking place on the surface of the capillaries, even during the drawing procedure. As the glass-ceramic contains 50% crystal phase, it demonstrates the properties of 80 GPa Young's modulus, and 3.0×10^{-6} K^{-1} coefficient of thermal expansion. The scratch resistance and bending strength are particularly noteworthy (Takeuchi et al., 1997, Nagase et al., 1997).

4.4 MEDICAL AND DENTAL GLASS-CERAMICS

In the development of biocompatible and bioactive glass-ceramics for medical uses, two different types of materials must be addressed that differ in their application environment and preferred properties: materials for use in implantology (medical prostheses) and materials used in restorative dentistry (dental prostheses). For reasons of clarity, the difference between the two different groups of biomaterials must be addressed at this stage. The first group is used to fabricate medical prostheses in dentistry or implants in human medicine. These products for human medicine are used in orthopedics and head and neck surgery, for example. Dental implants and root fillers also belong to this group, since they are inserted into the human body. The second and larger group of biomaterials, however, is that used for restorative purposes in dentistry. These materials are not implanted in the body (jaws of human beings). Rather they are used to restore natural teeth. These materials are used to fabricate, e.g., dental inlays, crowns, bridges, and veneers.

The following statistics show in which quantities the second group of materials is used in Germany. It must be noted that in contrast to the situation in the United States and Switzerland, social health insurance plans in Germany usually cover dental treatment. In 1994, for example, the German social health insurance paid for 9,532,600 single crowns and veneer crowns with bridges (KZBV 1996). This figure includes restorations made of metals, polymers, and increasingly ceramics, glass-ceramics, and glasses. Clearly, glass-ceramics are biomaterials with considerable potential in the future for dental restoration.

The requirements in the glass-ceramic development of these two types of materials (medical and dental prostheses) are entirely different. Implantology calls for biocompatible, and in most cases, bioactive properties. A bioactive glass-ceramic forms a biologically active hydroxycarbonate apatite layer that permits bonding with bone and even with soft tissue. Depending on the application of the glass-ceramic material, it may be load bearing or not and may need to fulfill special requirements with respect to the bending strength, toughness, and Young's modulus. Special optical properties such as translucency and color are not important in the development of this type of bioactive material.

The situation with regard to glass-ceramics for restorative dental applications is different. These materials must also fulfill the standards for biomaterial use, such as compatibility with the oral environment. Bioactivity on the surface of the dental restoration, however, must not occur. More importantly, the surface properties of the glass-ceramics, such as shade, translucency, toughness, and wear, must correspond to those of natural teeth. Even higher standards are placed on the chemical durability of the material compared with that of natural teeth, since cavities should not occur in the new glass-ceramics.

As a result of these different requirements, distinct chemical systems are used, and the development of the glass-ceramic materials is concentrated on different main crystal phases (see Sections 1.3, 2.1.1, 2.2.9, 2.3.1, 2.4, and 2.6.5) and, therefore, different properties.

4.4.1 Glass-Ceramics for Medical Applications

The following bioactive glass-ceramics are used for implants in human medicine: CERABONE® (apatite–wollastonite glass-ceramic), CERAVITAL® (apatite–devitrite glass-ceramic), and BIOVERIT® I (mica–apatite glass-ceramic).

Bioactive glasses such as BIOGLASS® are used in head and throat surgery in the form of middle-ear devices and implants for the orbital floor. In stomatology, they are used in the form of endosseous ridge maintenance devices (ERMI) and particles for injection (Wilson et al., 1993). Glass-ceramics with the same composition as bioactive glasses of this type are not used for such a wide range of applications.

Other apatite-containing glass-ceramics (Ilmaplant® L1 and Ap40) were reported by Steinborn et al. (1993).

4.4.1.1 CERABONE®

CERABONE® is the most widely and successfully used bioactive glass-ceramic for bone replacement in human medicine. Nippon Electrical Glass Co., Ltd. produces the apatite–wollastonite (A–W) glass-ceramic under the brand name of CERABONE® A-W. It is distributed as a biomaterial by Lederle (Japan), Ltd.

The glass-ceramic demonstrates particularly high bioactivity. The reaction mechanism that forms a bond between CERABONE® and the living bone, without connective tissue, is addressed in Section 2.4.2. In this section, it is shown that the release of Ca^{2+} ions and the reactivity of the $\equiv Si–OH$ surface groups of the glass-ceramic are of particular significance. The body fluids provide structural elements of phosphate, which allows apatite to form on the implant surface in a solid-state reaction. Kokubo (1993) showed that the formation of a thick layer of apatite proceeded very quickly. Therefore, it was possible for the apatite to form a 6-µm thick layer in 24 h. Furthermore, numerous clinical trials have shown intergrowth between glass-ceramics and human bone. Yamamuro (1993), for example, has used glass-ceramics in orthopedics in the form of vertebral prostheses, restorations of the iliac crest, and bone defect fillers.

The advantages provided by the combination of bioactivity and specific mechanical properties (Table 4-13), such as compressive strength and fracture toughness, are evidenced by the successful application of glass-ceramics as artificial vertebrae (Fig. 4-26). The CERABONE® A-W (1992) brochure presents a particularly effective example of how the material is used. The broken lumbar 2 vertebra in a 57-year-old patient was replaced by an implant made of CERABONE® A-W glass-ceramic. The implant was held in place with a Kaneda device. The X-ray examinations conducted three and six years

Table 4-13

Properties of Glass-Ceramics for Medical Application

Properties	CERABONE® A-W	BIOVERIT® I	BIOVERIT® II
Mechanical			
Density (g/cm³)	3.07	2.8	2.5
Flexural strength (MPa)	215	140–180	90–140
Compressive strength (MPa)	1080	500	450
Young's modulus (GPa)	118	70–88	70
Hardness (Vickers)	680 (HV)	5000 (HV10)	~8000 (HV10)
Fracture toughness (MPa·m$^{0.5}$)	2.0	1.2–2.1	1.2–1.8
Slow crack growth, n	33		
Roughness (after polishing) (µm)			0.1
Thermal			
CTE (20°C–400°C)		$8\text{–}10 \times 10^{-6}$ K^{-1}	$7.5\text{–}12 \times 10^{-6}$ K^{-1}
Chemical			
Hydrolytic class (DIN 12111)		2–3	1–2

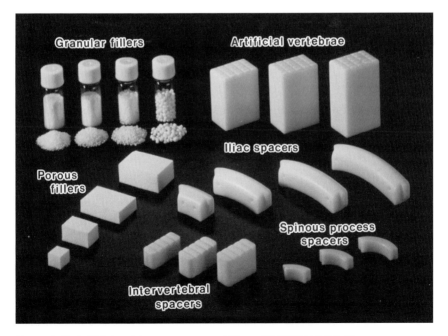

Figure 4-26 Artificial vertebrae, spacers, and fillers of CERABONE® A-W.

after the operation showed that the implant had not moved and that the formation of new bone structure was excellent. Between 1991 and 1996, glass-ceramics were successfully used for such applications in more than 10,000 patients (Kokubo 1996).

4.4.1.2 CERAVITAL®

Numerous in vitro and in vivo investigations of glass-ceramics of the apatite-type (Section 2.4.3) have proved the biocompatibility and bioactivity of these materials. The histological evaluations conducted with scanning electron micrographs and transmission electron micrographs demonstrated the different levels of surface reactivity of the glass-ceramics. Gross et al. (1993) have documented the most significant results. In their publication, they also showed that the surface reactivity was improved by special surface treatment. An example of glow discharge treatment is given.

In human medicine, this glass-ceramic has proved to be particularly useful for middle ear devices (Reck 1984). Since these implants carry only light weight, the mechanical properties of the material are suitable.

4.4.1.3 BIOVERIT®

Biomaterials such as mica–apatite and mica glass-ceramics are produced and distributed under the brand names of BIOVERIT® I and BIOVERIT® II by Vitron Spezialwerkstoffe GmbH, Germany. BIOVERIT® I and II are machinable glass-ceramics. In other words, they are biomaterials that are workable with standard metal tools and instruments. They also can be easily modified during the surgical procedure, e.g., middle ear implants of BIOVERIT® II (Beleites et al., 1988). The workability of the biomaterial depends on the mica content and on the morphology of the glass-ceramics, e.g., a glass-ceramic with high mica content permits excellent machinability. Therefore, the workability of BIOVERIT® II is improved over that of BIOVERIT® I. Main properties of BIOVERIT® I and II are demonstrated in Table 4-13.

Results of in vivo tests of BIOVERIT® I explain the reaction behavior of the biomaterial. One year after the operation, the reaction interface between the bone (tibia of a guinea pig) and glass-ceramic implants is less than 15 μm and shows very good intergrowth of the glass-ceramic and bone. Furthermore, optical microscopic investigations (histology) and the calculated bone connection show the bioreactive behavior of BIOVERIT® I glass-ceramic compared with corundum implants. These results of the bone connection of BIOVERIT® I and corundum can be compared directly because of the chosen model of implantation. The shearing strength of the implant–bone boundary has been determined by measuring the mechanical force necessary to push out the implant of BIOVERIT® I. The value of approximately 2.3 MPa found for glass-ceramic implants was eight times greater on average than the one for the sintered corundum implants, which was measured for comparison (Höland and Vogel, 1993).

BIOVERIT® II is a biocompatible glass-ceramic with lower bioreactivity than BIOVERIT® I. Animal experiments have demonstrated that intergrowth occurs without causing any adverse reactions and that the biocompatible implant is covered with epithelium. Intergrowth takes place as if the implant were part of the body.

The successful tests of BIOVERIT® I and II allowed the glass-ceramics to be applied as biomaterials for bone substitution in human medicine. More than 850 implants have been successfully applied (up to 1992) in orthopedic surgery, especially different types of spacers (Schubert et al., 1988) and in head and neck surgery, especially middle ear implants (Fig. 4-27) (Beleites and Rechenbach, 1992).

In addition to these applications, the material was also tested for use as a dental root substitute (Pinkert 1990).

4.4.2 Glass-Ceramics for Dental Restoration

4.4.2.1 Requirements for Their Development

While bioactivity is required of medical prostheses or implant materials that replace hard tissue in human medicine, other properties are required of the biomaterials used in restorative dentistry. This subject must be addressed initially to facilitate the understanding of the development of these materials as well as of their applications.

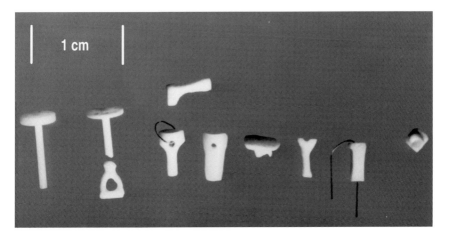

1 cm

Figure 4-27 Middle-ear implants of BIOVERIT® II.

Restorative dental materials are used to fabricate, e.g., dental crowns, bridges, inlays, and veneers. These materials, therefore, are not implanted in bone. Instead, they are bonded to the living tooth by bonding or cementation systems. The materials used to fabricate posts and core buildups in the replacement of nonvital teeth that have undergone root canal treatment also belong to this category. Furthermore, some glass-ceramic veneering materials can also be used to mask metal superstructures.

These desired applications determine the main requirements to be fulfilled in the development of glass-ceramics for dental applications. The main objective is to produce a new biomaterial, the properties of which correspond to those of natural teeth. The most important properties are mechanical properties, biochemical compatibility with the oral environment, and a degree of translucency, shade, opalescence, and fluorescence similar to that of natural teeth. An abrasion resistance similar to that of natural teeth must also be achieved. The new biomaterial must demonstrate higher chemical durability than natural teeth, to prevent it from being susceptible to decay.

The following glass-ceramics fulfill the above properties. Moreover, they even possess additional properties, such as favorable working techniques, that are important for both dentists and dental technicians:

DICOR®, IPS EMPRESS®, IPS EMPRESS® Cosmo, IPS EMPRESS® 2, IPS d.SIGN®, Pro CAD®, and IPS ERIS for E2.

These glass-ceramics represent highly specialized materials in which the different sophisticated properties mentioned have been developed. Therefore, they are highly refined tailor-made products. Furthermore, the fabrication technique has also been optimized. This measure was necessary because patients have different requirements, and dental restorations (e.g., inlay, crown, or bridge) must be suited exactly to the clinical situation. As a result, special techniques were developed that can be applied in the dental laboratory.

These biomaterials are an integral part of a system used to fabricate dental restorations. The system comprises the base material, the processing techniques, the methods for characterizing the materials, and the various applications. The properties, processing techniques, and applications will be described in detail for each of the following product groups to illustrate this system. Reference is made to the development of the materials and to the relevant parts in Chapters 1, 2, and 3.

This detailed description of glass-ceramics is intended for both dentists and dental technicians. In addition, it conveys general information about biomaterials for patients. At the same time, the description of this system of materials provides students, technicians, and engineers with information that will help them develop complex systems of materials for other technical and scientific fields.

4.4.2.2 DICOR®

Two types of DICOR® glass-ceramics are produced for dental applications. The first type is a castable glass-ceramic with which dental restorations are fabricated in the dental laboratory. The second type is a machinable glass-ceramic with which dental restorations are machined using CAD/CAM technology in the dental clinic.

A) Properties

DICOR® glass-ceramics are mica glass-ceramics the chemical composition and microstructure of which were discussed in Section 2.3.1. Adair and Grossmann (1984) demonstrated that the properties of the material could be controlled by controlling the crystallization of the base glass. Translucency, machinability, and a modulus of rupture of 153 MPa characterize DICOR® glass-ceramics. This strength is very good in comparison with 10.3 MPa of natural enamel unsupported by dentin. The modulus was obtained in bending or flexure tests. The compressive strength of 828 MPa is much higher than that of conventional dental porcelains (172 MPa). According to Table 4-14, the glass-ceramic demonstrates a linear thermal expansion coefficient of 7.2×10^{-6} K^{-1} and a Young's modulus of 70.3 GPa. Clinical investigations have shown good wear properties.

The optical quality produced by controlled translucency is based on the microstructure of mica crystals embedded in the glassy matrix. The crystals measure approximately 1μm in length after casting and heat treatment of the glass. The shading is performed using colorants premixed with a glazing material.

B) Processing

Dental technicians fabricate dental restorations, e.g., dental crowns or inlays, using the lost-wax technique. The pattern is prepared in the same manner as for cast metal crowns. The pattern is sprued and invested in a special investment material. The wax burnout takes place at 900°C . The material for fabricating the DICOR® glass-ceramic is supplied as a glassy ingot, which is placed in an electric muffle furnace of a specially designed casting machine. The ingot becomes viscous during heating, and following a 6-min hold at 1370°C, the material is forced into a mold by centrifugal casting for 4 min. The casting procedure is characterized by a high degree of accuracy. The mold is cooled to room temperature and the cast glass dental restoration is divested. The glass is transparent and quality control can be applied to possible defect locations.

Table 4-14

Characteristic Properties of DICOR® Glass-Ceramic

Properties	Parameter
Mechanical	
Density (g/cm^3)	2.7
Modulus of rupture (psi)	22,000
(MPa)	152
Compressive strength (psi)	120,000
(MPa)	828
Young's modulus (psi)	10.2
(GPa)	70.3
Hardness (Knoop) (MPa)	362
Thermal	
Coefficient of thermal expansion	
($\times 10^{-6}$ K^{-1})	7.2
Optical	
Refractive index	1.52
Translucency	0.56

The glassy dental restoration is embedded in an investment-like material and placed in an oven for the controlled crystallization process. During this heat treatment, the glass is converted into the DICOR® glass-ceramic. Crystallization takes place at 1075°C during 6 h of heat treatment (Abendrot 1985). After the completion of the heat treatment and cooling to room temperature, the dental restoration is grit blasted and cleaned. Subsequently, the dental technician applies a colored glaze to the surface of the restoration.

In 1989, the sintered ceramic DICOR® Plus was developed (DICOR® Plus 1989). This material allowed dental technicians to create more individually shaded restorations and to mask the cast DICOR® glass-ceramic. The heating temperature of this material was 950°C.

The second type of DICOR® glass-ceramic was developed for the purpose of machining dental restorations. This glass-ceramic carries the brand name DICOR® MGC. The material is machined using the Cerec 1 system (Siemens AG, Germany). Details of this machining system have been reported by Mörmann et al. (1987).

The microstructure and physical properties of the DICOR® MGC glass-ceramic were described by Grossman (1991). The microstructure of DICOR® MGC (Corning 9670) is characterized by a high crystallinity of the glass-ceramic. The crystal content is estimated to be approximately

70 vol%. The average diameter of the mica plates is approximately 2 μm, and their thickness is approximately 0.5 μm. DICOR® MGC is produced in two shades: light and dark. The crystals of the dark glass-ceramic measure 1 μm. The interlocking microstructure of the glass-ceramic allows the material to be machined. During machining, microscopic fractures proceed along the glass–mica interface and the cleavage planes of the mica. The induced fractures are repeatedly deflected, keeping the cut from propagating outside the localized cutting area. The cutting efficiency of diamond and carbide burs is very good. Gagauff et al. (1989) showed that the machining behavior of DICOR® MGC closely resembles that of natural enamel (Fig. 4-28).

Moreover, Grossman (1991) demonstrated the strength dependence on the crystal size of DICOR® MGC. He showed that mica crystals that were smaller than 4.5 μm increased by 1/5 power of the average mica-plate diameter (Fig. 4-29). The main physical properties are shown in Table 4-15. The flexural strength measured 127–147 MPa (biaxial testing) and the fracture toughness K_{IC} 1.4–1.5 MPa·m$^{0.5}$.

C) Applications

The preferred applications of DICOR® glass-ceramics in restorative dentistry are veneers, inlays, onlays, as well as dental crowns. Early clinical studies were conducted by Richter and Hertel (1987). They reported favorable aesthetic characteristics of the glass-ceramic. Long-term clinical functioning was established by Malament and Grossman (1987). The cast DICOR® glass-ceramic was predominantly used to fabricate complete crowns. The glass-ceramic was not recommended for fixed partial dentures, abutments, or dental bridges. The in vivo research revealed low plaque adherence. Moreover, bacterial plaque growth was seven times lower than on natural teeth.

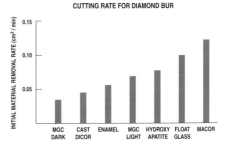

Figure 4-28 Machinability of DICOR® MGC compared with that of other materials and natural enamel.

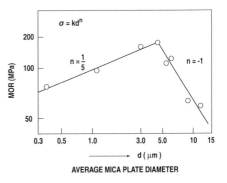

Figure 4-29 Strength dependence (modulus of rupture, MOR) on the crystal size of DICOR® MGC.

Table 4-15

Physical Properties of DICOR® MGC

Properties	Code 9670 (light)	Code 9671 (dark)
Mechanical		
Density (g/cm^3)	2.8	2.8
Machinability (cm^3/min)	0.068	0.034
Modulus of rupture (MPa)	138	122
Biaxial flexure strength (MPa)	147	127
Young's modulus (GPa)	68	68
Fracture toughness, K_{IC} (MPa·m$^{0.5}$)	1.5	1.4
Microhardness, KNH_{100} (MPa)	3300	3500
Thermal		
Coefficient of thermal expansion ($\times 10^{-6}$ K^{-1})	6.4	6.4
Optical		
Refractive index	1.52	1.52
Translucency	0.41	0.44
Chemical		
Chemical durability (95°C/24 h) (mg/cm^2)		
pH 6	−0.007	−0.007
pH 7	−0.006	−0.005
pH 8	−0.005	−0.003
Characteristics		
Crystal size (μm)	2	1

In a 14-year clinical study, Malament and Socransky (1998) reported an estimated risk of failure of 2.45% per year. They also showed that long-term survival significantly improved if the restorations were acid etched prior to luting. During this long-term study, no DICOR® glass-ceramic restorations on lateral incisors failed. The preferred application of DICOR® MGC glass-ceramic in restorative dentistry is dental inlays. DICOR® is produced by Corning Inc./Dentsply International (USA).

4.4.2.3 IPS EMPRESS® Glass-Ceramic

A) Properties

This glass-ceramic material is derived from the SiO_2–Al_2O_3–K_2O system. Its composition, controlled crystallization, and microstructure featuring the main crystal phase of leucite, $KAlSi_2O_6$, are described in Section 2.2.9.

Given its special microstructure, the IPS EMPRESS® glass-ceramic demonstrates translucent properties. Hence, it is highly suitable for dental applications. The leucite crystals cause dispersion strengthening of the glass-ceramic. As a result, a bending strength of 100–160 MPa is achieved. Additional post treatment of the surface, such as glazing and heat treatment, produces bending strengths of approximately 180 MPa (Dong et al., 1992). This improvement of properties by post treatment is particularly favorable for inlays and crowns fabricated in the dental laboratory. The suitability of the IPS EMPRESS® glass-ceramic for the fabrication of crowns was proven by Ludwig (1994). He discovered that the loading capacity of 500 N demonstrated by the crowns he had fabricated was more than twice the capacity required for this type of restoration.

The leucite crystals are responsible for the high linear thermal expansion of the glass-ceramic. Different forms of the IPS EMPRESS® glass-ceramic are offered: glass-ceramic for the staining technique and glass-ceramic for the layering technique (see Section 4.4.2.3B). The coefficient of linear thermal expansion of the specific materials may range between $15 \times 10^{-6} \text{ K}^{-1}$ and $18.25 \times 10^{-6} \text{ K}^{-1}$ in the temperature range of 100°C–500°C (Höland and Frank, 1993).

The chemical durability of the glass-ceramic in an acid environment is particularly important. It is a well-known fact that this type of environment may reduce the chemical durability of silicate materials. A special ISO standard for dentistry (ISO dental standard 1995) determines the test conditions to which glass-ceramics must be subjected: 16-h immersion in 4% hot (80°C) acetic acid. The weight loss of the glass-ceramic should not exceed the limiting values during this time. The IPS EMPRESS® glass-ceramic fulfills this standard with 13 $\mu g/cm^2$ (glass-ceramic for staining technique, see Section 4.4.2.3B) and 122 $\mu g/cm^2$ (for layering technique).

The biomaterial must also demonstrate special abrasion properties that stand up to the oral environment. The most favorable properties are those demonstrated by natural teeth. If the surface of a glass-ceramic is too hard for example, it could damage the antagonist tooth. However, if the biomaterial is too soft compared with natural dentition, it would wear too easily. Comprehensive tests simulating the actual oral conditions have been conducted in a mastication simulator. Heinzmann et al. (1990) and Krecji et al. (1993) were the first to show that the IPS EMPRESS® glass-ceramic demonstrates favorable abrasion properties corresponding to those of natural teeth. Subsequent studies have confirmed these findings. It is very likely that this characteristic has been achieved with the microstructure containing a glass matrix and crystals of a close to uniform size of 1–6 μm in length.

A retentive pattern on the surface of the glass-ceramic inlays and crowns enhances the adhesive cementation to the natural teeth of patients (Salz 1994). This pattern is achieved by etching the contact surface of the restorations with 4.5% hydrofluoric acid etching gel before they are seated. The etching process selectively starts at the interface between the leucite crystals and the glass matrix. From there, the etching process advances toward the leucite crystals. Depending on the etching time, for example, 30 or 60 seconds, the leucite crystals are dissolved in a surface layer measuring a few micrometers. An open-porous glass matrix results on the surface, as well as the retentive pattern required for the adhesive system.

The IPS EMPRESS® glass-ceramic demonstrates an additional property favorable for the processing of the material in the dental laboratory. It begins to flow at temperatures between 1000°C and 1200°C, despite its crystal content of approximately 34 vol% leucite. Hence, the glass matrix reaches a viscosity of approximately 10^{11} Pa·s at these temperatures. Once relatively low pressure of 200–300 N is applied, the glass-ceramic demonstrates viscous flow. Given this property, dental restorations of virtually every shape can be formed from the material.

B) Processing

The IPS EMPRESS® glass-ceramic is produced with a base glass using controlled surface crystallization (Section 1.5). The material also contains additives that influence its processing temperature and stability and produce fluorescence. A raw glass-ceramic is produced in this controlled crystallization process. At a later stage, this raw material, in the form of ingots (Fig. 4-30) is used to fabricate dental restorations, (e.g., crowns, inlays) in the laboratory (Fig. 4-31).

The procedure used by dental technicians to fabricate the restorations includes the following steps. First, a wax model is produced on the basis of the impression of the clinical situation provided by the dentist.

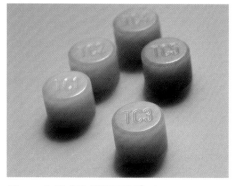

Figure 4-30 IPS EMPRESS® glass-ceramic ingots.

This wax model is invested in the corresponding material. Once this investment material has set into a cylinder, the wax is fired out (lost-wax technique) to produce a hollow mold of the restoration. Subsequently, the investment cylinder is placed in special furnaces called EP 500

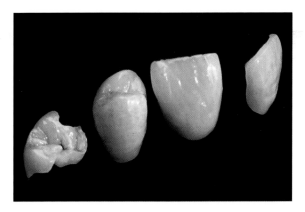

Figure 4-31 IPS EMPRESS® glass-ceramic dental products (inlay, crowns, veneer).

and EP 600 (Ivoclar Vivadent AG, Lichtenstein) (Fig. 4-32). Once the glass-ceramic ingot has become viscous, it is pressed into the hollow mold at temperatures ranging between 1000°C and 1200°C (1075°C for staining technique, 1180°C for layering technique). After the cylinder has cooled adequately, the investment material is removed from the glass-ceramic restoration by blasting it.

Subsequently, two methods are available for post treating the surface. A glaze is applied to inlay surfaces either alone or in combination with stains. This method is called *staining technique*. As a second method, an additional layer of sintered ceramic is applied to crowns. This method is called *layering technique*. The sintered ceramic layer enhances the optical properties of anterior teeth in particular. Translucency and opalescence are the most important factors. Opalescence, the ability to demonstrate a brownish-yellow color in transmitted light and a whitish-blue color in reflected light, is achieved by adding a layer of opal glass-ceramic. This ma-terial is added to the sintered ceramic in the incisal regions of the tooth (Fig. 4-33). This technique allows the successful simulation of the optical properties of natural teeth. The microstructure of the material is shown in Section 3.2.3.

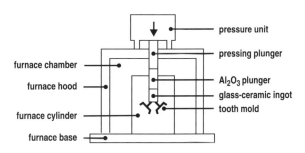

Figure 4-32 Scheme of EP 500 and EP 600 press furnaces.

In preparation for the cementation to the natural tooth, the contact surface of the restoration is etched to produce a retention pattern. A detailed description of the procedure is provided by Wohlwend and Schärer (1990) and Haller and Bischoff (1993).

C) Applications

The success of the IPS EMPRESS® glass-ceramic as a dental restorative has been proven in comprehensive clinical tests and successful application worldwide since 1991. This glass-ceramic fulfills new properties, which could not be achieved by metals. The IPS EMPRESS® glass-ceramic is used worldwide to fabricate inlays, onlays, anterior and posterior crowns, and veneers. The material is not recom-

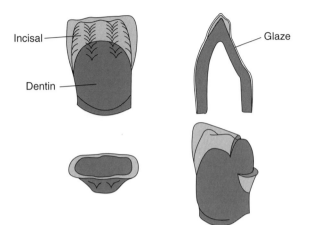

Figure 4-33 Schematic diagram of a glass-ceramic crown of IPS EMPRESS® leucite glass-ceramic. Pressed glass-ceramic (dentin), sintered ceramic with opal glass-ceramic (incisal), and glaze.

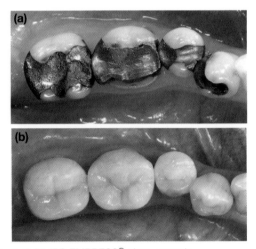

Figure 4-34 IPS EMPRESS® glass-ceramic
a) preoperative situation showing four amalgam fillings,
b) four glass-ceramic inlays/onlays of IPS EMPRESS® staining technique (Courtesy Brodbeck and Arteco Dentaltechnik, Zürich, CH)

mended for fabricating multi-unit dental bridges. Figure 4-34 shows four IPS EMPRESS® inlays/onlays in a clinical situation. The clinical results of a six-year study show that the failure rate is approximately 2–3%. Most of the problems experienced thus far have been traced to preparation errors. An outline of the clinical results is presented by Brodbeck (1996). IPS EMPRESS® is produced by Ivoclar Vivadent AG (Liechtenstein).

4.4.2.4 IPS EMPRESS® Cosmo Glass-Ceramic for Dental Core Buildups

The entire system of a dental core buildup comprises a product consisting of a ZrO_2-sintered ceramic (CosmoPost) and a ZrO_2-containing glass-ceramic (IPS EMPRESS® Cosmo). As a result, two entirely different materials are combined to form one product in a special process. The composition and the microstructure of the ZrO_2-containing glass-ceramic are discussed in detail in Section 2.6.5.

A) Properties

CosmoPost (ZrO₂ Ceramic)

The properties of both materials systems have been presented in detail by Schweiger et al. (1996, 1998) and Kakehashi et al. (1998). The ZrO_2 posts are cylindrical rods with a conical point tip. The conical angle is 6°. The posts were produced in two sizes with a diameter of either 1.4 or 1.7 mm. The post surface was roughened to optimize the strength of the bond between the ZrO_2 and the glass-ceramic dentin core buildup and between the ZrO_2 and the root. The surface roughness R_a measures 0.5–1.2 μm.

The ZrO_2 ceramic (ZrO_2–TZP ceramic, 3 mol% Y_2O_3) is distinguished for its outstanding mechanical properties (Rieger 1993). Its four-point flexural strength registers 900 MPa. The fracture toughness K_{IC} of 7.0 MPa·m$^{0.5}$ is exceptionally high for a monolithic ceramic. The high strength and toughness values were achieved with transformation toughening. This mechanism provided one technique for achieving high strength and toughness in materials. The tetragonal form of the ZrO_2 crystals was capable of transforming itself into a monoclinic form under stress. A volume increase in the crystals accompanies this reaction. The volume increase in turn induces compressive strain in the transformation zone. This reaction inhibits or even stopped the formation of a crack with tensile stress in the crack tip (Evans and Heuer, 1980). In addition, the microstructure of the ceramic is composed of densely sintered grains exhibiting an average diameter of 0.4 μm. The supplementary high-temperature isostatic pressing of the posts during the fabrication procedure eliminates porosity. The modulus of elasticity as a standard of rigidity was measured at 210 GPa. The modulus of elasticity values of most dental alloys are found in this range. Between 100°C and 500°C, the coefficient of linear thermal expansion is 10.7×10^{-6} K^{-1}.

IPS EMPRESS® Cosmo (ZrO₂-Containing Glass-Ceramic)

The properties of this glass-ceramic are also influenced by the growth and formation of crystalline phases during the hot-press procedure.

The three-point flexural strength of the hot-pressed glass-ceramic without the reinforcement of ZrO_2 ceramic registers 164 ± 26 MPa. This strength value is considerably higher than that of glass. The strength of the material may have resulted from the precipitation of crystals or by the large share of Zr^{4+} ions in the form of complex $(ZrO_4)^{4-}$ structural units in the glass network structure. The high chemical durability, another feature of the glass-ceramic, is achieved by incorporating $(ZrO_4)^{4-}$ structural elements. The core buildup featuring the high inherent stability of the glass-ceramic is additionally reinforced by using a ZrO_2 post.

The Vickers hardness of the glass-ceramic was 5340 MPa when an indentation force of 9.8 N was applied. The modulus of elasticity registered 55 GPa. The coefficient of linear thermal expansion of the glass-ceramic was established as 9.4×10^{-6} K^{-1} between 100°C and 500°C. The glass transition point occurred at 545°C. A lower coefficient of thermal expansion was required for the glass-ceramic than the ZrO_2 ceramic (100°C–500°C): 10.7×10^{-6} K^{-1} to obtain a stress- and crack-free bond between the two materials. The bond is shown in Fig. 4-35. Neither microcracks nor porosity is evident.

The bond strength was determined in a push-out test (Kakehashi et al., 1997). Good bond strength results were obtained for ZrO_2 posts that had been roughened. The bond strength registered 35 ± 9 MPa. Following an aging cycle in water for 333 h and additional thermocycling of 10,000 cycles between 5°C and 55°C, the bond strength registered 44 ± 10 MPa. Weakening of the bond was not observed. Bond strength values that are higher than 30 MPa are considered to be very good in dental adhesive technology.

Contrast ratio measurements (British Standards Institution 1978) on samples that were 1.10 mm thick produced a value of 0.72 (1 corresponds to 100% opacity; 0 to

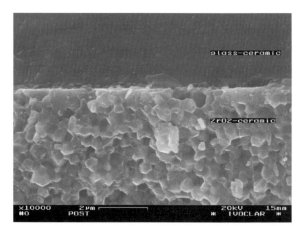

Figure 4-35 Microscopic image of a defect-free bond between the ZrO_2-containing glass-ceramic and the ZrO_2 ceramic.

100% transparency, Schweiger et al., 1998). Highly aesthetic dental restorations were produced as a result of these translucency values. These restorations were used in the anterior region in particular.

Another glass-ceramic containing 20 wt% ZrO_2 was also produced in the P_2O_5–SiO_2–Li_2O–ZrO_2 system (Section 2.6.5). In this case considerably more ZrO_2 crystals were precipitated. This glass-ceramic demonstrated 280 MPa and was white-opaque in appearance due to the high ZrO_2 content, thus the translucency was low.

B) Processing

As the object of the process is the fabrication of a core buildup using a glass-ceramic and a ZrO_2 sintered ceramic post, a two-step procedure is required. The first step involves the fabrication of a base glass-ceramic in the form of ingots; the second step involves the fabrication of the final product, consisting of ZrO_2 post and ZrO_2 containing glass-ceramic. The glass-ceramic ingots, ZrO_2 posts, and the final product are shown in Fig. 4-36.

The first step in the development of the glass-ceramic was the production of a base glass. This glass was melted with raw materials including oxides, carbonates, phosphates, and fluorides and homogenized at temperatures between 1400°C and 1600°C. The melt was subsequently poured into water. The glass grains were ground to a fine powder. The cylinders were densely sintered. Nucleation and crystallization took place at the same time as the densification process. An intermediate product in the form of a glass-ceramic ingot was produced in this manner.

In a second step, the glass-ceramic intermediate product was hot-pressed in the dental laboratory. During this process, the bond with the ZrO_2 ceramic was produced and the glass-ceramic shaped according to the individual requirements. The glass-ceramic was pressed into a mold at high temperatures (900°C) to produce dental restorations with individualized shapes. The hot-press furnaces (EP 500, EP 600, Ivoclar, Ltd.) used in the

Figure 4-36 ZrO_2-containing glass-ceramic ingot, ZrO_2 post and two products of glass-ceramic and ZrO_2 post.

process are shown in Fig. 4-32. The glass-ceramic ingots were heated to 900°C in the preheated mold. They were held in a vacuum at this temperature for 10 min. Subsequently, the automatic molding cycle with a pressure of 200–300 N began. The high temperature and the pressure buildup forced the glass-ceramic into the mold. During the hot-press procedure and the cooling phase, the final microstructure of the glass-ceramic was formed. Once the molding cycle had ended, the mold was left to cool to room temperature. Subsequently, the pressed part was divested from the mold by blasting the mold material with corundum powder and glass beads at 1–2 bar pressure.

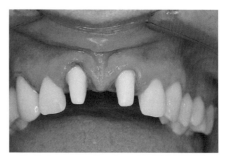

Figure 4-37 Clinical situation of cemented combination of CosmoPost® and IPS EMPRESS® Post.

C) Application

To date, metal root posts predominantly have been used to restore nonvital teeth. In the process, the dentist prepares the dental root canal and cements a metal post in the tooth. On this post, the tooth is restored with an organic composite or a ceramic material.

The possibility of using metal-free root post systems has been presented in the past (Meyenberg et al., 1995). Based on these results, the development of a ZrO_2 post in conjunction with a glass-ceramic was initiated. The clinical result of the cemented post and ZrO_2-containing glass-ceramic is shown in Figs. 4-37 and 4-38. This product provides a highly aesthetic metal-free solution for core buildups. The final clinical result with all-ceramic restoration is shown in Fig. 4-38 b. Sorensen and Mito (1998) have discussed the special properties and the clinical

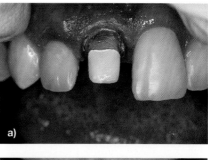

a)

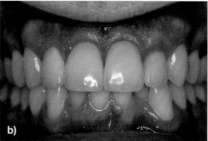

b)

Figure 4-38 Final clinical result of CosmoPost® and IPS EMPRESS® Post. a) core buildup, b) final result (Sorensen and Mito, 1998).

success of the materials. As a result they concluded that the CosmoPost® and IPS EMPRESS® Post system provide the following features for restoration of endodontically treated teeth:

- improved aesthetics
- unified core and post
- inert material to avoid discoloration from corrosion by-products
- stable core material, unlike resin composite core material
- high stiffness to resist bending forces from functional fatigue loading
- simple method for fabrication

IPS Empress® Cosmo is produced by Ivoclar Vivadent AG (Liechtenstein).

4.4.2.5 IPS EMPRESS®2 Glass-Ceramic

The special properties and favorable uses of IPS EMPRESS® glass-ceramics provided the impetus for further development of this material with the objective of making it suitable for the demanding application of dental bridges. As a result, a new system of materials had to be developed for the layering technique (see Section 4.4.2.3). The objective was to increase the material's mechanical parameters, such as flexural strength and fracture toughness. On the basis of these requirements, a new glass-ceramic framework material had to be developed. In addition, a new layering glass-ceramic was also required. The materials shown in Table 4-16 were specifically developed for the system. In Table 4-16, the glass-ceramic, which is used as the framework material, is called a pressed glass-ceramic, while those that are applied to the pressed glass-ceramic in layers are called sintered glass-ceramics. As these names imply, the framework is fabricated by pressing a base glass-ceramic into a mold (as described in Section 4.4.2.3) and the layering glass-ceramic powder is sintered on this framework. The glass-ceramics, their processing, and properties were reported by Frank et al. (1998), Schweiger et al. (1999), and Höland et al. (2000b).

Table 4-16

Glass-Ceramics of the IPS EMPRESS®2 System

Glass-ceramic	Chemical nature
pressed glass-ceramic	lithium disilicate glass-ceramic
sintered glass-ceramics	apatite-containing glass-ceramics

A) Properties

Pressed Glass-Ceramic

The pressed glass-ceramic is derived from the basic SiO_2–Li_2O system. It contains a lithium disilicate, $Li_2Si_2O_5$, as main crystal phase. The most important additions are oxides, such as ZnO, P_2O_5, K_2O, and La_2O_3. The composition of the glass-ceramic and the mechanisms of nucleation and crystallization that are involved in the formation of the microstructure are described in Section 2.1.1 (composition: see Table 2-3 in Section 2.1.1). The special characteristic of the microstructure is also discussed in this section. That is, the ingot (base glass-ceramic) as well as the pressed dental framework contains 70 ± 5 vol% crystal phase. Furthermore, these crystals are tightly crosslinked (see Fig. 2-3 in Section 2.1.1 and Fig. 3-21 in Section 3.2.11).

This high crystal content in conjunction with the coordinated residual glass matrix is responsible for a series of special properties. Flexural strength of 400 ± 40 MPa were determined by three-point bending. This strength value provides a considerable improvement compared with the IPS EMPRESS® leucite glass-ceramic. An evaluation of the strength values of IPS EMPRESS®2 according to Kappert (1998) in Fig. 4-39, clearly shows that these pressed glass-ceramics belong to a new generation of restorative dental materials that can be used to fabricate both crowns and bridges.

The strength or load-bearing capability of a three-unit bridge, for example, is of particular interest for clinical applications. Kappert (1998) has conducted in vitro investigations of this topic. In these investigations, anatomical three-unit bridges (two abutments and one pontic) and a connector cross section of 4 mm × 4 mm were fabricated and placed on metal dies, without bonding them. These bridges were subjected to fatigue loading tests. During cyclic loading, they were simultaneously exposed to changes in

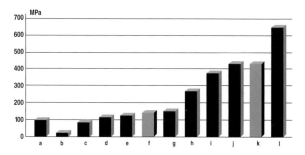

Figure 4-39 Flexural strength of pressed glass-ceramic IPS EMPRESS®2 in comparison to other dental materials, natural tooth, and technical ceramics (Kappert 1998). a) ISO limit, b) tooth enamel, c) tooth dentin, d), e) leucite-based ceramics, f) IPS EMPRESS®, g) feldspar ceramic, h) mica glass-ceramic, i) sintered spinel and infiltrated glass, j) sintered Al_2O_3 and infiltrated glass, k) IPS EMPRESS®2, l) dense sintered Al_2O_3.

temperature (water with a temperature of 35°C and 55°C). The applied load during the test cycles was 75% of the initially measured fracture load. Following preliminary loading, force was applied to rupture the specimen.

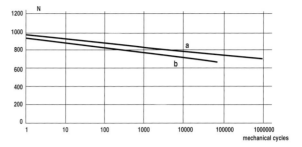

Figure 4-40 Experimental results of fatigue loading test of IPS EMPRESS®2 bridge. a) in comparison to another dental product (sintered Al_2O_3 and infiltrated glass) b) of the same geometry (Kappert 1998).

The results of this test are shown in Fig. 4-40. Compared with another dental product (sintered Al_2O_3 and infiltrated glass), the IPS EMPRESS® material shows similar results.

The fracture toughness (determined as K_{IC}), translucency, coefficient of linear thermal expansion, and the chemical durability of the IPS EMPRESS®2 glass-ceramic are compared with those of IPS EMPRESS® in Table 4-17. It must be noted that standardized test methods were used to produce the test results. Therefore, K_{IC} was determined using the three-point bending test with a 20-mm span between the support points on test specimens with dimensions 3 mm × 6 mm × 28 mm. The notch depth was approximately 1.5–2.0 mm. The fracture load, notch depth, support width, and the cross section of the specimen were used to calculate K_{IC}. The K_{IC} value of 3.3 MPa·m$^{0.5}$ is approximately 2.5 times higher than that of the IPS EMPRESS® leucite glass-ceramic and is achieved with a new microstructure of the glass-ceramic and high crystal content of 70 vol%. This degree of toughness was only previously achieved with high-strength sintered ceramics, such as B_4C, Al_2O_3, or ZrO_2 and certain glass-ceramics based on chain silicates (Sections 2.3.3, 2.3.4). ZrO_2 sintered material is characterized by a very high toughness, but the material cannot be processed by pressing in a dental laboratory. Also, the optical transmission of a glass-ceramic is significantly higher than that of ZrO_2.

The coefficient of linear thermal expansion was measured in specimens according to dental standards. The value of $10.6 × 10^{-6}$ K^{-1} recorded for IPS EMPRESS®2 compared with $15.0 × 10^{-6}$ K^{-1} for IPS EMPRESS®, clearly shows that the two materials are incompatible (Table 4-17). As a result, two different sintered ceramics must be used for the two systems of materials.

Table 4-17

Properties of Pressed Glass-Ceramic (Lithium Disilicate) of IPS EMPRESS®2 in Comparison to IPS EMPRESS® (Layering Technique) (Leucite Glass-Ceramic)

Properties	IPS EMPRESS®	IPS EMPRESS®2
Mechanical		
Flexural strength (MPa)	112 ± 10	400 ± 40
Fracture toughness (MPa·m$^{0.5}$)	1.3 ± 0.1	3.3 ± 0.3
Abrasion behavior	similar to that of natural teeth	like IPS EMPRESS®
Optical		
Translucency	similar to that of natural teeth	similar to that of natural teeth
(British Standards Institution 1978)	0.58	0.55
Thermal		
Coefficient of linear thermal expansion (α), ($\times 10^{-6}$ K^{-1})	15.0 ± 0.25 (25°C–500°C)	10.6 ± 0.25 (100°C–400°C)
Chemical		
Durability (solubility in acetic acid) (µg/cm^2)	100–200	50
Technical		
Pressing temperature (°C)	1180	920
Application of the sintered glass-ceramics (dentin and incisal materials) (°C)	910	800

The chemical durability of the material was determined according to the relevant ISO dental standard (1995). Following this test, the loss of substance on the surface of the specimens was established. After the treatment, a framework material that must be veneered with a chemically durable glass-ceramic must achieve a minimum of weight loss of <1000 µg/cm^2, while the chemically durable veneering material must achieve <100 µg/cm^2. As shown in Table 4-17, the IPS EMPRESS®2 framework material also meets the standard for the veneering material. That is, the framework can be used without a veneering material. Other dental framework materials show much lower chemical durability (Geis-Gerstorfer 1997).

Wear characteristics are a particularly important property of dental materials, as these materials should not damage natural teeth. Some dental ceramics have been shown to be highly abrasive to antagonist natural tooth structure. Sorensen et al. (1999) examined this issue in the course of a six-month clinical study. Eight three-unit premolar bridges (of pressed glass-ceramic and sintered glass-ceramic) and 30 natural antagonist teeth were studied. An examination of the depth and volume of wear showed wear of 0.0268 mm^3 with a mean maximum depth of 23 μm on the surface of the glass-ceramic and of 0.0701 mm^3 with a mean maximum depth of 32 μm in the antagonist tooth. Fifty percent of antagonist teeth evidenced no wear facets. A novel observation was that 42% of glass-ceramic surfaces had small wear facets, demonstrating high biocompatible properties.

These in vivo investigations focused on the surface characteristics of the sintered glass-ceramic (Sorensen et al., 1999). A three-body *in vitro* investigation was carried out to evaluate the wear of antagonist tooth structure of a variety of dental ceramics. These examinations showed that the high-strength lithium disilicate glass-ceramic demonstrated wear characteristics similar to those of natural teeth. Obviously, different wear mechanisms of substructure and veneer glass-ceramic produce the same wear results. These results confirm the suitability of IPS EMPRESS®2 glass-ceramics for dental crowns and bridges.

The translucency, brightness, and fluorescence of the sintered glass-ceramics, available in the form of dentin and incisal materials, are particularly noteworthy. Comparable translucency is achieved in the pressed glass-ceramic microstructure. Light transmission was compared against a white and a black background, according to the British Standards Institution (1978), to determine translucency. When layer thicknesses of 1.0 were used, typical standard values of 0.55 were recorded. A simple experiment is useful in illustrating the material's translucency. Ubassy (1998) showed that light is effectively transmitted throughout the framework of a three-unit bridge if it is illuminated with white light. The optical effect is comparable to that achieved in natural teeth.

Sintered Glass-Ceramics

As shown in Section 2.4.7, it is possible to precipitate apatite in the $SiO_2-Al_2O_3-CaO-Na_2O-K_2O-P_2O_5$ system without crystallization of leucite. The microstructure is characterized by needlelike fluoroapatite crystals. This glass-ceramic can be bonded to lithium disilicate glass-ceramic. An optimum bond between the lithium disilicate glass-ceramic and sintered apatite glass-ceramics (Dentin, Incisal, Effect, and Transpa materials) is achieved when the two materials are sintered at 800°C. After this procedure, the sintered apatite glass-ceramics demonstrate a coefficient of thermal expansion of $9.7 \pm 0.5 \times 10^{-6} \text{ K}^{-1}$ (100°C–400°C). Therefore, these glass-ceramics are compatible with the lithium disilicate glass-ceramic, but not with the leucite glass-ceramic. According to British Standards Institution (1978), these materials demonstrated translucency of 0.46. Furthermore, the translucency of the glass-ceramics can be controlled, depending on the application of the materials.

Other special properties of these sintered glass-ceramics are their chemical durability and abrasion behavior. After treatment with 4% acetic acid, the materials exhibited a very low loss in mass of less than 100 $\mu g/cm^2$. The abrasion characteristics of the sintered glass-ceramics are also very good. The abrasion of the materials is similar to that of the pressed glass-ceramic. This surprising result was presented by Sorensen et al. (1998). Therefore, the materials have been shown to be highly compatible with natural antagonist teeth. Table 4-18 shows the properties of the sintered glass-ceramic.

B) Processing

Dental crowns and bridges are fabricated according to the viscous flow process described in Section 4.4.2.3.B, in which the biomaterial IPS EMPRESS® is introduced. The basic fabrication principles, therefore, remain unchanged. As a result, dental technicians can fabricate precision restorations quickly and easily. The quality of the IPS EMPRESS®2 material, however, provides several benefits as far as processing is concerned. The special features of the material are illustrated in the following example of the fabrication of a three-unit bridge.

Table 4-18

Properties of IPS EMPRESS®2 Sintered Apatite-Containing Glass-Ceramic

Properties	Parameter
Mechanical	
Flexural strength (MPa)	80 ± 25
Abrasion behavior	like IPS EMPRESS®
Optical	
Translucency	similar to that of natural teeth
(British Standards Institution 1978)	0.46
Thermal	
Coefficient of linear thermal expansion (α), (100°C–400°C) ($\times 10^{-6}$ K^{-1})	9.7 ± 0.25
Chemical	
Durability (solubility in acetic acid) ($\mu g/cm^2$)	20
Technical	
Sintering temperature (°C)	800

First, the dentist prepares the teeth and takes an impression. Next, the dental technician fabricates a wax model of the framework (Fig. 4-41). Two sprues are attached to the model. All edges of the wax pattern must be rounded. Sharp edges must be avoided. Specific wall thicknesses must be observed when designing the crown framework; crowns (abutments) must measure a minimum

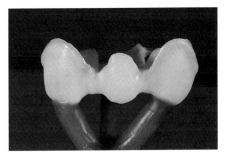

Figure 4-41 Wax object for the three-unit bridge.

thickness of 0.8 mm. An analysis of a load-bearing bridge conducted during the research and development work provided important information about the bridge design. The determination of the weak points of a loaded three-unit bridge is illustrated in Fig. 4-42 (Wintermantel 1998). Clearly, the con-

nectors play an important role, since load peaks occur in these areas when the restoration is loaded. As a result, Sorensen (1999) established that the cross section of the connector should measure at least 4 mm × 4 mm in the anterior region and 4 mm × 5 mm in the posterior region, because the masticatory force is higher (Fig. 4-43). This thickness can be achieved in the posterior region. In the anterior region, however, a special solution is needed. As the pressed glass-ceramic demonstrates a high degree of translucency and good chemical durability, the load-bearing part of the restoration (in this case the lingual region) can be fabricated with the pressed glass-ceramic exclusively, without having to apply the apatite glass-ceramic in a second step. Therefore, crowns can be

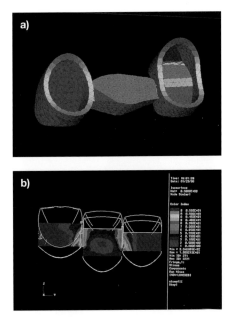

Figure 4-42 Finite element calculation of a loaded three-unit bridge. a) Scheme of a three-unit bridge consisting of 30,000 tetrahedra elements (PATRAN/P3). b) When the bridge is subjected to loading of 600 N, the maximum tensile stress in the connector area is approximately 110 MPa.

fabricated with lingual walls that are 1 mm thick and connectors that measure 4 mm × 4 mm. As a result, new ways of designing the material have become available. Furthermore, the high-strength core should be maximized in the design of the prosthesis (Fig. 4-44(a)).

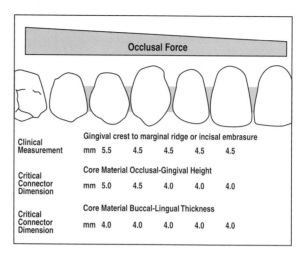

Posterior restorations can be designed in a similar way. For example, load-bearing

Figure 4-43 Minimum occlusal-gingival and buccal-lingual framework connector dimension as a function of position of bridge connector and occlusal forces (Sorensen 1999).

cusps in the lingual region can be created entirely out of core material without layering material (Fig. 4-44(b)). When using these special designs of framework fabrication, the walls must be prepared of uniform thickness. Therefore, the buccal side must not be too thin or the lingual side too thick.

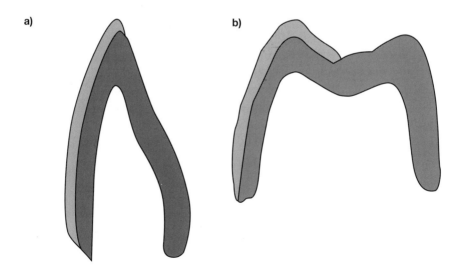

Figure 4-44 Special design of bridges in stress-bearing areas:
a) cross section of anterior tooth
b) cross section of posterior tooth.

It is important to note that the connectors should not be adjusted with rotary instruments to avoid introduction of flaws. Any grinding work should be carried out with water.

An apatite glass-ceramic is applied in layers to the three-unit bridge to achieve an appearance that matches natural teeth. In addition to the translucency, the brightness of the restoration's shade is enhanced. Very small apatite crystals are responsible for the unique way in which light is scattered in the material mimicking tooth structure. The microstructure of the framework and layering materials are shown in two cross sections of the bridge in Fig. 4-45. The microstructure of the pressed glass-ceramic is also shown in Fig. 3-21 and the microstructure of the sintered glass-ceramic in Fig. 4-45 (a). The adaptation of the interface between the two glass-ceramics is visible in Fig. 4-45 (b). Before applying the apatite glass-ceramic on the surface of the lithium disilicate glass-ceramic, the framework (minimum thickness of the framework: 0.8 mm) was cleaned by etching with an acid (see Section 4.4.2.5.C). The clean surface enhanced the contact between the two types of glass-ceramics.

The microstructure and composition of materials in the class of sintered glass-ceramics help produce optical properties,

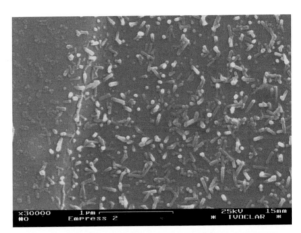

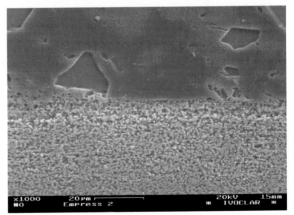

Figure 4-45 Microstructure of a three-unit bridge of IPS EMPRESS®2

a) sintered glass-ceramic

b) interface between pressed glass-ceramic and sintered glass-ceramic.

such as fluorescence and opalescence. Fluorescence makes the restoration glow like natural teeth when it is exposed to UV light, for example, at a discotheque. Opalescence makes natural teeth look red/yellow in transmitted light and blue/white in incident light.

C) Application

As shown in Sections 4.4.2.5.A and 4.4.2.5.B, lithium disilicate and apatite glass-ceramics can be combined to fabricate three-unit bridges. The apatite glass-ceramic is the layering material for the lithium disilicate glass-ceramic. The basic prerequisites for any technical process in which two different materials are fused with each other is a clean interface. Therefore, before applying the apatite glass-ceramic on the surface of the lithium disilicate glass-ceramic, the framework has to be cleaned by etching with a special acid (Invex, Ivoclar Vivadent AG). It is critical to remove all residual investment material with the liquid and blast the substructure surface with type 100 Al_2O_3. This procedure produces a very clean surface and enhances the contact between the two types of glass-ceramics. The strength of these ceramics allows fabrication of bridges for the anterior and posterior regions up to the second premolar. Because of their special design, frameworks made of the high-strength glass-ceramic can absorb load peaks. Sorensen (1999), Edelhoff et al. (1999), and Pospiech et al. (2000) conducted clinical tests on both crowns and bridges to determine clinical indications. The restorations that failed were shown to have inadequate framework dimensions resulting in catastrophic failure or defects in the sintered glass-ceramic causing minor chipping. Therefore, the sintered glass-ceramic must be properly sintered. Healing of surface flaws of the dental restoration is achieved by the application of a glaze.

Adhesive luting is recommended for the cementation of crowns and bridges on the tooth. The pressed glass-ceramic is etched with a hydrofluoric acid gel (approximately 4.5% HF) to achieve a micromechanical retentive surface microstructure as shown in Fig. 4-46. Following silanization with Monobond-S (Ivoclar Vivadent AG) and cementation with a resin composite (Variolink II, Ivoclar Vivadent AG, Liechtenstein), a high bond strength of 35 MPa is achieved. In clinical situations in which adhesive cementation is not possible, dentists can use a non-adhesive luting agent (ProTec CEM, Ivoclar Vivadent AG). The application of glass ionomer cements was tested with high success (Pospiech 2000). Figure 4-47 demonstrates that restorations made of IPS EMPRESS®2 appear natural. IPS EMPRESS®2 is produced by Ivoclar Vivadent AG (Liechtenstein).

4.4.2.6. IPS d.SIGN®
Glass-Ceramic

The materials system containing the apatite–leucite glass-ceramic called IPS d.SIGN® is described as a composite from a materials science point of view. This restorative dental product is produced by joining and combining three types of materials, that is, metal

Figure 4-46 Micromechanical retention of the pressed glass-ceramic IPS EMPRESS®2 created after etching in 4.5% HF gel for 20 sec.

(the framework material), a composite containing a high content of ZrO_2, called an opaquer, and various apatite–leucite glass-ceramics. The opaquer masks the metal, so that the framework is no longer visible. Apatite–leucite

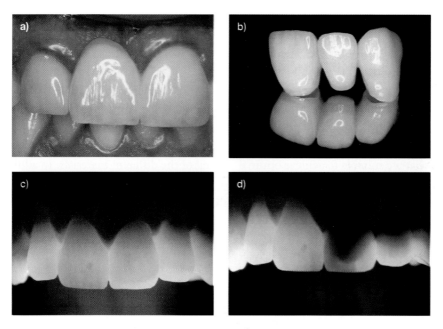

Figure 4-47 Postoperative view of IPS EMPRESS®2 restoration (three-unit bridge).
a) Clinical situation (courtesy Sorensen). b) Three-unit bridge on a mirror (courtesy Edelhoff).
c) and d) Dental bridges of IPS EMPRESS®2 (c) in comparison to a bridge of a metal framework (d) (courtesy Edelhoff).

glass-ceramics are applied to the opaquer in powder form. Because of this underlying layer, the biomaterial gives a natural appearance, which could not be achieved if the metal restoration were to remain unmasked.

A typical apatite–leucite glass-ceramic is presented in Section 2.4.6. This section shows that the characteristic part of phase formation in the glass-ceramic is the controlled crystallization of the leucite and apatite crystal phases. The formation of the crystals and the typical microstructure of the glass-ceramic with the needlelike apatite crystals are described in Sections 2.4.6 and 3.2.12.

Because of the different materials used in this system, the joining technology represents a very important part of the fabrication process. Metals comprise one of the three main groups of materials. A large variety of dental metal alloys (more than 600) is available throughout the world. These alloys are distributed by a variety of manufacturers and suppliers. As a result, the object in the development of IPS d.SIGN® was to produce a glass-ceramic that could be used with different types of metal alloys. The metals that can be used to fabricate frameworks for IPS d.SIGN® can be divided into five basic groups: 1) high-gold alloys, 2) low-gold alloys, 3) palladium-based alloys, 4) palladium–silver alloys, and 5) Co–Cr alloys. The fifth largest group of dental alloys comprises the nonprecious metal alloy. This alloy is very popular in Eastern Europe and Asia. The five most important alloys of types 1–5 are IPS d.SIGN® 98, 91, 84, 67, and 20 (Ivoclar Vivadent AG, Liechtenstein) with CTE of 13.9–14.6 $\times$ 10^{-6} K^{-1} (25°C–500°C). The biocompatibility of all metal alloys was tested and confirmed by specific methods. In the United States, for example, the tests are conducted according to Food and Drug Administration criteria.

Selected examples in Section 4.4.2.6.B are used to describe the processes that are suitable to achieve optimal joining of the different materials. In addition, measuring methods for evaluating the results were carried out. These glass–ceramic–metal composites for dental applications demonstrate that a multicomponent glass-ceramic and metals can be used for a variety of applications in medical and technical fields.

A) Properties

The IPS d.SIGN® glass-ceramics can be divided into three different materials according to their use in the restoration of different parts of the tooth:

- Sintered glass-ceramic to simulate dentin
- Material to simulate incisal area
- Special materials that produce specific optical effects (e.g., opalescence)

The most important properties of the dentin and incisal materials are shown in Table 4-19. The coefficient of linear thermal expansion plays an important role in the optimal joining of various types of apatite–leucite glass-ceramics and the ZrO_2-rich opaquer, which are applied to the different metals. Therefore, CTE of the opaquer has been included as a comparative value in Table 4-19. A comparison of CTE of glass-ceramics and of the opaquer with that of metals clearly shows that the application of the glass-ceramic to the metal framework systematically builds up compressive strain. As a result, the finished dental product demonstrates surface tension and a controlled increase in strength, ensuring retention on the substructure.

It must be noted that all the materials exhibit very good chemical durability. The weight loss of the glass-ceramics following immersion in 4% acetic acid (80°C, 16 h) was established according to ISO standards (see Table 4-19). *In vitro* wear tests have shown that this glass-ceramic is similar in wear to natural tooth structure (Sorensen et al., 1999). Section 4.4.2.6.C. describes the special optical characteristics, which can be achieved with additional assorted materials. Their use in the finished product is discussed.

B) Processing

The fabrication of dental restorations comprising a metal framework and various glass-ceramics is characterized by the fact that the glass-ceramics must be sintered on the framework in powder form. This process allows dental technicians to reproduce the natural shape and shade of the natural tooth as

Table 4-19

Physical Properties of IPS® d.SIGN, Incisal, and Dentin Material

Properties	Incisal and dentin material	Opaquer
Mechanical		
Flexural strength (MPa)	80	
Optical		
Translucency	See Table 4-20	
Thermal		
Coefficient of linear thermal expansion, (α), (25°C–500°C, $\times 10^{-6}$ K^{-1})		
after two heat treatments	12.0 ± 0.5	13.6 ± 0.5
after four heat treatments	12.6 ± 0.5	13.8 ± 0.5
Chemical		
Durability (solubility in acetic acid) (μg/cm^2)	<50	<100

accurately as possible. The first step in the dental laboratory following the fabrication of the model and the casting of the metal alloy is the conditioning of the metal framework. This step must be conducted according to the instructions of the alloy manufacturer. In general, the surface resulting from firing under oxidation is coated by a layer of opaquer. High-gold alloys must be conditioned with acid. Subsequently, a thin initial layer of opaquer (approximately 10–30 µm) is applied. This step, called wash firing in technical terms, optimally seals the metal oxide surface and ensures the best conditions for further sintering processes. In the subsequent working step, the opaquer layer is applied at 900°C with a holding time of 1 min. Figure 4-48 is an SEM image of the opaquer layer.

The dentin and incisal glass-ceramic materials are sintered at 860°C. Figures 4-49 and 4-50 reveal the microstructures of the glass-ceramics. The SEM micrograph in Fig. 4-51 shows the total thickness of the glass-ceramics compared with that of the opaquer and metal. The microstructure of the dentin and incisal glass-ceramics is characterized by the existence of two types of crystals: leucite and apatite. Figure 4-49 shows the microstructure of the dentin glass-ceramic containing needlelike apatite and leucite. Moreover, two forms of needlelike apatite are clearly

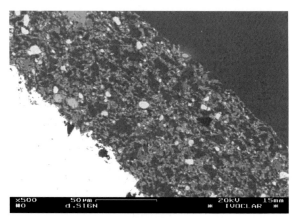

Figure 4-48 SEM image showing the microstructure of the opaquer layer of IPS® d.SIGN. Etched in 3% HF for 10 sec.

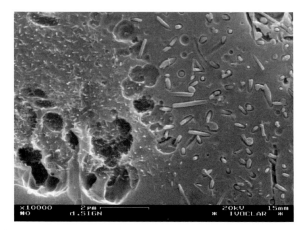

Figure 4-49 SEM image showing the microstructure of apatite–leucite glass-ceramic (dentin). Etched in 3% HF for 10 sec.

discernible (in a different morphology): large and small needlelike apatite. The incisal material (Fig. 4-50) contains only fine needlelike apatite, which enhances brightness (value in technical terms).

Some dental technicians choose to apply an additional veneer to the shoulder region of dental crowns and bridges. As a result, metal is no longer visible in the gingival region. Thus the metal is optimally masked. This technique requires that the shoulder material from the additional assortment of glass-ceramics be applied prior to the sintering of dentin and incisal materials.

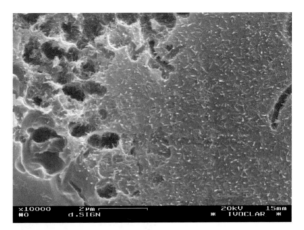

Figure 4-50 SEM image showing the microstructure of apatite–leucite glass-ceramic (incisal). Etched in 3% HF for 10 sec.

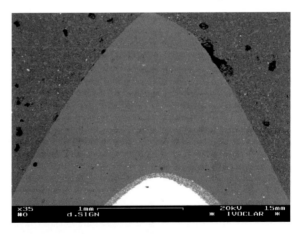

Figure 4-51 SEM image showing the microstructure of dental crown (cross section). Etched in 3% HF for 10 sec.

C) Application

In terms of quantity, the incisal and dentin materials comprise the largest part of the apatite–leucite glass-ceramics of the IPS d.SIGN® system. Their microstructure and properties are described in Section 2.4.6. The main applications of these glass-ceramics include dental crowns and multi-unit bridges.

Apart from the dentin and incisal materials, the IPS d.SIGN® system includes further glass-ceramics of the apatite–leucite type as well as special glasses. The additional assortment of materials comprises base-kit, deep-dentin-kit, margin kit, impulse kit (with Effect and Transpa), gingiva kit, essence kit, stains kit (with glaze), and add-on materials. These system components help dental technicians create special optical effects that make the artificial tooth look like its natural counterpart. The Transpa and Effect materials, as well as the staining and glazing materials, are particularly effective. The Transpa material heightens the transmission properties of the dental product. These materials are either applied with the incisal material or layered on the incisal glass-ceramic. The Effect materials of IPS d.SIGN® are opal glass-ceramics that produce an opalescent appearance (see Section 2.4.6). These materials are available in different degrees of opalescence for the incisal region. Figure 4-52 shows several typical applications of opal materials in dentistry.

Staining materials are used to characterize the restoration. Margin materials are used to mask the metal framework in the gingival region. Glazing materials are highly transparent. They are sin-

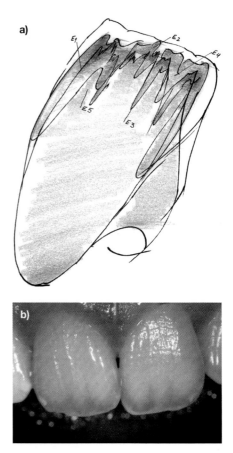

Figure 4-52 Application of opal glass and opal glass-ceramic (E1–E5) in restorative dentistry (dental crown). a) scheme, b) clinical situation.

tered on the restoration by the dental technician in a final firing process at 830°C. Glaze firing produces a smooth, glossy surface. In addition to enhancing the appearance of the glass-ceramic, the glazed surface also keeps plaque accumulation to a minimum.

As modifications of the dental restorations may be necessary once the restorations have been completed, the assortment collage includes a low-sintering temperature add-on material, which can be sintered on the glass-ceramic at 750°C. Because of the low sintering temperature, the material exhibits a T_g of 460°C and a coefficient of linear thermal expansion of 11.8 × 10^{-6} K^{-1} (100°C–400°C). Moreover, the material demonstrates a high degree of translucency.

Table 4-20 provides a clear comparison of the optical properties of the dentin and incisal materials with those of the materials in the additional assortment. The table includes one representative each from the different

Table 4-20

Absorption and Transmission of IPS® d.SIGN Products Thickness of the Samples: 1.0 mm		
Product	**Absorption (%)**	**Transmission (%)**
Transpa	10	90
Incisal	20	80
Dentin	40	60
Margin	45	55
Deep dentin	60	40
Brilliant dentin	90	10

ranges of materials together with their absorption and transmission values. This table shows that the Transpa materials demonstrate the highest, and the Brilliant dentin materials the lowest transmission value. Both are bright and have wear characteristics of natural teeth. Figure 4-53 shows a dental restoration that is composed of a metal framework, which has been veneered with IPS d.SIGN® glass-ceramics. The development of different shades was carried out by Kerschbaumer and Foser (Ivoclar Vivadent AG, Liechtenstein) and Winter and Cornell (USA). Clinical tests were successfully applied by Winter, and Cornell (USA), as well as Reitemeier and Walter (Germany). IPS d.SIGN® is manufactured by Ivoclar Vivadent AG (Liechtenstein).

4.4.2.7 ProCAD®

The ProCAD® (professional computer assisted design) system is directed to the preparation of dental restoration by machinability using the CEREC® 2 or CEREC® 3 apparatus. The ProCAD® system comprises ProCAD® glass-ceramic blocks; shade/stains kit; the furnaces of Programat type, e.g., P100 or Programat X1 (Ivoclar Vivadent AG); and a Cem kit.

A) Properties

ProCAD® is a leucite-reinforced glass-ceramic derived from the system $SiO_2–Al_2O_3–K_2O–Na_2O$. Its chemical composition and microstructure are comparable to that reported in Section 2.2.9.

ProCAD® glass-ceramic is characterized by a three-point flexural

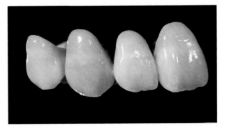

Figure 4-53 Final dental restoration of different types of leucite–apatite glass-ceramics.

strength of 135–160 MPa. The fracture toughness as K_{IC} was determined as 1.3 MPa. After machining, the surface of the glass-ceramic can be improved by polishing or applying a specially developed ProCAD® glaze. The application of the glaze improves the flexural strength of the material to 180–240 MPa.

The chemical durability of the glass-ceramic is very good and measured 49 µg/cm² according to ISO dental standard 6872. The chemical durability of the glaze was characterized as 36 µg/cm². It is possible to etch the glass-ceramic by applying the HF-etch gel and to form a retentive layer for a preferable adhesive cementation of the dental restoration. ProCAD® is produced in different degrees of translucency and color for high aesthetic applications.

B) Processing

The processing of the glass-ceramic dental products is carried out by machines of the CEREC® 2 or CEREC® 3 system. The CEREC® system is a CAD process reported by Mörmann and Krejci (1992) and Pfeiffer (1996). This system allows the dentist to prepare a dental restoration within approximately 15 minutes. Thus a chair-side preparation of a dental restoration is possible. The CAD processing of ProCAD® dental restorations using the CEREC® 3 system is reported by Mörmann and Bindl (2000).

C) Application

The main indication of glass-ceramic restorations by the CEREC system was the preparation of inlays and veneers; crowns also were prepared. The ProCAD® glass-ceramic is characterized by an improved mechanical strength and allows the application for dental posterior and anterior crowns, inlays, and onlays. ProCAD® is produced by Ivoclar Vivadent AG (Liechtenstein).

4.4.2.8 IPS ERIS for E2

The sintered glass-ceramic IPS ERIS for E2 was specially developed for the veneering of a lithium disilicate core in dental restorative procedures. IPS ERIS for E2 consists of glasses derived from the SiO_2–Li_2O–K_2O–ZnO system and glass-ceramics of the SiO_2–Li_2O–K_2O–ZnO–CaO–P_2O_5–F system. Needlelike fluoroapatite represents the main crystal phase of the glass-ceramic. By means of a precipitation of fluoroapatite crystals, optical properties similar to that of the natural tooth could be achieved. IPS ERIS for E2 veneering material is characterized by a biaxial flexural strength of 106 ± 17 MPa (requirement according to dental ISO standard 6872: >50 MPa), chemical durability in 4% acetic acid/80°C: weight loss of 25 µg/cm² (requirement according to ISO 6872: <100 µg/cm²), coefficient of thermal expansion (100°C–400°C) : 9.75 ± 0.25 × 10⁻⁶ K⁻¹ and Vickers hardness HV5 of 557 ± 14 kg·mm⁻².

IPS ERIS for E2 can be very homogeneously sintered in a conventional dental furnace at approximately 750°C. It shows a low residual porosity after firing. More than 99% of the theoretical density can be achieved. The low porosity content has considerable advantages with regard to the mechanical properties of the sintered product. IPS ERIS for E2 was developed by Ivoclar Vivadent AG (Liechtenstein).

4.5 ELECTRICAL AND ELECTRONIC APPLICATIONS

4.5.1. Insulators

Mica-type glass-ceramics are very good insulators. MACOR® is a preferred glass-ceramic for this application. Its properties and microstructure were reported in Sections 2.3.1 and 3.2.6.

Glass-ceramics to which other materials have been added and that are referred to as composites in this section for reasons of simplicity have been developed by the Nippon Electric Glass Co., Ltd., Japan, for the electronics industry. These glass-ceramics are produced in thin sheets and are suitable for low-temperature firing multilayer substrates. Furthermore, they exhibit high dielectric constants. This glass-ceramic composite material with the product name MLS-1000 is composed of a glass-ceramic that is derived from the SiO_2–Al_2O_3–PbO base system. As shown in Section 2.7.1, glass-ceramics with high dielectric constants have been produced in this glass-ceramic system. In addition to this type of glass-ceramic, the composite material also contains a sintered ceramic in the form of Al_2O_3. Selected properties of this composite are shown in Table 4-21 (Electric Glass Materials 1996).

Another glass-ceramic composite with a higher dielectric constant than that of MLS-1000 has been achieved with a product called MLS-40. The composite contains a glass-ceramic derived from the SiO_2–TiO_2–Nd_2O_5 system as well as a sintered ceramic. The properties of the product (composite) are also shown in Table 4-21.

The procedure described in this paragraph is recommended when the glass-ceramic is used as a substrate for electrical conductors. The material in the form of green sheets contains a binder resin (butylmethacrylate), a solvent (toluene), and a plasticizer (dibutylphthalate). The 0.1–0.3-mm thick glass-ceramic composite sheets are produced according to the doctor blading procedure. In the electronics industry, these sheets are cut to the desired size and circuit patterns are screen printed on them. Subsequently, these products are subjected to heat treatment at 400°C to burn out the organic components. Furthermore, they are densely sintered at 900°C. The end product is a glass-ceramic composite substrate on which an electrical conductor has been applied.

Table 4-21

Properties of Glass-Ceramic Composites Used as Multilayer Substrates (Nippon Electric Glass Co., Japan)

Properties	Glass-Ceramics	
	MLS-1000	MLS-40
Mechanical		
Density (g/cm^3)	3.39	4.33
Bending strength (MPa)	280	230
Thermal		
Coefficient of thermal expansion ($\times$ 10^{-6} K^{-1}) (~30°C–380°C)	6.0	9.9
Coefficient of thermal conductivity (W·m^{-1}·K^{-1})	3.1	1.7
Firing temperature (°C)	900	900
Transformation point (°C)	565	685
Electrical		
Dielectric constant (25°C)		
1 MHz	7.8	28.0
2.4 gHz	8.0	30.0
tan δ ($\times$ 10^{-4}) (25°C)		
1 MHz	16	30
2.4 gHz	20	30
Volume resistivity (150°C)		
log ρ (Ω·cm)	13.5	–
Characteristics		
Particle size (µm)		
D_{50}	1.8	2.2
D_{max}	15	15

4.5.2 Electronic Packaging

4.5.2.1 Requirements for Their Development

Until the 1970s, Al_2O_3 sintered ceramics represented the most popular ceramic substrate for encapsulating strip conductors. The application of Al_2O_3 as the substrate necessitated sintering temperatures of 1500°C for cofiring. The conducting metal was preferably molybdenum. Significant developments in the electronics industry, however, demanded new substrate materials.

As early as in the1970s, there was a demand for lowering the cofiring temperature and using metals demonstrating increased conductivity. The objective of heightening the performance was paramount in the development of high-capacity computers. On the basis of these requirements, new applications for glass-ceramics were discovered.

4.5.2.2 Properties and Processing

Beall (1993) and MacDowell and Beall (1990) showed that the most important advantage of glass-ceramic substrates is their low dielectric constant (K) of 4–6 as compared to 9 for alumina. Glass-ceramics can be cofired below 1000°C and metallized with copper, silver, or gold. Alumina must be cofired at above 1500°C using molybdenum or tungsten. Cordierite glass-ceramics fulfil these advantages (MacDowell and Beall, 1990). The thermal expansion behavior of cordierite glass-ceramics can best be controlled by tailoring the bulk composition to produce a secondary phase. Additives to the stoichiometric composition of cordierite allowed the glass powder compact to sinter to high density before the onset of cordierite crystallization. P_2O_5 and B_2O_3 alone or in combination with lead–zinc or alkali-earth oxides have been used for this purpose.

Tummala (1991) and Knickerbocker (1992) reported that cordierite glass-ceramics are suitable for replacing Al_2O_3 as the substrate. Tummala (1991) describes the formation of cordierite glass-ceramics with the composition of 50–55 wt% SiO_2, 18–23 wt% Al_2O_3, 18–25 wt% MgO, 0–3 wt% P_2O_5, and 0–3 wt% B_2O_3. According to his findings, nucleation and binder burn-off take place during an initial reaction at temperatures up to 800°C. Subsequently, sintering and complete densification take place at temperatures up to 900°C. Crystallization of α-cordierite and clino-enstatite proceeds at temperatures above 900°C. It is interesting to note that optimum processing properties (sintering) and coefficients of linear thermal expansion can be achieved by adding B_2O_3 and P_2O_5. Table 4-22 shows the properties of the glass-ceramic/copper substrate produced according to this general process.

Similarly, Pannhorst (1995) describes a special composite material that is suitable as an alternative to Al_2O_3. This material is produced by controlling the surface crystallization of a cordierite glass of an almost stoichiometric composition. In the sintering process, powdered ZrO_2 is added to the base glass to produce a composite made of glass-ceramic and ZrO_2 sintered ceramic. The controlled surface crystallization of cordierite glass is described in Section 2.2.5.

The fact that this composite material can be cofired at temperatures of approximately 950°C represents a favorable characteristic. Hence, highly conductive metals such as copper and silver can be used to increase the conductivity compared with molybdenum.

The composite material also demonstrates an additional improvement with regard to its properties. Its dielectric constant has been decreased to 5. This dielectric constant is favorably lower than the 9.4 demonstrated by Al_2O_3.

Based on the developments by Budd (1993) and Partridge et al. (1989), cordierite glass-ceramics have been produced as advanced microwave

Table 4-22

Properties of Cordierite Glass-Ceramic/Copper Substrates According to Tummala (1991) for IBM System 390/ES9000 in Comparison to Alumina-Molybdenum

Properties	IBM System 390/ES9000 with glass-ceramic/copper	IBM System 3090 with alumina-molybdenum
Mechanical		
Shrinkage control (%)	± 0.1	± 0.15
Thermal		
Coefficient of thermal expansion (RT to 200°C) ($\times 10^{-6}$ K^{-1})	3.0	6.0
Electrical		
Dielectric constant	5.0	9.4
Volume resistivity ($\mu\Omega$·cm)	3.5	11
Substrate characteristics		
Size (mm)	127.5 × 127.5	110.5 × 117.5
Layers	63	45
Number of vias (total)	2×10^6	4.7×10^5
Wiring density (cm/cm^2)	844	450
Line width (mm)	75	100
Via diameter (mm)	90 and 100	125

substrates. The glass-ceramics have been produced either as a bulk or powdered material. Their properties are shown in Table 4-23.

4.5.2.3 Applications

The composite material made from glass-ceramic and ZrO_2, according to Pannhorst (1995), is manufactured by Schott, Germany. Since 1991, it has been used for multilayer chip carriers in the 390 mainframe computers from IBM. MEMIX® cordierite glass-ceramics are produced by GEC Alsthom, UK. The material is used for microwave- and millimeter-wave integrated circuits (MICs), amplifiers, and range-finding and communication equipment.

In addition to cordierite-based glass-ceramics, GEC Alsthom also manufactures glass-ceramics demonstrating a "stabilized" dielectric constant and low electrical loss. Apart from SiO_2–Al_2O_3–MgO glass-ceramics, SiO_2–Al_2O_3–ZnO glass-ceramics are also suitable for this purpose. Donald (1993) has reported that the main crystal phases of these glass-ceramics may contain willemite, $2ZnO·SiO_2$, β-quartz solid solution ($nSiO_2·ZnAl_2O_4$), or in the case of glass-ceramics with high coefficients of thermal expansion,

Table 4-23

Properties of Cordierite Glass-Ceramics (GEC Alsthom, UK)

Properties	Bulk Material Mexim™ (3404, 3447, 3449)	Powder Materials Mexim™ (4041, 4060, 4070)
Mechanical		
Flexural strength (MPa)	200–348	148–174
Thermal		
Coefficient of thermal expansion (20°C–200°C) ($\times 10^{-6}$ K^{-1})	2.6–5.2	3.9–4.6
Electrical		
Dielectric constant (9.4 GHz; 20°C) (9.4 GHz; 400°C)	5.49–5.72 5.47–5.81	4.32–4.60
Loss tangent ($\times 10^{-4}$) (9.4 GHz; 20°C) (9.4 GHz; 400°C)	2–4 17–29	1–3

α-quartz, tridymite, and Li_2ZnSiO_4. Glass-ceramics with thermal expansion coefficients that are suitable for fusing the material to molybdenum, copper, and some iron- and nickel-based alloys are already commercially available. These materials are manufactured by GEC Alsthom. Apart from the development of cordierite glass-ceramics, glass-ceramics of the SiO_2–Al_2O_3–MgO system containing a BPO_4 main crystal phase were fabricated for microelectronic packaging. These glass-ceramics are characterized by a favorable dielectric constant of 3.8–4.5 and a coefficient of thermal expansion of 40 $\times$ 10^{-7} K^{-1}. Compared with cordierite glass-ceramics, BPO_4 glass-ceramics exhibit a lower modulus of rupture. Thus, it is possible to produce dense multilayer substrates using a tape-casting process (Beall 1993).

A glass-ceramic from the P_2O_5–B_2O_3–SiO_2 system has also been developed for microelectronic packaging. The controlled reaction described in Section 2.6.4 produces a special type of glass-ceramic known as a "gas-ceramic." This particular glass-ceramic exhibits a very low density as well as other favorable properties compared with other materials. Its properties are shown in Table 4-24.

4.6 ARCHITECTURAL APPLICATIONS

The most significant glass-ceramic for building applications is manufactured by Nippon Electric Glass under the brand name Neoparies™ (Neoparies 1995). This glass-ceramic is produced on a large scale as a building material

Table 4-24

Properties of Glass-Ceramics from the P_2O_5–B_2O_3–SiO_2 System in the Form of a "Gas-Ceramic" (Hydrogen Microfoams), Corning Inc., USA

Properties	
Mechanical	
Low density	as low as 0.5 g/cm^3
High strength/weight ratio	2 to 5 Ksi (abraded) on <1 g/cm^3 foams 136–345 MPa
Small bubble size	1–100 µm
Smooth, durable glassy skin	10 µm–1 mm thick
Machinable/field workable	can be sawed, scored, drilled
Thermal	
Low to moderate thermal expansion	0–70 × 10^{-7} K^{-1}
Low thermal conductivity	heat retention/insulation value from closed cell porosity
Electrical	
Low dielectric constant	as low as 2 (and potentially lower)
High d.c. resistivity (at 250°C)	10^{16} (Ω·cm)
Low dielectric loss	<.001 in non-alkali compositions

for interior and exterior walls in particular. Large flat or curved sheets of the material are produced for facing buildings (Fig. 4-54). The Neoparies glass-ceramic, for example, is produced in panels measuring up to 900 × 1200 mm, with a minimum width of 50 mm being possible. Furthermore, glass-ceramics for exterior applications are produced in different colors and textures. The product catalog of Nippon Electric Glass (Neoparies 1995) contains a total of 24 different color and texture samples.

The glass-ceramic is produced in the following manner: A base glass (see Section 2.2.6) is produced in the form of granular glass particles. In a fully automatic process, the granular glass is formed into flat or curved panels in a tunnel furnace at temperatures up to approximately 1100°C. During the heat treatment, sintering of the granular glass takes place at 850°C. The crystallization of wollastonite begins at temperatures above 950°C.

The properties of Neoparies, which are particularly important for building applications, were shown in Section 2.2.6. These main properties can be summarized as follows:

- high resistance to weathering
- zero water absorption rate
- harder than natural stone

Figure 4-54 Neoparies™ glass-ceramic as a building material.

- 30% lighter in weight than natural stone building materials
- easy re-forming of curved panels
- special texture of the glass-ceramic

Furthermore, this glass-ceramic is easy to shape. The panels made of this material acquire their curved shape directly in the sintering process. As a result, they do not require mechanical post processing as is the case for panels made of natural stone.

Asahi Glass Co., Ltd., Japan produces another wollastonite glass-ceramic (Cryston®) for architectural application as cladding material (Hatta and Kamei, 1987). The crystal size of wollastonite measures 0.5–2 mm. The thermal and mechanical properties are shown in Table 4-25. The chemical durability of Cryston® glass ceramic is also very good. Therefore it is widely used as cladding materials for architectural application. Asahi Glass Co., Ltd., Japan, also produces opal glasses/opal glass-ceramics as walling materials for architectural application. The material contains NaF and CaF_2 as crystal phases. The crystallization was carried out in glasses of the composition 71–77 wt% SiO_2, 1.9–4.9 wt% Al_2O_3, 3–5.4 wt% CaO, 0.1 wt% MgO, 4.0 wt% F, 0–0.2 wt% Sb_2O_3, 11.8–15.2 wt% Na_2O, and 0.7–3.8 wt% K_2O. The chemical properties are nearly the same as float glass.

Table 4-25

Properties of Cryston® Glass-Ceramic

Properties	
Mechanical	
Density (g/cm³)	2.76
Young's modulus (GPa)	97
Flexural	
Strength (MPa)	50
Compressive strength (MPa)	450
Fracture toughness (MPa·m$^{0.5}$)	1.8
Hardness (Vickers)	6000
Thermal	
Coefficient of thermal expansion (~30°C–380°C, ($\times 10^{-6}$ K^{-1}))	8
Coefficient of thermal conductivity at 20°C (W·m^{-1}·K^{-1})	0.9
Coefficient of specific heat at 50°C (J·g^{-1}·K^{-1})	0.8

Therefore, the application as a textureless opal material in architecture is used (Hatta et al., 1986).

Eurokera (Corning Inc./St. Gobain, USA/France) produces transparent Eclair® glass-ceramics for architectural application. The Eclair® glass-ceramic is similar to Keraglas® (Section 4.2), but much lower in FeO content (much less than 200 ppm).

In addition to the development of the Neoparies™, Cryston®, and Eclair® glass-ceramics, the 1970s and 1980s also saw the manufacturing of slag sital for building applications in the Ukraine and Hungary. Panels for facing large building surfaces were produced of this glass-ceramic (see Section 2.2.8).

In Japan, transparent β-quartz glass-ceramics are used for architectural purposes. The transparent glass-ceramic known as Firelite™ is used to make fire-resistant windows. The material is used to fabricate large scale windows in high-tech buildings, windows in department stores and other public buildings, as well as doors and windows in banks and even kindergarten facilities. The safety glass demonstrates almost zero thermal expansion and a temperature resistance up to approximately 800°C. The architectural application of this type of window is shown in Fig. 4-55. The production methods and properties of the glass-ceramic are discussed in Section 4.2. The glass-ceramic is manufactured and distributed by Nippon Electric Glass Co., Ltd., Japan.

Figure 4-55 Fire resistant glass-ceramic Firelite™.

4.7 COATINGS AND SOLDERS

In today's microelectronic industry, the technology that is used to produce thin or thick films of materials on special substrates plays an increasingly important role. A speciality of thin-film production was discussed in Sections 2.7.1 and 2.7.3. These sections demonstrated that thin films with high dielectric constants could be produced. Furthermore, sol–gel processing was shown to allow economical manufacturing.

Boron phosphate materials demonstrate the best dielectric properties observed in glass-ceramics to date. The glass-ceramic can be produced as multi-layer glass-ceramic sheets. This type of glass-ceramic is discussed in Section 2.6.4. A BPO_4 glass-ceramic from the P_2O_5–B_2O_3–SiO_2 system (900°C/2 h) is characterized by dc resistivity of 10^{16} $\Omega \cdot$cm at 250°C. This parameter is higher than that of a sintered ceramic consisting of 99.9% Al_2O_3. Above 1 kHz and up to 200°C, loss tangents are generally below 10^{-3} (MacDowell and Beall, 1990). The dielectric constants of these glass-ceramics are between 3.8 and 4.5, depending on the SiO_2 content. Processing of BPO_4 glass-ceramics involves internal nucleation of the bulk glass or surface crystallization of glass powder compacts. These powders can be shaped via tape casting to form multilayer glass-ceramic sheets. This processing method, which is in widespread use in the microelectronics industry, combined with the aforementioned properties and a favorable coefficient of thermal expansion (4.0×10^{-6} K^{-1}) make this a viable material for microelectronic applications. The properties are well matched with those of silicon, but the modulus of rupture is over 15,000 psi (103.5 MPa) less than that of Al_2O_3. Metallization using gold, silver, or copper appears feasible. The glass-ceramic was developed by Corning, Inc.

Mica glass-ceramic films can be produced as fluorhectorite, $LiMg_2LiSi_4O_{10}F_2$ by sol–gel processing (Beall et al., 1980, 1981). Also the fluorhectorite glass-ceramic spontaneously gelatinized on contact with water forming fine sol–gel dispersion. This itself could be dried to a film, but this film was readily attacked by water to reform the gel. To make a stable film, the fluorhectorite gel was pumped through a slot into KCl solution, and the K$^+$ for Li$^+$ ion exchange in the interlayer produced a stable film of $KLiMg_2Si_4O_{10}F_2$ (fluortaeniolite). This film could be used directly or beat as a pulp to make paper by the Fourdrinier process. The paper could then be used with epoxy to make a good dielectric circuit board (MacDowell and Beall 1990).

Glass-ceramics seals for the microelectronics industry were developed from the P_2O_5–Na_2O–BaO system (Section 2.6.2) and from the SiO_2–Li_2O system (Section 2.1.1). The phosphate base glass-ceramics are preferably used for hermetic seal applications in conjunction with aluminum or copper. High-strength seals have been developed as lithium disilicate glass-ceramics. This glass-ceramic with the special composition of

67.1 mol% SiO_2, 2.6 mol% B_2O_3, 1.0 mol% P_2O_5, 23.7 mol% Li_2O, 2.8 mol% K_2O, 2.8 mol% Al_2O_3

is produced as "S-glass" by Sandia National Laboratories, USA (Brow et al., 1995). The glass-ceramic fulfills the requirement of withstanding high pressure upward of 90,000 psi (621 MPa). The base glass of the glass-ceramic yielded a glass-ceramic with a coefficient of thermal expansion between 11 and 13×10^{-6} K^{-1} after heat treatment. It was also possible to yield a glass-ceramic (Headley and Loehman, 1984) with 14–15×10^{-6} K^{-1} to match the Ni-based super alloy (Inconel), using a special heat treatment method for the crystallization of additional cristobalite. The sealing yielded an interface between the glass-ceramic and Inconel of 20 μm with the phosphide $C_{17}P_7$. The "S-glass" is used for industrial applications such as high-voltage connectors, high-pressure actuators and detonators, and components demonstrating complicated geometries.

Future Directions

S ince the discovery of glass-ceramics in the 1950s, the major applications have been in fields where thermomechanical properties (strength, low CTE, thermal stability) are most critical. These include missile nose cones (radomes), the first application; then cookware, tableware, stovetops, and electronic packaging. Each of these applications also required secondary properties of considerable diversity. For example, radomes must be transparent to microwaves, kitchenware must be chemically durable, stovetops must be transparent to near-infrared radiation, and electronic packaging materials must have low dielectric constants and losses.

In recent years, there has been an increasing interest in glass-ceramic applications where optical properties are key. A parallel but unrelated trend involves the use of glass-ceramics as dental and surgical prostheses. In the optical area, the most significant properties are luminescence in the near infrared range in combination with excellent transparency. Efficient broadband luminescence in crystallites is the basis of applications such as tunable lasers and optical amplifiers, both of which can be made in both bulk and fiber form as glass-ceramics.

Dental biomaterials and surgical implants require different properties: aesthetic appearance, good durability, and good mechanical properties at ambient temperatures are critical to the former, while biocompatibility and flexural strength are essential to the latter. Dental biomaterials are continually developed to satisfy the demands of patients, dentists, and dental technicians.

We foresee a dramatic increase in technical interest and applications in both optical and biological areas over the next few decades. There will also be continual application of glass-ceramics in traditional areas, although with less growth. Then, there are always unexpected applications, which may surface and require an entirely new combination of material properties. In any event, the wide range of potential properties combined with the flexibility of high-speed hot glass forming and the intricacy of shape associated with powder and extrusion processing will ensure continued growth of glass-ceramic technology.

Appendix 1

Twenty figures of crystal structures, calculated with atoms V 4.0 for Windows (Dowty 1997).

Crystal Structures

	Name and Formula	Structure Type	Reference
1	α-Quartz (SiO_2)	Framework Silicate	Levien et al., (1980)
2	β-Quartz (SiO_2)	Framework Silicate	Wright and Lehmann, (1981)
3	Cristobalite (SiO_2)	Framework Silicate	Dowty (1999a)
4	Tridymite (SiO_2)	Framework Silicate	Kihara (1977)
5	β-Eucryptite ($LiAlSiO_4$)	Framework Silicate	Guth and Heger, (1979)
6	β-Spodumene ($LiAlSi_2O_6$-SiO_2)	Framework Silicate	Li and Peacor, (1968)
7	Enstatite ($MgSiO_3$)	Chain Silicate	Ghose et al., (1969)
8	Wollastonite ($CaSiO_3$)	Chain Silicate	Ohashiy and Finger, (1978)
9	Diopside ($CaMgSi_2O_6$)	Chain Silicate	Clark et al., (1969)
10	K-fluorrichterite ($KNaCaMg_5Si_8O_{22}F_2$)	Chain Silicate	Cameron et al., (1983)
11	Cordierite ($Mg_2Al_4Si_5O_{18}$)	Ring Silicate	Predecks et al., (1987)
12	Lithium Disilicate ($Li_2Si_2O_5$)	Layer Silicate	Liebau (1961)
13	Fluorophlogopite ($KMg_3AlSi_3O_{10}F_2$)	Layer Silicate	Mc Cauley et al., (1973)
14	Leucite ($KAlSi_2O_6$)	Framework Silicate	Mazzi et al. (1976)
15	Nepheline ($KNa_3\,[AlSiO_4]_4$)	Framework Silicate	Simmons and Peacor, (1972)
16	Mullite ($3Al_2O_3 \cdot 2SiO_2$)	Chain Aluminosilicate	Sadanaga et al., (1962)
17	Spinel ($MgAl_2O_4$)	Oxide	Dowty (1999 b)
18	Rutile (TiO_2)	Oxide	Dowty (1999 c)
19	Fluoroapatite ($Ca_{10}(PO_4)_6F_2$)	Phosphate	Sänger and Kuhs, (1992)
20	Monazite ($CePO_4$) for ($LaPO_4$)	Phosphate	Ueda (1953)

Remark:

$[SiO_4]$-tetrahedra and $[AlO_4]$-tetrahedra show the same color in structures of leucite and phlogopite.

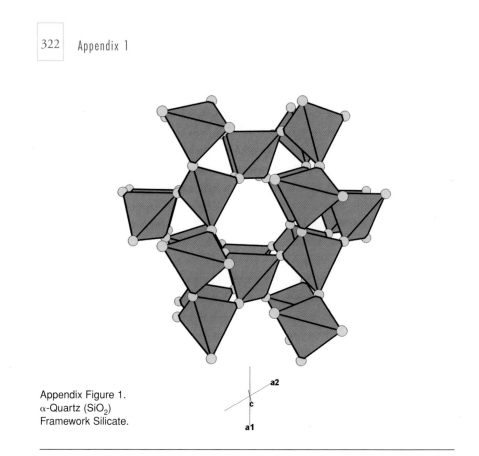

Appendix Figure 1.
α-Quartz (SiO$_2$)
Framework Silicate.

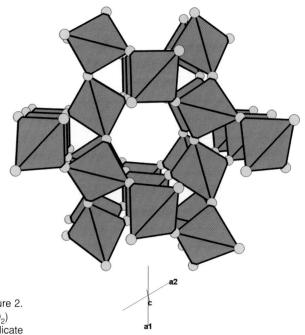

Appendix Figure 2.
β-Quartz (SiO$_2$)
Framework Silicate

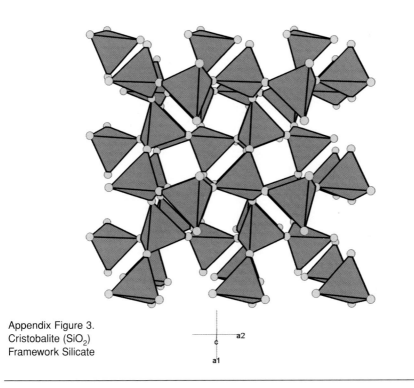

Appendix Figure 3.
Cristobalite (SiO$_2$)
Framework Silicate

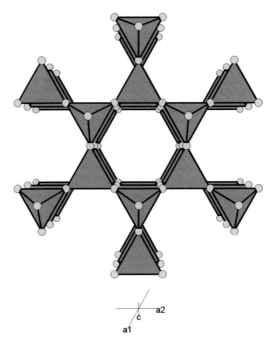

Appendix Figure 4.
Tridymite (SiO$_2$)
Framework Silicate

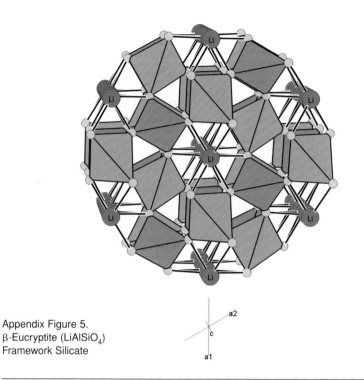

Appendix Figure 5.
β-Eucryptite (LiAlSiO$_4$)
Framework Silicate

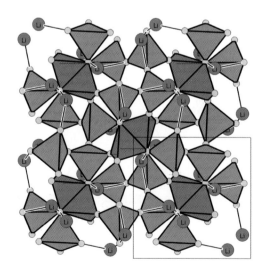

Appendix Figure 6.
β-Spodumene (LiAlSi$_2$O$_6$-SiO$_2$)
Framework Silicate

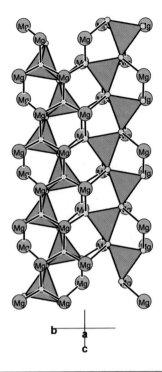

Appendix Figure 7.
Enstatite (MgSiO₃)
Chain Silicate

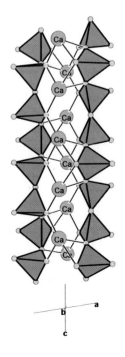

Appendix Figure 8.
Wollastonite (CaSiO₃)
Chain Silicate

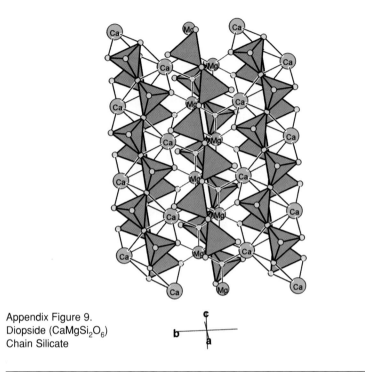

Appendix Figure 9.
Diopside (CaMgSi$_2$O$_6$)
Chain Silicate

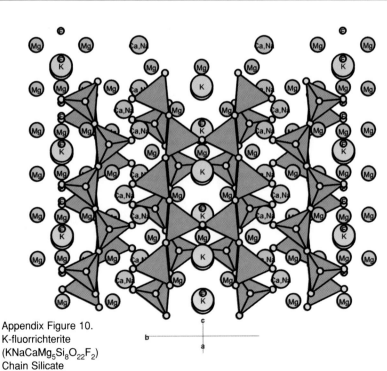

Appendix Figure 10.
K-fluorrichterite
(KNaCaMg$_5$Si$_8$O$_{22}$F$_2$)
Chain Silicate

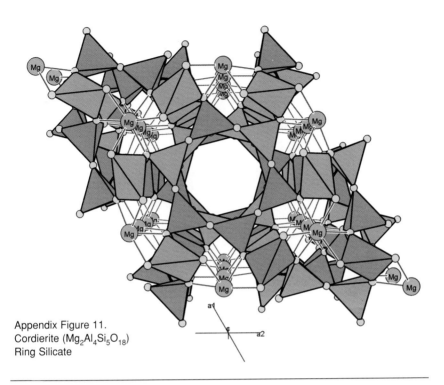

Appendix Figure 11.
Cordierite (Mg$_2$Al$_4$Si$_5$O$_{18}$)
Ring Silicate

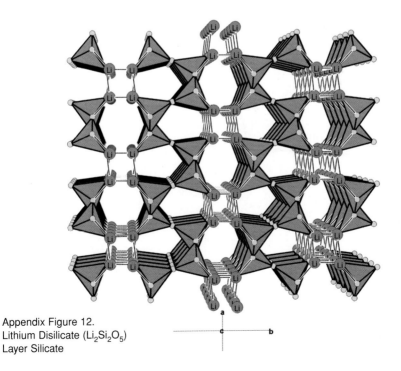

Appendix Figure 12.
Lithium Disilicate (Li$_2$Si$_2$O$_5$)
Layer Silicate

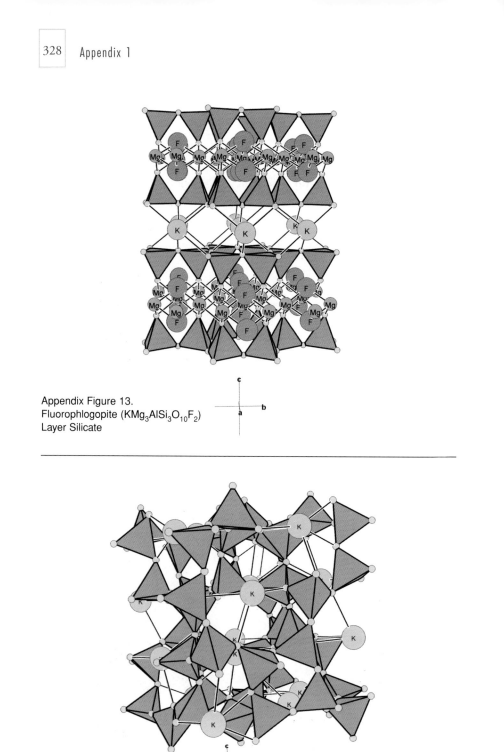

Appendix Figure 13.
Fluorophlogopite (KMg$_3$AlSi$_3$O$_{10}$F$_2$)
Layer Silicate

Appendix Figure 14.
Leucite (KAlSi$_2$O$_6$)
Framework Silicate

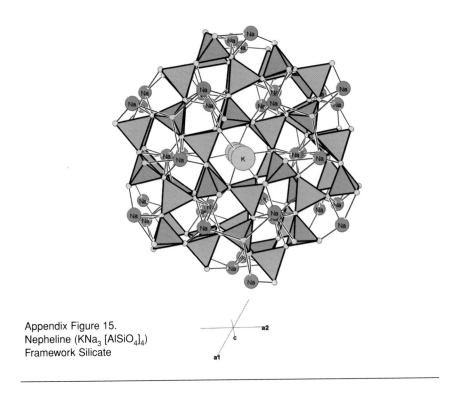

Appendix Figure 15.
Nepheline (KNa$_3$ [AlSiO$_4$]$_4$)
Framework Silicate

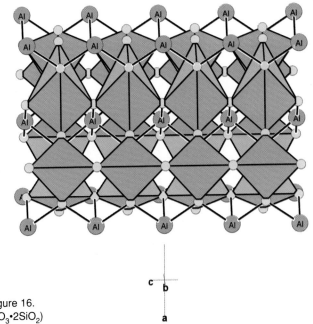

Appendix Figure 16.
Mullite (3Al$_2$O$_3$•2SiO$_2$)
Chain Aluminosilicate

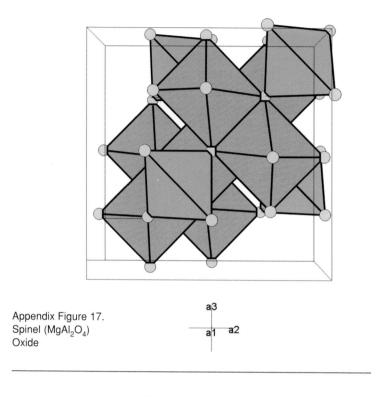

Appendix Figure 17.
Spinel (MgAl$_2$O$_4$)
Oxide

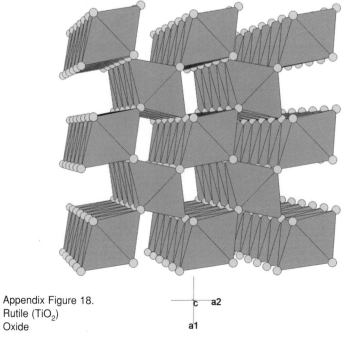

Appendix Figure 18.
Rutile (TiO$_2$)
Oxide

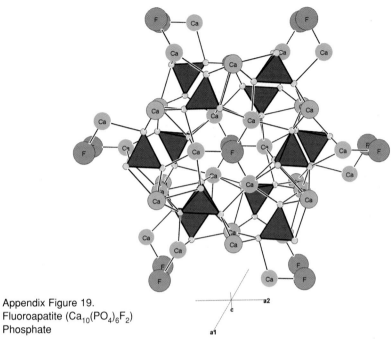

Appendix Figure 19.
Fluoroapatite (Ca$_{10}$(PO$_4$)$_6$F$_2$)
Phosphate

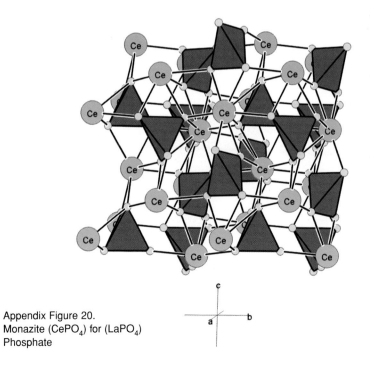

Appendix Figure 20.
Monazite (CePO$_4$) for (LaPO$_4$)
Phosphate

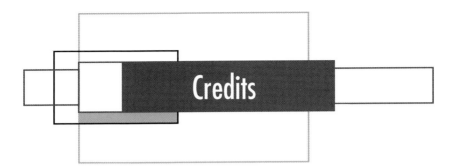

Credits

1-32

Reprinted from *Journal of Non-Crystalline Solids,* 180, W. Höland, M. Frank, and V. Rheinberger, "Surface Crystallization of Leucite in Glass," 292–307, 1995, with permission from Elsevier Science.

1-44

Reprinted from *Journal of Non-Crystalline Solids,* 180, W. Höland, M. Frank, and V. Rheinberger, "Surface Crystallization of Leucite in Glass," 292–307, 1995, with permission from Elsevier Science.

1-45

Reprinted from *Journal of Non-Crystalline Solids,* 180, W. Höland, M. Frank, and V. Rheinberger, "Surface Crystallization of Leucite in Glass," 292–307, 1995, with permission from Elsevier Science.

2-13

Reprinted with permission from Masamichi Wada.

2-17

Reprinted from *Journal of Non-Crystalline Solids,* 180, W. Höland, M. Frank, and V. Rheinberger, "Surface Crystallization of Leucite in Glass," 292–307, 1995, with permission from Elsevier Science.

2-18A

Reprinted, with permission, from W. Höland, M. Frank, and V. Rheinberger, "Realstruktur und Gefüge der Empress Glaskeramik nach Aetzung" *Quintessenz,*" 44, 761–73, 1993.

2-36

Reprinted from *Journal of Non-Crystalline Solids,* 253, W. Höland, V. Rheinberger, and M. Frank, "Mechanism of Nucleation and Controlled Crystallization," 170–177, 1999, with permission from Elsevier Science.

2-39

Reprinted from *Journal of Non-Crystalline Solids,* 253, W. Höland, V. Rheinberger, and M. Frank, "Mechanism of Nucleation and Controlled Crystallization," 170–177, 1999, with permission from Elsevier Science.

2-41

Reprinted, with permission, from W. Höland, V. Rheinberger, S. Wegner, and M. Frank, "Needlelike Apatite–Leucite Glass-Ceramic as a Base Material for the Veneering of Metal Restorations in Dentistry." *Journal of Mater. Sci.: Mater. Med.,* 11, 1–7, Plenum Publishing, 2000.

2-44

Reprinted, with permission, from W. Höland, A.D. Nguyen, E. Heidenreich, E. Tkalcec, and W. Vogel, "Einfluess von Eisenoxiden auf Kristallisationskinetik und Eigenschaften Glimmerhaltiger Maschinellberbeitbarer Glaskeramiken." Teil 1 Phasentrennung, Keimbildung und Kristallisation. *Deutsche Glastechnische Gesellschaft,* 55, 41-49, 1982, Fig. 11.

2-47

Reprinted from *Journal of Non-Crystalline Solids,* 190, H. Hosono and Y. Abe "Porous Glass-Ceramics Composed of a Titanium Phosphate," 185–197, 1995, with permission from Elsevier Science.

2-50

Reprinted, with permission, from M. Schweiger, et al., "Microstructure and Properties of a Pressed Glass-Ceramic Core to a Zirconia Post." *Quintessence of Dental Technology 1998,* Chicago Quintessence, 1998, Fig. 5.

2-51

Reprinted, with permission, from W. Höland, M. Frank, M. Schweiger, S. Wegner, and V. Rheinberger, "Glass Development and Controlled Crystallization in the SiO_2–Li_2O–ZrO_2–P_2O_5 System." *Glastechn. Ber. Glass Sci. Technol,* 69, 25–33, 1996.

2-52A

Reprinted, with permission, from W. Höland, M. Frank, M. Schweiger, S. Wegner, and V. Rheinberger, "Glass Development and Controlled Crystallization in the SiO_2–Li_2O–ZrO_2–P_2O_5 System." *Glastechn. Ber. Glass Sci. Technol,* 69, 25–33, 1996.

2-52B

Reprinted, with permission, from W. Höland, M. Frank, M. Schweiger, S. Wegner, and V. Rheinberger, "Glass Development and Controlled Crystallization in the SiO_2–Li_2O–ZrO_2–P_2O_5 System." *Glastechn. Ber. Glass Sci. Technol,* 69, 25–33, 1996.

3-3

Reprinted from *Thermochimica Acta,* 280/281, W. Höland, M. Frank, and V. Rheinberger, "Opalescence in Dental Products," 491–499, 1996, with permission from Elsevier Science.

3-4

Reprinted from *Thermochimica Acta,* 280/281, W. Höland, M. Frank, and V. Rheinberger, "Opalescence in Dental Products," 491–499, 1996, with permission from Elsevier Science.

3-5

Reprinted from *Thermochimica Acta,* 280/281, W. Höland, M. Frank, and V. Rheinberger, "Opalescence in Dental Products," 491–499, 1996, with permission from Elsevier Science.

3-6

Reprinted from *Thermochimica Acta,* 280/281, W. Höland, M. Frank, and V. Rheinberger, "Opalescence in Dental Products," 491–499, 1996, with permission from Elsevier Science.

3-8

Reprinted, with permission, from W. Höland, V. Rheinberger, S. Wegner, and M. Frank, "Needlelike Apatite–Leucite Glass-Ceramic as a Base Material for the Veneering of Metal Restorations in Dentistry." *Journal of Mater. Sci.: Mater. Med.,* 11, 1–7, Plenum Publishing, 2000.

4-13

Used by permission of Schott Glass.

4-26

Reprinted with permission from Tadashi Kokubo.

4-27

Reprinted from *Journal of Non-Crystalline Solids,* 129, W. Höland, P. Wange, K. Naumann et al., "Control of Phase Formation Process in Glass-Ceramics," 152–162, 1991, with permission from Elsevier Science.

4-36

Reprinted, with permission, from J.A. Sorensen/W.T. Mito, "Rationale and Clinical Technique for Esthetic Restoration of Endodontically Treated Teeth with the CosmoPost and IPS EmpressPost System." *Quintessence of Dental Technology 1998,* Chicago Quintessence, 1998, Fig. 3, Fig. 21.

4-38A

Reprinted, with permission, from J.A. Sorensen/W.T. Mito, "Rationale and Clinical Technique for Esthetic Restoration of Endodontically Treated Teeth with the CosmoPost and IPS EmpressPost System." *Quintessence of Dental Technology 1998,* Chicago Quintessence, 1998, Fig. 3.

4-38B

Reprinted, with permission, from J.A. Sorensen/W.T. Mito, "Rationale and Clinical Technique for Esthetic Restoration of Endodontically Treated Teeth with the CosmoPost and IPS EmpressPost System." *Quintessence of Dental Technology 1998,* Chicago Quintessence, 1998, Fig. 21.

4-47

Reprinted with permission from John A. Sorensen (4-47a), from Daniel Edelhoff (4-47 b, c, d).

4-54

Reprinted with permission from Masamichi Wada.

4-55

Reprinted with permission from Masamichi Wada.

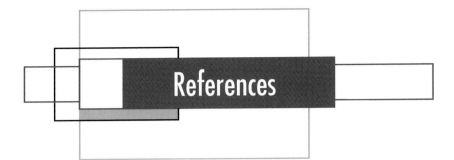

References

Abe Y., "Abnormal Characteristics in Crystallization of $Ca(PO_3)_2$ Glass," *Nature (London)*, **282,** 55–56 (1979).

Abe Y., Hosoe M., Kasuga T., Ishikawa H., Shinkai N., Suzuki N., and Nakayama J. "High-Strength $Ca(PO_3)_2$ Glass-Ceramic Prepared by Unidirectional Crystallization," *J. Am. Ceram. Soc.*, **65**, C-189–C-190 (1982).

Abe Y., Kasuga T., Hosono H., and de Groot K., "Preparation of High-Strength Calcium Phosphate Glass-Ceramic by Unidirectional Crystallization," *J. Am. Ceram. Soc.*, **67**, C-142–C-144 (1984).

Abe Y., Hosono H., Hosoe M., Iwase J., and Kub Y., "Superconducting Glass-Ceramic Rods in $BiCaSrCu_2O_x$ Prepared by Crystallization under a Temperature Gradient," *Appl. Phys. Lett.*, **53**, 1341–42 (1988).

Abe Y., Arakawa H., Hosoe M., Hikich Y., Iwase J., Hosono H., and Kubo Y., *Jpn. J. Appl. Phys.*, **28**, 929–931 (1989).

Abe Y., Hosono H., Nogami M., Kasuga T. and Nagase M., "Development of Porous Glass-Ceramics in Ag-Titanium Phosphates and its Antibacterial Effect," *Bioceramics*, **8**, 247–251 (1995).

Abendroth U., "Die Herstellung von Kronen aus Giessbarer Dicor-Glaskeramik," *Dentallabor*, **33**, 1281–86 (1985).

Adair P.J. and Grossman D., "The Castable Ceramic Crown," *The Int. J. Periodontics Restorative Dent.*, 33–45 (1984).

Aitken B.C.,"Perovskite Glass-Ceramics," *Bol. Soc. Esp. Ceram. VID*, 31–C {5} 33–38 (1992).

Alkemper J., Paulus H., and Fuess H., "Crystal Structure of Aluminium Pentasodium Tetrakis (Phosphate), $Na_5Ca_2Al(PO_4)_4$," *Z. Kristallogr.*, **209**, 76 (1994).

Alkemper J., Paulus H., and Fuess H., "Crystal Structure of Aluminum Calcium Sodium Diphosphate, $Na_{27}Ca_3Al_5(P_2O_7)_{12}$." *Z. Kristallogr.*, **210**, 298–99 (1995).

Allen and Herczog, "Transparent High Dielectric Constant Material, Method and Electroluminescent Device." U.S. Pat. No. 3 114 066, 1962.

Andrews L.J., Beall G.H., and Lempicki A., "Luminescence of Cr^{3+} in Mullite Transparent Glass-Ceramics," *J. Lumin.*, **36**, 65–74 (1986).

Bapna M.S., Mueller H.J., and Campbell S.D., "Kinetic Parameters for Crystallization in Dicor® Glass," *Thermochim Acta*, **275**, 287–93 (1996).

Barrett J.M., Clark D.E., and Hench L.L., "Glass-Ceramic Dental Restoration," U.S. Pat. No. 4 189 325, 1980.

Barry T.I., "The Crystalliation of Glasses Based on the Eutectic Compositions in the System $Li_2O–Al_2O_3–SiO_2$," *J. Mater. Sci.*, **5**, 117–26 (1970).

337

Beall G.H., "Structure, Properties, and Nucleation of Glass-Ceramics"; pp. 251–61 in *Advances in Nucleation and Crystallization in Glasses*, Special Publication. No. 5, Edited by L.L. Hench and S.W. Freiman, American Ceramic Society, Columbus, OH, 1971a.

Beall G.H., "Ta_2O_5-Nucleated Glass-Ceramic Articles, U.S. Pat. No. 3 573 939, 1971b.

Beall G.H., "Alkali Metal, Calcium Fluorosilicate Glass-Ceramic Article," U.S. Pat. No. 4 386 162, 1983.

Beall G.H., "Glass-Ceramics"; pp. 157–72 Advances in Ceramics, Vol. 18, *Commercial Glass*. Edited by D.C. Boyd and J.F. MacDowell. American Ceramic Society, Columbus, OH, 1986.

Beall G.H., "Refractory Glass-Ceramic Containing Enstatite," U.S. Pat. No. 4 687 749, 1987.

Beall G.H., "Design of Glass-Ceramics," *Solid State Sci.*, **3**, 333–354 (1989).

Beall G.H., "Chain Silicate Glass-Ceramics," *J. Non-Cryst. Solids*, **129**, 163–173 (1991).

Beall G.H., "Design and Properties of Glass-Ceramics," *Annu. Rev. Mater. Sci.*, **22**, 91–119 (1992).

Beall G.H., "Glass-Ceramics: Recent Development and Application"; pp. 241–66 in *Nucleation and Crystallization in Glasses and Liquids*. Ceramic Transactions, Vol 30, Edited by M.C. Weinberg. American Ceramic Society, Westerville, OH, 1993

Beall G.H., "Magnetic Memory Storage Device and Disk Having a Glass-Ceramic Substrate," U.S. Pat. No. 5 744 208, 1998.

Beall G.H., and Duke D.A., "Glass-Ceramic Technology"; pp. 404–45 in *Glass Science and Technology*, Vol. 1. Edited by D.R. Uhlmann, N.J. Kreidl. Academic Press, Orlando, FL, 1983.

Beall G.H. and Pinckney L.R., "Variably Translucent Glass-Ceramic Article and Method for Making," Eur. Pat. No. EP 0 536 478 02.07.92, 1992.

Beall G.H. and Pinckney L.R., "High Modulus Glass-Ceramics Containing Fine Grained Spinel-Type Crystals," Eur. Pat. No. EP 0 710 627, 1995.

Beall G.H. and Pinckney L.R., "Nanophase Glass-Ceramics," *J. Am. Ceram. Soc.*, **82**, 5–16 (1999).

Beall G.H. and Reade R.F., "Glass and Glass-Ceramic for Induction Heating," U.S. Pat. No. 4 140 645, 1979.

Beall G.H. and Rittler H.L., "Basalt Glass Ceramics," *Am. Ceram. Soc. Bull.*, **55** [6] 579–82 (1976).

Beall G.H. and Rittler H.L., "Glass-Ceramics Based on Pollucite"; pp. 301–12 in Advances in Ceramics, Vol. 4, *Nucleation and Crystallization in Glasses*. Edited by J.H. Simmons, D.R. Uhlmann, and G.H. Beall. American Ceramic Society, Columbus, OH, 1982.

Beall G.H. and Weidman D.L., "A Thermal Optical Device," U.S. Pat. No. Appl. 60/010058, 1996.

Beall G.H., Karstetter B.R., and Rittler H.L., "Crystallization and Chemical Strength of Stuffed β-quartz Glass-Ceramic," *J. Am. Ceram. Soc.*, **50**, 67–74 (1967).

Beall G.H., Montierth M.R., Smith G.P., "Bearbeitbare Glaskeramik," *Glass-Keram. Technol.* **11**, 409–415 (1971).

Beall G.H., Chirino A.M., Chuyung K., Martin F.W., and Taylor M.P., "Glass-Ceramic Articles Containing Osumilite," U.S. Pat. No. 4 464 474, 1984.

Beall G.H., Chyung K., Stewart R.L., Donaldson K.Y., Lee H.L., Baskaran S. and Hasselman D.P.H., "Effect of Test Method and Crack Size on the Fracture Toughness of Chain-Silicate Glass-Ceramics," *J. Mater. Sci.*, **21**, 2365–72 (1986).

Beall G.H., Chyung K., and Pierson J.E., "Negative CTE β-eucryptite Glass-Ceramics for Fiber Bragg Gratings"; in *Proceedings of the International Congress on Glass XVIII* (CD-ROM) (San Francisco, CA), 1998.

Beevers C.A. and McIntyre D.B., "The Atomic Structure of Fluorapatite and Its Relation to that of Tooth and Bone Mineral," *Miner. Mag.* **27**, 254–259 (1956).

Beleites E., Gudziol H., and Höland W., "Maschinell Bearbeitbare Glaskeramik für die Kopf-Hals-Chirurgie," *HNO-Praxis*, **13**, 121–125 (1988).

Beleites E. and Rechenbach G., "Implantologie in der Kopf-Hals-Chirurgie Gegenwärtiger Stand," *HNO Praxis*, **12**, 170–199 (1992).

Blume R.D. and Drummond III C.H., "Crystallization in Synthetic Basaltic Glass-Ceramics," *Glastech. Ber. Glass Sci. Technol.*, **73** [C1] 43–50 (2000).

Boccaccini A.R., Petitmermet M., and Wintermantel E., "Glass-Ceramic from Municipal Incinerator Fly Ash," *Am. Ceram. Soc. Bull.*, **76**, 75–78 (1997).

Borom M.P., Turkalo A.M., and Doremus R.H., "Strength and Microstructure in Lithium Disilicate Glass-Ceramics," *J. Am. Ceram. Soc.*, **58**, 385–91 (1975).

Borrelli N.F., Herczog A., and Maurer R.D., "Electro Optic Effect of Ferroelectronic Microcrystals in a Glass Matrix," *Appl. Phys. Lett.*, **7**, 117–18 (1965).

Borelli NF, Morse F.B., Bellman R.H., and Morgan W.L., "Photolytic Technique for Producing Microlenses in Photosensitive Glass," *Appl. Opt.*, **24** [16] 250–52 (1985).

Bradt R.C., Newnham R.E. and Biggens J.V. (1973) "The Toughness of Jade,"*Am Mineral*, **58**, 727–32

Brodbeck U., "Six Years of Clinical Experience with an All-Ceramic System," *Signature Int.*, 8–14 (1996).

(a) Brömer H., Pfeil E., and Käs H., Ger. Pat. No. 2 326 100, 1973. (b) Cao W. and Hench L.L., "Bioactive Materials," *Ceram. Int.*, **22**, 493–507 (1996).

Brömer H., Deutscher K., Blencke B., Pfeil E., and Strunz V., "Properties of the Bioactive Implant Material 'CERAVITAL'," *Sci. Ceram.*, **9**, 219–225 (1977).

Brow R.K., "Ion Beam Effects on the Composition and Structure of Glass Surfaces," *J. Vac. Sci. Technol.* A, **7**, 1673–76 (1989).

Brow R.K., Tallant D.R., Sharon T.M., and Phifer C.C., "The Short Range Structure of Zinc Polyphosphate Glass," *J. Non-Cryst. Solids*, **1991**, 45–55 (1995).

Budd M.I., "Sintering and Crystallization of a Glass Powder in the $MgO–Al_2O_3–SiO_2–ZrO_2$ System," *J. Mater. Sci.*, **28**, 1007–14 (1993).

Buerger M.J., "The Stuffed Derivatives of Silica Structures," *Am. Mineral.*, **39**, 600–14 (1954).

Buessem W.R., Thielke N.R., and Sarakauskas R.V., "Thermal Expansion Hysteresis of Aluminum Titanate," *Ceram. Age*, **60** [5] 38–40 (1952). (Also p. 185 in *Ceram. Abstr.*, 1954.)

Bürke H., Durschang B., Meinhardt J., and Müller G., "Nucleation and Crystal Growth Kinetics in the ZrO_2-Strengthened Mica Glass-Ceramic for Dental Application," *Glastech Ber. Glass Sci. Technol.*, **73** [C1] 270–77 (1971).

Burnett D.G., "Nucleation and Crystallization in the Soda–Baria–Silica System," *Phys. Chem. Glasses*, **12**, 117–124 (1971).

Cahn J.W. "The Metastable Liquidus and its Effect on the Crystallization of Glass," *J. Am. Ceram. Soc.*, **52**, 118–121 (1969).

Cameron M., Sueno S., Papike J.J., and Prewitt C.T., "High Temperature Crystal Chemistry of K and Na Fluorrichterites," *Am. Mineral.*, **68**, 924–43 (1983).

Carr S.M. and Subramanian K.N., "Spherulitic Crystal Growth in P_2O_5-Nucleated Lead Silicate Glasses," *J. Cryst. Growth*, **60**, 307–12 (1982).

Caroli B., Caroli C., Roulet B., and Faivre G., "Viscosity-Induced Stabilization of the Spherical Mode of Growth from an Undercooled Liquid," *J. Cryst. Growth*, **94**, 253–60 (1989).

Ceran® Technical Specification TL1001-01 for CERAN®-Cooking Surface, SCHOTT Glass, 1996.

Charles R.J., "Immiscibility and its Role in Glass Processing," *Am. Ceram. Soc. Bull.*, **52**, 673–80 (1973).

Chen F.P.H., "Kinetic Studies of Crystallization of Synthetic Mica Glass," *J. Am. Ceram. Soc.*, **46**, 476–85 (1963).

Chen X., Hench L.L., Greenspan D., Zhong J., and Zhang X., "Investigation on Phase Separation, Nucleation and Crystallization in Bioactive Glass-Ceramics Containing Fluorophlogopite and Fluorapatite," *Ceram. Int.*, **24**, 401–10 (1998).

Chyung C.K., "Secondary Grain Growth of $Li_2O–Al_2O_3–SiO_2–TiO_2$ Glass-Ceramics," *J. Am. Ceram. Soc.*, **52**, 61–64 (1969).

Chyung K., Beall G.H., and Grossman D.G., "Fluorophlogopite Mica Glass-Ceramic"; pp. 122–29 in *Proceedings of the Tenth International Congress on Glass*, Part II. Edited by M. Kunugi, M. Tashiro, and N. Saga. The Ceramic Society of Japan, Kyoto, Japan, 1974.

Clark J.R., Appleman D.E., and Papike J.J., "Chrystal-Chemical Characterization of Clinopyroxenes Based on Eight New Structure Refinements," *Miner. Soc. Am. Spec. Paper*, **2**, 31–50 (1996).

Clifford A. and Hill R., "Apatite-Mullite Glass-Ceramic," *J. Non-Cryst. Solids*, **196**, 346–51 (1996).

Cramer von Clausbruch S., Schweiger M., Höland W., and Rheinberger V., "The Effect of P_2O_5 on the Crystallization and Microstructure of Glass-Ceramics in the $SiO_2–Li_2O–K_2O–ZnO–P_2O_5$ System," *J. Non-Cryst. Solids*, **263 & 264**, 388–94 (2000).

Davis M.J., Phillip D.I., and Lasaga A.C., "Influence of Water on Nucleation Kinetics in Silicate Melt," *J. Mater. Sci.*, **219**, 62–69 (1997).

Dejneka M.J., "The Luminescence and Structure of Novel Transparent Oxyfluoride Glass-Ceramics," *J. Non-Cryst. Solids*, **239**, 149–55 (1998a).

Dejneka M.J., "Transparent Oxyfluoride Glass-Ceramics," *Mater. Res. Bull.*, **23** [11] 57–62 (1998b).

Dental Ceramic, ISO Dental Standard ISO 6872, 1995-09.01, 1995.

"Dental Porcelains for Jacket Crowns," BS No. 5612, British Standards Institution 1978.

Deubener J., Brückner R., and Sternitzke M.. "Induction Time Analysis of Nucleation and Crystal Growth in Di- and Metasilicate Glasses," *J. Non-Cryst. Solids*, **163**, 1–12 (1993).

Dicor®Plus Technique Manual, pp. 1–8 Dentsply 1989.

Dickinson Jr. J.E., Jong B.H.W.S., and Schramm M., "Hydrogen Gas and Gas-Ceramic Microfoams: Raman, XPS, and MASNMR Results on the Structure of Precusor SiO_2–B_2O_3–P_2O_5 Glasses," *J. Non-Cryst. Solids*, **102**, 196–204 (1988).

Dietrich T.R., Ehrtfeld W., Lacher M., Krämer M., and Speit B., "Fabrication Technologies for Microsystems Utilizing Photoetchable Glass," *Microelect. Eng.*, **30**, 497–504 (1996).

Donald I.W., "Glass-Ceramics: An Update;" pp. 1689–95 in *Encyclopedia of Materials Science and Engineering*, Edited by R.W. Cahn. Pergamon, New York, 1993.

Donald I.W., "The Crystallization Kinetics of a Glass Based on the Cordierite Composition Studied by DTA and DSC," *J. Mater. Sci.*, **30**, 904–15 (1995).

Donald I.W., "Crystallization of Iron Containing Glass"; presented at the TC 7 Meeting of ICG, San Francisco, CA, 1998.

Dong J.K., Lüthy H., Wohlwend A., and Schärer P., "Heat-Pressed Ceramics—Technology and Strength," *Quintessenz*, **43**, 1373–85 (1992).

Dowty E., "Atoms V4.0. Computer Program for Windows," Shape Software U.S.A., 1997.

Dowty E., "Idealized Structure of Cristobalite"; personal information, 1999a.

Dowty E., "Structure of spinel"; personal information, 1999b.

Dowty E., "Structure of Rutile"; personal information, 1999c.

Duan R.G., Liang K.M., and Gu S.R., "A Study on the Mechanism of Crystal Growth in the Process of Crystallization of Glasses," *Mater. Res. Bull.*, **38**, 1143–49 (1968).

Duke D.A., Megles J.E., MacDowell J.F., and Bropp H.F., "Strengthening Glass-Ceramics by Application of Compressive Glazes," *J. Am. Ceram. Soc.*, **52**, 98–102 (1968).

Duke D.A., MacDowell J.F., and Karstetter B.R., "Crystallization and Chemical Strengthening of Nepheline Glass-Ceramic," *J. Am. Ceram. Soc.*, **50**, 67–74 (1967).

Dupree R., Holland D., and Mortuza M.G., "A MAS-NMR Investigation of Lithium Silicate Glasses and Glass-Ceramics," *J. Non-Cryst. Solids*, **116**, 148–60 (1990).

Ebisawa Y., Sugimato Y., Hayashi T., Kokubo T., Ohura K., and Yamamuro T., "Crystallization of (FeO, Fe_2O_3)–CaO–SiO_2 Glasses and Magnetic Properties of their Crystallized Products," *Nippon Seramikkusu Kyokai Gakujutsu Ronbunshi*, **99**, 7–13 (1991).

Echeverría L.M., "New Lithium Disilicate Glass-Ceramic," *Bol. Soc. Esp. Ceram. VID*, **5**, 183–88 (1992).

Echeverría L.M. and Beall G.H., "Enstatite Ceramics; Glass and Gel Routes;" pp. 235–44 in Ceramic Transactions, Vol. 20; *Glasses for Electronic Applications*. Edited by K.M. Nair. American Ceramic Society, Westerville, OH, 1991.

Edelhoff D., Spiekermann H., Rübben A., and Yildirim M., "Kronen- und Brückengerüste aus Hochfester Presskeramik," *Quintessenz*, **50**, 177–89 (1999).

Electronic Glass Materials, 13th Ed. Nippon Electric Glass Co., Ltd, Sales Division, 1996.

Elliott J.C., "Structure and Chemistry of the Apatites and Other Calcium Orthophosphates", in *Studies in Inorganic Chemistry*, Vol. 18. Elsevier, Amsterdam, The Netherlands, 1994.

Elsen J., King G.S.D., Höland W., Vogel W., and Carl G., "Crystal Structure of a Fluorophlogopite Synthesized in a Glass-Ceramic," *J. Chem. Res.*, **(M)**, 1253–63 (1989).

El-Shennawi A.W.A., Mandour M.A., Morsi M.M., and Abdel-Hameed S.A.M., "Monopyroxenic Basalt-Based Glass-Ceramics," *J. Am. Ceram. Soc.*, **82**, 1181-86 (1999).

Eurokera, Keraglas® product information, 1995.

Evans A.G. and Heuer A.H., "Review-Transformation Toughening in Ceramics: Martensitic Transformation in Crack-Tip Stress Fields," *J. Am. Ceram. Soc.*, **63**, 241–48 (1980).

Evans E.G., "Toughening Mechanisms in Zirconia Alloys"; pp. 193–212 in Advances in Ceramics, Vol. 12, *Science and Technology of Zirconia II*, Edited by N. Claussen, and A.H. Heuer. American Ceramic Society, Columbus, OH, 1984.

Firelite brochure, Nippon Electric Glass, Tokyo, Japan, 1998.

Frank M., Schweiger M., Rheinberger V., and Höland W., "High-Strength Translucent Sintered Glass-Ceramic for Dental Application," *Glastech Ber. Glass Sci. Technol.*, **71C**, 345–48 (1998).

Gaber M., Müller R., and Höland W., "Degasing Phenomena during Sintering and Crystallization of Glass Powders," *Glastech Ber. Glass Sci. Technol.*, **71C**, 353–56 (1998).

Gee B. and Eckert H., "Cation Distribution in Mixed-Alkali Silicate Glasses; NMR Studies by ^{23}Na–(^{7}Li) and ^{23}Na–(^{6}Li) Spin Echo Double Resonance," *J. Phys. Chem.*, **100**, 3705–12 (1996).

Gegauff A.G., Rosenstiel S.F., Bleiholder R.F., and McCafferty, "Substrates in Rotary Instruments Testing" (Abstr. No 1745), *J. Dent. Res.*, **68**, 400 (1989).

Geis-Gerstorfer J. and Schille C., "Influence of Surface Grinding on Chemical Solubility of Dental Ceramics, (IADR abstract), *J. Dent. Res.*, 3090 (1997).

Gillery F.H. and Bush E.A., "Thermal Contraction of β-eucryptite ($Li_2O \cdot Al_2O_3 \cdot 2SiO_2$) by X-ray and Dilatometer Methods," *J. Amer. Ceram. Soc.*, **42**, pp. 175–77 (1959).

Ghose S., Schomaker V., and McMullan R.K., "Enstatite, $Mg_2Si_2O_6$: A Neutron Diffraction Refinement of the Crystal Structure and a Rigid-Body Analysis of the Thermal Vibration," *Z. Kristallogr.*, **176**, 159–75 (1986).

Goto N., "Glass-Ceramic for Magnetic Disks," *New Glass*, **10**, 56–60 (1995).

Goto N. and Yamaguchi K., "Magnetic Disk Substrate and Method for Manufacturing the Same," U.S. Pat. No. 5 626 935, 1997.

Griffith E.J., *Phosphate Fibers*. Plenum, New York, 1995.

Gross U.M. and Strunz V., "The Anchoring of Glass-Ceramics of Different Solubility in the Femur of the Rat," *J. Biomed. Mater. Res.*, **14**, 607–18 (1981).

Gross U.M., Müller-Mai C. and Voigt C., pp. 105–23 in *An Introduction to Bioceramics*, Edited by L.L. Hench and J. Wilson. World Scientific, Singapore, 1993.

Grossman D.G., "Machinable Glass-Ceramics Based on Tetrasilicic Mica," *J. Am. Ceram. Soc.*, **55**, 446–49 (1972).

Grossman D.G., "Glass-Ceramic Application"; pp. 249–60 in *Nucleation and Crystallization in Glasses*. Edited by J.H. Simmons, D.R. Uhlmann, and G.H. Beall. Advances in Ceramics, Vol. 4, American Ceramic Society, Columbus, OH, 1982.

Grossman D.G., "Der Werkstoff Gussglaskeramik"; pp. 117–34 in *Perspektiven der Dentalkeramik*. Edited by J.D. Preston. Quintessenz Verlags GmbH, Berlin, Germany, 1989.

Grossman D.G., "Structure and Physical Properties of Dicor/MGC Glass-Ceramic"; pp. 103–15 in *International Symposium on Computer Restorations*. Edited by W.H. Mörmann. Quintessence, Chicago, IL, 1991.

Guth H. and Heger G., "Temperature Dependence of the Crystal Structure of the One-Dimensional Li^+ Conductor β-eucryptite, $LiAlSiO_4$"; pp. 499–502 in *Fast Ion Transport in Solids*, Elsevier, Amsterdam, The Netherlands, 1979.

Gutzow I., "Induced Crystallization of Glass-Forming Systems: A Case of Transient Heterogeneous Nucleation, Part 1," *Contemp. Phys.*, **21**, 121–137 (1980).

Gutzow I. and Schmelzer J., *The Vitreous State*. Springer, Berlin, Germany, 1995.

Gutzow I., Zlateva K., Alyakov S., and Kovatscheva T., "The Kinetics and Mechanism of Crystallization in Enstatite-Type Glass-Ceramic Materials," *J. Mater. Sci.*, **12**, 1190–202 (1977).

Gutzow I. and Penkov I., "Nucleation Catalysis in Glass-Forming Melts: Principal Methods, Their Possibilities and Limitations," Wiss Ztschr Friedr–Schiller–Uni Jena, Naturwiss R **36**, 907–19 (1987).

Haller B. and Bischoff H., *Metallfreie Restaurationen aus Presskeramik*. Edited by B. Haller and H. Bischoff. Quintessenz Verlags-GmbH, Berlin, Germany, 1993.

Halliyal A., Bhalla A.S., Newnham R.E., Cross L.E., and Lewis M.H., *Glasses and Glass-Ceramics*; pp. 272–315. Chapman and Hall, London, U.K., 1989.

Hase H., Nasu H., Mito A., Hashimoto T., Matsuoka J., and Kamiya K., "Second Harmonic Generation from Surface Crystallized $Li_2O–Ta_2O_5–SiO_2$ Glass," *Jpn. J. Appl. Phys.*, **30**, 5355–66 (1996).

Hatta G. and Kamei F., "Properties of Glass-Ceramic CRYSTON for Walling Materials," *Rep. Res. Lab. Asahi Glass*, **37**, 149–56 (1987).

Hatta G. Isakai K., Manabe C., Sakai K., and Ichikura E., "Crystallization of Fluorides in Opal Glass," *Rep. Res. Lab Asahi Glass*, **36**, 181–92 (1986).

Headley T.J. and Loehman R.E., "Crystallization of a Glass-Ceramic by Epitaxial Growth," *J. Am. Ceram. Soc.*, **67**, 620–25 (1984).

Heide K., Völksch G., and Hanay C., "Characterisation of Crystallization in Cordierite Glasses by Means of Optical and Electron Microscopy," *Bol. Soc. Ceram. VID*, **31-C** [5] 111–16 (1992).

Heinzmann J.L., Krejci I., and Lutz F., "Wear and Marginal Adaptation of Glass-Ceramic Inlays, Amalgams, and Enamel" (Abstr. No. 423), *J. Dent. Res.*, **69**, 161 (1990).

Hench L.L., "Bioceramics: From Concept to Clinic," *J. Am. Ceram. Soc.*, **74**, 1487–510 (1991).

Hench L.L., "Summary and Future Directions"; pp. 365–74 in *An Introduction to Bioceramics*. Edited by L.L. Hench and J. Wilson. World Scientific, Singapore, 1993.

Hench L.L., Splinter R.J., Allen W.C., and Greenlee Jr. T.K., "Bonding Mechanism at the Interface of Ceramic Prosthetic Materials," *J. Biomed. Mater. Res. Symp.*, **2**, 117–41 (1972).

Hench L.L., and Andersson Ö., "Bioactive Glass Coatings"; pp. 239–59 in *An Introduction to Bioceramics*. Edited by L.L. Hench and J. Wilson. World Scientific, Singapore, 1993.

Herczog A., "Microcrystalline $BaTiO_3$ by Crystallization from Glasses," *J. Am. Ceram. Soc.*, **47**, 107–55 (1964).

Herczog A. and Stookey S.D., U.S. Pat. No. 30,413, 1960.

Heslin M.R. and Shelby J.E., "The Effect of Hydroxyl Content on the Nucleation and Crystallization of Li$_2$O·2SiO$_2$ Glass"; pp. 19–96 in Ceramic Transactions, Vol. 30, *Nucleation and Crystallization in Liquids and Glasses*, Edited by M.C. Weinberg. American Ceramic Society, Westerville, OH, 1993.

Hing P. and McMillan P.W., "The Strength and Fracture Properties of Glass-Ceramics," *J. Mater. Sci.*, **8**, 1041–48 (1973).

Hinz W., *Silikate*, Vol. 2; pp. 399–403. Verlag für Bauwesen, Berlin, Germany, 1970.

Hobo S. and Takoe I., "Castable Apatite Ceramic as a New Biocompatible Restorative Material, I. Theoretical Consideration," *Quint. Intl.* **2/1985**, 135–41 (1985).

Hoda S.N. and Beall G.H., "Alkaline Earth Mica Glass-Ceramics"; pp. 287–300 in Advances in Ceramics, Vol. 4, *Nucleation and Crystallization in Glasses*. Edited by J.H. Simmons, D.R. Uhlmann, and G.H. Beall. American Ceramic Society, Columbus, OH, 1982.

Höland W., "Phase Formation and Properties of Dental Glass-Ceramics in the SiO$_2$–Al$_2$O$_3$–K$_2$O–CaO–P$_2$O$_5$ and SiO$_2$–Li$_2$O–ZrO$_2$–P$_2$O$_5$ Systems," *J. Inorg. Phosphors Chem.*, **6**, 111–14 (1996).

Höland W., "Biocompatible and Bioactive Glass-Ceramic-State of the Art and New Directions," *J. Non-Cryst. Solids*, **219**, 192–97 (1997).

Höland W. and Frank M., "IPS-Empress Glaskeramik"; pp. 147–60 in *Metallfreie Restaurationen aus Presskeramik*. Edited by B. Haller and H. Bischoff. Quintessenz Verlags-GmbH, Berlin, Germany, 1993.

Höland W., Frank M., and Rheinberger V., "Realstruktur und Gefüge der Empress Glaskeramik nach Aetzung," *Quintessenz*, **44**, 761–73 (1993).

Höland W., Rheinberger V., Frank M., and Wegner S., "Glass-Ceramic Containing Needle-like Apatite for Dental Restorations"; pp. 445–48 in *Bioceramics 9* (Otsu, Japan) Pergamon, New York, 1996.

Höland W. and Vogel W., "Fundamentals of Controlled Formation of Glass-Ceramic"; pp. 116–53 in *International Meeting on New Glass Technology*. New Glass Forum, Tokyo, Japan, 1992.

Höland W. and Vogel W., "Machinable and Phosphate Glass-Ceramics"; pp. 125–37 in *An Introduction to Bioceramics*. Edited by L.L. Hench and J. Wilson. World Scientific, Singapore, 1993.

Höland W., Naumann K., Seifert H.G., and Vogel W., "Neuartige Erscheinungsform von Phlogopitkristallen in Maschinell Bearbeitbaren Glaskeramiken," *Z. Chem.*, **21**, 108–109 (1981).

Höland W., Zlateva K., Vogel W., and Gutzow I., "Kinetik der Phasenbildung in Phlogopitglaskermaiken," *Z. Chem.*, **22**, 197–202 (1982a).

Höland W., Nguyen A.D., Heidenreich E., Tkalcec E., and Vogel W., "Einfluss von Eisenoxiden auf Kristallisationskinetik und Eigenschaften Glimmerhaltiger Maschinellbearbeitbarer Glaskeramiken," Teil 1 Phasentrennung, Keimbildung und Kristallisation. *Glastech. Ber.*, **55**, 41–49 and "Einfluss von Eisenoxiden auf Kristallisationskinetik und Eigenschaften Glimmerhaltiger Maschinellbearbeitbarer Glaskeramiken," Teil 2 Ferrimagnetische Eigenschaften, *Glastech. Ber.*, **55**, 70–74 (1982b).

Höland W., Vogel W., Mortier W.J., Duvigneaud P.H., Naessens and Plumat E., "A New Type of Phlogopite Crystal in Machinable Glass-Ceramics," *Glass Technol.*, **24**, 318–22 (1983a).

Höland W., Naumann K., Vogel W., and Gummel J., "Maschinell Bearbeitbare Bioaktive Glaskeramik," Wiss Z Uni Jena, mat-nat wiss Reihe **32**, 571–80 (1983b).

Höland W., Vogel W., Naumann K., and Gummel J., "Interface Reactions between Machinable Bioactive Glass-Ceramics and Bone," *J. Biomed. Mater. Res.*, **9**, 303–12 (1985).

Höland W., Wange P., Naumann K., Vogel J., Carl G., Jana C., and Götz W., "Control of Phase Formation in Glass-Ceramics for Medicine and Technology," *J. Non-Cryst. Solids*, **129**, 152–62 (1991a).

Höland W., Wange P., Carl G., Jana C., Götz W., and Vogel W., "Fundamentals of Controlled Formation of Glass-Ceramics"; pp. 57–63 in *Fundamentals of the Glass Manufacturing Process 1991*. Society of Glass Technology, Sheffield, England, 1991b.

Höland W., Götz W., Carl G., and Vogel W., "Microstructure of Mica Glass-Ceramics and Interface Reactions between Mica Glass-Ceramics and Bone," *Cells Mater.*, **2**, 105–12 (1992).

Höland W., Frank M., Schweiger M., and Rheinberger V., "Development of Translucent Glass-Ceramics for Dental Application," *Glastech. Ber. Glass Sci. Technol.*, **67C**, 117–22 (1994).

Höland W., Frank M., and Rheinberger V., "Surface Crystallization of Leucite in Glass," *J. Non-Cryst. Solids*, **180**, 292–307 (1995a).

Höland W., Rheinberger V., Frank M., and Schweiger M., "Glass-Ceramics for Dental Restoration," *Bioceramics*, **8**, 299–301 (1995b).

Höland W., Frank M., Schweiger M., Wegner S., and Rheinberger V., "Glass Development and Controlled Crystallization in the SiO_2–Li_2O–ZrO_2–P_2O_5 System," *Glastech. Ber. Glass Sci. Technol.*, **69**, 25–33 (1996a).

Höland W., Frank M., and Rheinberger V., "Opalescence in Dental Products," *Thermochim. Acta*, **280/281**, 491–99 (1996b).

Höland W., Rheinberger V., and Frank M., "Mechanism of Nucleation and Controlled Crystallization of Needlelike Apatite in Glass-Ceramics of the SiO_2–Al_2O_3–K_2O-CaO–P_2O_5 Systems," *J. Non-Cryst. Solids*, **253**, 170–77 (1999).

Höland W., Schweiger M., Cramer von Clausbruch S., and Rheinberger V., "Complex Nucleation and Crystal Growth Mechanisms in Applied Multi-Component Glass-Ceramics," *Glastech. Ber. Glass Sci. Technol.*, **73** [C1[12–19 (2000a).

Höland W., Schweiger M., Frank M., and Rheinberger V., "A Comparison of the Microstructure and Properties of the IPS Empress®2 and the IPS Empress® Glass-Ceramic," *J. Biomed. Mater. Res. (Appl. Biomater.)*, **53**, 297–303 (2000b).

Höland W., Rheinberger V., Wegner S. and Frank M., "Needlelike Apatite–Leucite Glass-Ceramic as a Base Material for the Veneering of Metal Restorations in Dentistry," *J. Mater. Sci.: Mater. Med.*, **11**, 1–7 (2000c).

Holland D., Iqbal Y., James P., and Lee B., "Early Stages of Lithium Disilicate Glasses Containing P_2O_5—An NMR Study," *J. Non-Cryst. Solids*, **232-234**, 140–46 (1998).

Höness H., Jacobson A., Knapp K., Marx T., Morian H., Müller R., Reisert N. and Thomas A., "Production of ZERODUR® in Special Shapes"; pp. 143–83 in *Low Thermal Expansion Glass-Ceramics*. Edited by H. Bach. Springer, Berlin, Heidelberg, Germany, 1995.

Hopper R.W., "Stochastic Theory of Scattering from Idealized Spinodal Structures: II. Scattering in General and for the Basic Late Stage Model," *J. Non-Cryst. Solids*, **70**, 111–42 (1985).

Hosono H., Kazunari K., and Abe Y., "Integrated Microporous Glass-Ceramics with Skeleton of $LiTi_2(PO_4)_3$ with Three-Dimensional Network Structure and of $Ti(HPO_4)_2 \cdot 2H_2O$ with Two-Dimensional Layered Structure," *J. Non-Cryst. Solids*, **162**, 287–93 (1993).

Hosono H., Tsuchitani F., Kazunari K., and Abe Y., "Porous Glass-Ceramic Cation Exchangers: Cation Exchange Properties of Porous Glass-Ceramics with Skeleton of Fast Li Ion-Conducting $LiTi_2(PO_4)_3$ Crystals," *J. Mater. Res.*, **9**, 755–61 (1994).

Hummel F.A., "Thermal Expansion Properties of Some Synthetic Lithium Minerals," *J. Am. Ceram. Soc.*, **34**, 235–39 (1951).

Ito S., Kokubo T., and Tashiro M., "Transparency of $LiTaO_3$–SiO_2–TiO_2–Al_2O_3 Glass-Ceramics in Relation to their Microstructure," *J. Mater. Sci.*, **13**, 930–38 (1978) and *J. Am. Ceram. Soc.*, **55** [9] 446–49 (1978).

Jacquin J.R. and Tomozawa M., "Crystallization of Lithium Metasilicate from Lithium Disilicate Glass," *J. Non-Cryst. Solids*, **190**, 233–37 (1995).

Jaha L.J., Best S.M., Knowles J.C., Rehman I., Santos J.D., and Bonfield W., "Preparation and Characterization of Fluoride-Substituted Apatites," *J. Mater. Sci.:Mater. Med.*, **8**, 185–91 (1997).

James P.F., "Nucleation in Glass Forming Systems: A Review"; pp. 1–48 in Advances in Ceramics, Vol. 4, Edited by J.H. Simmons, D.R. Uhlmann, and G.H. Beall. American Ceramic Society, Columbus, OH, 1982.

James P.F., "Kinetics of Crystal Nucleation in Silicate Glasses," *J. Non-Cryst. Solids*, **73**, 517–40 (1985).

James P.F. and McMillan P.W., "Transmission Electron Microscopy of Partially Crystallized Glasses," *J. Mater. Sci.*, **6**, 1345–49 (1971).

Jana C., Wange P., Grimm G., and Götz W., "Bioactive Coatings of Glass-Ceramics on Metals," *Glastech. Ber. Glass Sci. Technol.*, **68**, 117–22 (1995).

Jebsen-Marwedel H. and Brueckner R., *Glastechnische Fabrikationsfehler*; pp. 132–33. Springer–Verlag, Berlin, Germany, 1980.

Jona F., Shirane G., and Pepinsky R., "Dielectric, X-ray, and Optical Study of Ferroelectric $Cd_2Nb_2O_7$ and Related Crystals," *Phys. Rev.*, **98**, 903–909 (1955).

Kakehashi Y., Lüthy H., Naef R., Wohlwend A., and Schärer P., "A New All Ceramic Post and Core System: Clinical, Technical and *In Vitro* Results," *Int. J. of Periodontics & Restorative Dent.*, **18**, 587-93 (1998).

Kappert H., "Impressive Properties—IPS Empress®2," presented at Dental Competence 2000, Berlin, Germany, 1998.

Kasten C., Carl G., and Rüssel C., "The Behavior of Polyvalent Ions in the Glass Melt and Their Influence on the Crystallization of Mica Glass-Ceramic BIOVERIT II"; pp. 298–304 in *Fundamentals of Glass Science and Technology* (Glato, Sweden) 1997.

Kasuga T., Ichino A., and Abe Y., "Preparation of Calcium Phosphate Fibers for Applications to Biomedical Fields," *J. Ceram. Soc. Jpn.*, **100**, 1089 (1992).

Kasuga T., Hosono H., and Abe Y., "Bioceramics Composed of Calcium Polyphosphate Fibers," *Phosphorus Sulfur, Silicon*, **76**, 247–50 (1993).

Kasuga T., Ota Y., Tsuji K., and Abe Y., "Preparation of High-Strength Calcium Phosphate Ceramics with Low Modulus of Elasticity Containing β-Ca(PO$_3$)$_2$ Fibres," *J. Am. Ceram. Soc.*, **79**, 1821–24 (1996a).

Kasuga T., Nakamura K., Inukai E., and Abe Y., "Direct Joining of BSCCO Superconducting Glass-Ceramics Using a Flame-Melting Method," *J. Am. Ceram. Soc.*, **79**, 885–88 (1996b).

Kawamura S., Yamanaka T., Toya F., Nakamura S., and Ninomiya M., "β-Wollastonite Glass-Ceramic"; pp. 68–74 in *Tenth International Congress on Glass* (Kyoto, Japan). The Ceramic Society of Japan, 1974.

Kay J.F., "Calcium Phosphate Coatings for Dental Implants," *Dent. Clin. Nortn Am.*, **36**, 1–18 (1992).

Keding R. and Rüssel C., "The Influence of Electric Fields on the Crystallization of Glass"; pp. 313–19 in *Fundamentals of the Glass Science and Technology Proceedings* (Sweden), 1997.

Keefer C., Mighell A., Mauer F., Swanson H., and Block S., "The Crystal Structure of Twinned Low-Temperature Lithium Phosphate," *Inorg. Chem.*, **6**, 119–25 (1967).

Keith H.D., and Padden F.J., "A Phenomenological Theory of Spherulitic Crystallization," *J. Appl. Phys.*, **34**, 2409–21 (1963).

Keith M.L., and Schairer J., "The Stability Field of Sapphirine in the System MgO–Al$_2$O$_3$–SiO$_2$," *J. Geol.*, **60**, 181–86 (1952).

Keraglas® product information, Eurokera, 1995.

Kerker M., *The Scattering of Light*. Academic Press, New York, 1969.

Kihara K., "An Orthorhombic Superstructure of Tridymite Existing between about 105° and 180°C," *Z. Kristallogr.*, **146**, 185–203 (1977).

Kingery W.D., Bowen H.K., and Uhlmann D.R., *Introduction to Ceramics*. Wiley, New York, 1975.

Kirkpatrick R.J. and Brow R.K., "Nuclear Magnetic Resonance Investigation of the Structures of Phosphate-Containing Glass: A Review," *Solid State Magn. Reson.*, **5**, 9–21 (1995).

Kiselev A., Reisfeld R., Greenberg E., Buch A.N., and Ish-Shalom M., "Spectroscopy of Cr(III) in β-quartz and Petalite-like Transparent Glass-Ceramics: Ligand Field Strength of Chromium(III)," *Chem. Phys. Lett.*, **105**, 405–408 (1984).

Kleiber, *Einführung in die Kristallographie*, pp. 223–24. Verlag Technik, Berlin, Germany, 1968.

Knickerbocker J.U., "Overview of Glass-Ceramic/Copper Substrate—A High Performance Multilayer Package for the 1990s," *Am. Ceram. Soc. Bull.*, **71**, 1393–401 (1992).

Kokubo T., "Crystallization of BaOTiO$_2$–SiO$_2$–Al$_2$O$_3$ Glasses and Dielectric Properties of Their Crystallized Products," *Bull. Inst. Chem. Res., Kyoto Univ.*, **47**, 572–83 (1969).

Kokubo T. "Bioactive Glass-Ceramics Properties and Application," *Biomaterials*, **12**, 155–63 (1991).

Kokubo T., "A/W Glass-Ceramic: Processing and Properties"; pp. 75–88 in *An Introduction to Bioceramics*. Edited by L.L. Hench and J. Wilson. World Scientific, Singapore, 1993.

Kokubo T., Personal information, 1996.

Kokubo T. and Tashiro M., "Dielectric Properties of Fine–Grained PbTiO$_3$ Crystals Precipitated in a Glass," *J. Non-Cryst. Solids*, **13**, 328–40 (1973).

Kokubo T. and Tashiro M., "Fabrication of Transparent PbTiO$_3$ Glass-Ceramics," *Bull. Inst. Chem. Res., Kyoto Univ.*, **54**, 301–306 (1976).

Kokubo T., Kung Ch., and Tashiro M., "Preparation of Thin Films of BaTiO$_3$ Glass-Ceramics and Their Dielectric Properties," *J. Ceram. Assoc. Jpn.*, **76**, 89–94 (1968).

Kokubo T., Kung Ch., and Tashiro M., "Crystallization Process of a BaOTiO$_2$–Al$_2$O$_3$–SiO$_2$ Glass," *Yogyo Kyokaishi*, **77**, 367–71 (1969a).

Kokubo T., Nagao H., and Tashiro M., "Crystallization of PbO–TiO$_2$–Al$_2$O$_3$–SiO$_2$ Glasses and Dielectric Properties of Their Crystallized Products," *Yogyo Kyokaishi*, **77**, 293–301 (1969b).

Kokubo T., Yamashita K., and Tashiro M., "Effect of Al$_2$O$_3$ Addition on Glassy Phase Separation and Crystallization of a PbO–TiO$_2$–SiO$_2$ Glass," *Bull. Inst. Chem. Res., Kyoto Univ.*, **50**, 608–20 (1972).

Kokubo T., Setsuro I., and Tashiro M., "Formation of Metastable Pyrochlore-Type Crystals in Glasses," *Bull. Inst. Chem. Res., Kyoto Univ.*, **51**, 315–28 (1973).

Kokubo T., Shigamatsu M., Nagashima Y., Tashiro M., Nakamura T., Yamamuro Y., and Higashi S., "Apatite and Wollastonite-Containing Glass-Ceramics for Prosthetic Application," *Bull. Inst. Chem. Res., Kyoto Univ.*, **60**, 260–68 (1982).

Kokubo T., Arioka M., and Tashiro M., "Preparation of Li$_2$O·2SiO$_2$ Ceramics with Oriented Microstructure by Unidirectional Solidification of their Melts," *Bull. Inst. Chem. Res., Kyoto Univ.*, **57**, 355–75 (1979).

Kokubo T., Sakka S., Sako W., and Ikejiri S., "Preparation of Glass-Ceramic Containing Crystalline Apatite and Magnesium Titanate for Dental Crown," *J. Ceram. Soc. Jpn. Int. Ed.*, **97**, 236–40 (1989).

Kreidl N., "Inorganic Glass-Forming Systems"; pp. 107–299 in *Glass Science and Technology*, Vol. 1, *Glass-Forming Systems*. Edited by D.R. Uhlmann and N.J. Kreidl. Academic Press, Orlando, FL, 1983.

Krejci F., Lutz F., Reimer M., and Heinzmann J.L., "Wear of Ceramic Inlays, Their Enamel Antagonists and Luting Cements," *J. Prosthetic Dent.*, **69**, 425–30 (1993).

KZBV Jahrbuch 1995; pp. 94–95. Kassenärztliche Bundesvereinigung, Köln, Germany, 1996.

Lacy E.D., "Aluminium in Glasses and in Melts," *Phys. Chem. Glasses*, **4**, 234–38 (1963).

Le Bras E., "Vitrokeramische Erzeugnisse mit hohem Eisenoxid-Gehalt und Verfahren zu ihrer Herstellung," DE Pat. No. OS 26 33 744, 1976.

LeGeros R.Z. and LeGeros J.P., "Dense Hydroxyapatite"; pp. 139–80 in *An Introduction to Bioceramics*. Edited by L.L. Hench and J. Wilson. World Scientific, Singapore, 1993.

Levien L., Prewitt, and Weidner D.J., "Structure and Elastic Properties of Quartz at Pressure," *Am. Mineral*, **65**, 920–30 (1980).

Levin E.M., Robbins C.R., and McMurdie H.F., *Phase Diagrams for Ceramists*. American Ceramic Society, Columbus, OH, 1964.

Levin E.M., Robbins C.R., and McMurdie H.F., *Phase Diagrams for Ceramists*. American Ceramic Society, Columbus, OH, 1969.

Levin E.M., Robbins C.R., and McMurdie H.F., *Phase Diagrams for Ceramists*. American Ceramic Society, Columbus, OH, 1975.

Lewis M.H., and Smith G., "Sperulitic Growth and Recrystallization in Barium Silicate Glasses," *J. Mater. Sci.*, **11**, 2015–26 (1976).

Li C.T., and Peacor D.R.C., "The Crystal Structure of $LiAlSi_2O_6$-II (β-spodumene)." *Z. Kristallogr.*, **126**, 46–65 (1968).

Liebau F., "Untersuchungen an Schichtsilikaten des Formeltyps $A_m(Si_2O_5)_n$. I. Die Kristallstruktur der Zimmertemperaturform des $Li_2Si_2O_5$. *Acta Crystallogr.*, **14**, 389–95 (1961).

Liebau F., *Structural Chemistry of Silicates*, Springer–Verlag, Berlin, Germany, 1985.

Likivanichkul S. and Lacourse W.C., "Effect of Fluorine Content on Crystallization of Canasite Glass-Ceramic," *J. Mater. Sci.*, **30**, 6151–55 (1995).

Lindemann W., "Kristalline Phasen in Keramischen Verblendungen," *Dent. Lab.*, **33**, 993–94 (1985).

Ludwig K., "Untersuchnugen zur Bruchfestigkeit von IPS Empress—Kronen in Abhängigkeit von den Zementiermodalitäten," *Quint. Zahntech.*, **20**, 247–56 (1994).

Lynch S.M. and Shelby, "Crystal Clamping in Lead Titanate Glass-Ceramics," *J. Am. Ceram. Soc.*, **67**, 424–27 (1984).

MacDowell J.F., "Composition, Microstructure versus Heat Treatment and Properties Given," *Proc. Br. Ceram. Soc.*, **3**, 229–40 (1965a).

MacDowell J.F., "Nucleation and Crystallization of Barium Silicate Glasses," *Proc. Brit Ceram. Soc.*, **3**, 229-40 (1965b).

MacDowell J.F., "Boron Phosphate Glass-Ceramic," *Proceedings of the International Congress on Glass*, Vol. 3a. Edited by O.V. Mazurin, 1989.

MacDowell J.F. and Beall G.H., "Immiscibility and Crystallization in Al_2O_3-SiO_2 Glasses," *J. Am. Ceram. Soc.*, **52**, 17–25 (1969).

MacDowell J.F. and Beall G.H., "Low *K* Glass-Ceramics for Microelectronic Packaging"; pp. 259–77 in Ceramic Transactions, Vol. 15, *Materials and Processes for Microelectronic Systems*. Edited by K.M. Nair, R. Pohanka, and R.C. Buchanan, **15**, American Ceramic Society, Westerville, OH 1990.

Maeda K., Ichikura E., Nakao Y., and Ito S., "Nucleation of β-Wollastonite Crystal by Noble Metal Particles," *Bol. Soc. Esp. Ceram. VID*, **31-C** [5] 15–20 (1992).

Maier V. and Müller G., "Mechanism of Oxide Nucleation in Lithium Aluminosilicate Glass-Ceramics," *J. Am. Ceram. Soc.*, **70 C**, 176–178 (1989).

Malament K.A. and Socransjky S.S., "Survival of Dicor Glass-Ceramic Dental Restoration Over Fourteen Years"; presented at the Academy of Prosthodontics Annual Meeting, Newport Beach, CA.

Malament K.A. and Grossman D., "The Cast Glass-Ceramic Restoration," *J. Prosthetic Dent.*, **57**, 674–83 (1987).

Mazza D. and Lucca-Borlero M., "Effect of Substitution of Boron for Aluminum in the β-Eucryptite $LiAlSiO_4$ Structure," *J. Eur. Ceram. Soc.*, **13**, pp. 61–65 (1994).

Mazzi F., Galli E., and Gottardi G., "The Crystal Structure of Tetragonal Leucite," *Am. Mineral.*, **61**, 108–15 (1976).

McCauley J.W., Newnham R.E., and Gibbs G.V., "Crystal Structure Analysis of Synthetic Fluorophlogopite," *Am. Mineral.*, **58**, 249–54 (1973).

McCauley D., Newnham R.E., and Randall, "Intrinsic Size Effects in a Barium Titanate Glass-Ceramic," *J. Am. Ceram. Soc.*, **81**, 979–87 (1998).

McCracken W.J., Clark D.E., and Hench L.L., "Aqueous Durability of Lithium Disilicate Glass-Ceramics," *Ceram. Soc. Bull.*, **61**, 1218–29 (1982).

McLean J.W., "Dental Porcelains"; pp. 77–83 in *Dental Materials Research*, NBS Publication 354. Edited by G. Dickson and JM Cassels. National Bureau of Standards, Washington, DC, 1972.

McMillan P.W., *Glass-Ceramics*, 2nd Ed., Academic Press, New York, 1979.

McMillan P.W., Philips S.V., and Partridge G., "The Structure and Properties of a Lithium Zinc Silicate Glass-Ceramic," *J. Mater. Sci.*, **1**, 269–79 (1966).

Mecholsky J.J., "Fracture Mechanics Analysis of Glass-Ceramics"; pp. 261–76 in Advances in Ceramics, Vol. 4, American Ceramic Society, Columbus, OH, 1982.

Metoxit product information, Metoxit AG, Switzerland, 1998.

Meyenberg K.H., Lüthy H., and Schärer P., "A New All-Ceramic Concept for Nonvital Abutment Teeth," *J. Esth. Dent.*, , 73–80 (1995).

Meyer K., *Physikalisch-Chemische Kristallographie*; pp. 222–63. Verlag für Grundstoffindustrie, Leipzig, Germany, 1968.

Moisescu C., Carl G., and Rüssel C., "Glass-Ceramics with Different Morphology of Fluorapatite Crystals," *Phosphorus Res. Bull.*, **10**, 515–20 (1999).

Mora N.D., Ziemeth E.C., Zanotto E.D., "Heterogeneous Crystallization in Cordierite Glasses," *Bol. Soc. Esp. Ceram. VID*, **31-C** [5] 117–18 (1992).

Morena R. and Francis G.L., "Bonding Frits for Near-Zero and Negative Expansion Substrates;" *Proceedings of the International Congress on Glass XVIII* (CD-ROM) (San Francisco, CA, 1998).

Mörmann W.H. and Krejci I., "Computer-Designed Inlays after 5 Years *in Situ*: Clinical Performance and Scanning Electron Microscopic Evaluation," *Quint. Int.*, **23**, 109–15 (1992).

Mörmann W.H., Brandestini M., and Lutz F., "Das Cerec®-System; Computergestützte Herstellung direkter Keramikinlays in einer Sitzung," *Konservierende Zahnheilkunde*, **3**, 1–14 (1987).

Müller G., "Zur Wirkungsweise von Gemischen Oxidischer Keimbildner in Glaskeramiken des Hochquarz-Mischkristalltyps," *Glastech. Ber.*, **45**, 189–94 (1972).

Müller G., "Structure, Composition, Stability, and Thermal Expansion of High-Quartz and Keatite-Type Alumino-Silicate"; pp. 17–24 in *Low Thermal Expansion Glass-Ceramics*. Edited by H. Bach. Springer, 1995.

Müller R., Abu-Hilal L.A., Reinisch S., and Höland W., "Coarsening of Needle-Shaped Apatite Crystals in $SiO_2 \cdot Al_2O_3 \cdot Na_2O \cdot K_2O \cdot CaO \cdot P_2O_5 \cdot F$ Glass," *J. Mater. Sci.*, **34**, 65–69 (1999).

Müller R., Thamm D., and Pannhorst W., "On the Nature of Nucleation Sites at Cordierite Glass Surfaces," *Bol. Soc. Esp. Ceram. VID*, **31-C** [5] 105–10 (1992).

Murthy M.K., "Glass and Glass-Ceramic Materials in the System Eucryptite-SiO_2–$AlPO_4$," Ontario Research Foundation Rept. No. ORF-62-2, May 1–Sept. 3, 1962.

Nagase R., Takeuchi Y., and Mitachi S., "Optical Connector with Glass-Ceramic Ferrule," *Electr. Lett.*, **33**, 1243–44 (1997).

Nass P., Rodeck E.W., Schildt H., and Weinberg W., "Development and Production of Transparent Colorless and Tinted Glass-Ceramic"; pp. 60–79 in *Low Thermal Expansion Glass-Ceramics*. Edited by H. Bach. Springer, Berlin, Heidelberg, 1995.

"Neoceram, Zero-Expansion Glass-Ceramics for Innovative Application," Nippon Electric Glass Co. Ltd., 11 (1992).

"Neoceram, Low-Expansion Glass-Ceramic," Nippon Electric Glass Co. Ltd., 12 (1995).

"Neoceram, brochure, Nippon Electric Glass, 1998.

"Neoparies, Crystallized Glass Neoparies, Building Material for Interior and Exterior Walls," Nippon Electric Glass Co. Ltd., 9 (1995).

Newesely H., "Mechanism and Action of Trace Elements in the Mineralisation of Dental Hard Tissue," *Zyma SA*, 1972.

Norton product information, Norton Desmarquest, France, 1998.

O'Brien W.J., "Dental Porcelains"; pp. 180–94 in *An Outline of Dental Materials and Their Selection*. Edited by W.J. O'Brien, G. Ryge, W.B. Saunders, Philadelphia, PA, 1978.

Ohashi Y. and Finger L.W., "The Role of Octahedral Cations in Pyroxenoid Crystal Chemistry. I. Bustamite, Wollastonite, and the Pectolite–Schizolite–Serandite Series," *Am. Mineral.*, **63**, 274–88 (1978).

Onyiriuka E.C., "AM 2001 Lubricant Film on Canasite Glass-Ceramic Magnetic Memory Disk," *Chem. Mater.*, **5**, 798–801 (1993).

Ostertag W., Fischer G.R., and Williams J.P., "Thermal Expansion of Synthetic β-spodumene and β-spodumene Silica Solid Solutions," *J. Am. Ceram. Soc.*, **51**, 651- 54 (1968).

Ota R., Mashima N., Wakasugi T., and Fukunaga, "Nucleation of $Li_2O–SiO_2$ Glasses and Its Interpretation Based on a New Liquid Model," *J. Non-Cryst. Solids*, **219**, 70–74 (1997).

Palmer D.C., "Stuffed Derivatives of the Polymorphs"; pp. 83–122 in *Silica, Physical Behavior, Geochemistry and Materials Applications*. Edited by P.J. Heaney, C.T. Prewitt, and G.V. Gibbs. Mineral Society Of America, Washington D.C., 1994.

Pannhorst W., "Low-Expansion Glass-Ceramics—Review of the Glass-Ceramic Ceran and Zerodur and Their Application"; pp. 267–76 in Ceramic Transactions, Vol. 30, *Nucleation and Crystallization in Liquids and Glasses*. Edited by M.C. Weinberg. American Ceramic Society, Westerville, OH, 1993.

Pannhorst W., "Overview"; p. 9 in *Low Thermal Expansion Glass-Ceramic*. Edited by H. Bach. Springer, Berlin and Heidelberg, Germany, 1995.

Pannhorst W., "Development of the Optical Glass-Ceramic ZERODUR®"; pp. 107–30 in *Low Thermal Expansion Glass-Ceramics*. Edited by H. Bach. Springer, Berlin and Heidelberg, Germany, 1995.

Pantano C.G., Clark A.E., and Hench L.L., "Multilayer Corrosion Films on Bioglass Surfaces," *J. Am. Ceram. Soc.*, **57**, 412–13 (1974).

Partridge G. and Budd M.I., "Toughened Glass-Ceramic," U.K. Pat. Appl. No. GB 2172282, 1986.

Partridge G., Elyard C.A., and Budd M.I., "Glass-Ceramics in Substrate Applications"; pp. 226–71 in *Glasses and Glass-Ceramics*. Edited by M.H. Lewis. Chapman and Hall, London, U.K., 1989.

Pavluskin N.M., *Vitrokeramik Deutscher Verlag für Grundstoffindustrie*; 172–86, 263. Leipzig, Germany, 1986.

Pelino M., Cantalini C., Veglio F., and Plescia P.P., "Crystallization of Glasses Obtained by Recycling Goethite Industrial Wastes to Produce Glass-Ceramic Materials," *J. Mater. Sci.*, **29**, 2087–94 (1994).

Pernot F. and Rogier R., "Phosphate Glass-Ceramic–Cobalt–Chromium Composite Materials," *J. Mater. Sci.*, **27**, 2914–21 (1992).

Pernot F. and Rogier R., "Mechanical Properties of Phosphate Glass-Ceramic—316 L Stainless Steel Composites," *J. Mater. Sci.*, **28**, 6676–82 (1993).

Petzoldt J., "Metastabile Mischkristalle mit Quarzstruktur im Oxidsystem $Li_2O–MgO–ZnO–Al_2O_3–SiO_2$," *Glastech. Ber.*, **40**, 385–96 (1967).

Petzoldt J., "Transparente Glaskeramiken mit Einem Thermischen Ausdehnungskoeffizienten von 0 + 1.5 mal 10^{-7}/Grad C, der im Bereich von -30 bis +70°C Wenig temperaturabhängig ist," DE Pat. No. 1902432, 1977.

Petzoldt J. and Pannhorst W., "Chemistry and Structure of Glass-Ceramic Materials for High Precision Optical Application," *J. Non-Cryst. Solids*, **129**, 191–98 (1991).

Pfeiffer J., "The Character of CEREC 2"; pp. 255–67 in *CAD/CIM in Aesthetic Dentistry*. Edited by W.H. Mörmann. Quintessence, Berlin, Germany, 1996.

Pillars W.W. and Peacor D.R., "The Crystal Structure of β-eucryptite as a Function of Temperature," *Am. Mineral.*, **58**, pp. 681–90 (1973).

Pinckney L.R., "Transparent Glass-Ceramics Containing Gahnite," U.S. Pat. No. 4 687 750, 1987.

Pinckney L., "Microstructure of Leucite-Apatite Glass-Ceramic"; presented at the TC 7 Meeting of ICG Congress, San Francisco, CA, 1998.

Pinckney L.R. and Beall G.H., "Nanocrystalline Non-Alkali Glass-Ceramics," *J. Non-Cryst. Solids*, **219**, 219–28 (1997).

Pinckney L.R., "Phase Separated Glass and Glass-Ceramics"; pp. 433–38 in *Engineered Materials Handbook*, Vol. 4, American Society for Metals, Columbus, OH, 1993.

Pinckney L.R., "Transparent, High Strain Point Spinel Glass-Ceramic," *J. Non-Cryst. Solids*, **255**, 171–77 (1999).

Pinckney L.R., and Beall G.H., "Strong Sintered Miserite Glass-Ceramics," *J. Am. Ceram. Soc.*, **82**, 2523–28 (1999).

Pincus A.G., "Application of Glass-Ceramics"; pp. 210–33 in *Advances in Nucleation and Crystallization in Glasses*, Special Publication 5. Edited by L.L. Hench and S.W. Frieman. American Ceramic Society, Columbus, OH, 1971.

Pinkert E., "Individuell Hergestellte Enossale Offene Zahnimplantate aus Glaskeramik BIOVERIT," *Zahn-Mund-Kieferheilk*, **78**, 411–16 (1990).

Pirooz P.P., "Glass-Ceramic and Method for Making Same," U.S. Pat. No. 3 779 856, 1973.

Pospiech P., Kistler S., and Frasch C., "Clinical Success of Empress2 Glass-Ceramic as a Bridge Material," *Glastech. Ber. Glass Sci. Technol.*, **73** [C1] 310–17 (2000).

Predecki P., Haas J., Faber J., and Hittermann R.L., "Structural Aspects of the Lattice Thermal Expansion of Hexagonal Cordierite," *Acta Crystallogr.*, **70**, 175–82 (1987).

Prewo K.M., "Fibre Reinforced Glass-Ceramics"; pp. 336–68 in *Glasses and Glass-Ceramics*, Edited by M.H. Lewis. Chapman and Hall, Cambridge, U.K., 1989.

Rädlein E., and Frischat G.H., "Atomic Force Microscopy as a Tool to Correlate Nanostructure to Properties of Glasses," *J. Non-Cryst. Solids*, **222**, 69–82 (1997).

Raj R. and Chyung C.K., "Solution-Precipitation Creep in Glass-Ceramics," *Acta Metall.*, **29**, 159–66 (1981).

Ray C.S. and Day D.E., "Determining the Nucleation Rate Curve for Lithium Disilicate Glass by Differential Thermal Analysis," *J. Am. Ceram. Soc.*, **73**, 439–42 (1990).

Ray C.S. and Day D.E., "An Analysis of Nucleation Rate Type of Curves in Glass as Determined by Differential Thermal Analysis," *J. Am. Ceram. Soc.*, **80**, 3100–108 (1997).

Ray C.S., Yang Q., Huang W., Day D.E., "Surface and Internal Crystallization in Glasses as Determined by Differential Thermal Analysis," *J. Am. Ceram. Soc.*, **79**, 3155–160 (1996).

Reade R.F., "Method for Making Glass-Ceramics with Ferrimagnetic Surfaces," U.S. Pat. No. 4 083 709, 1977.

Reamur M., "The Art of Matching a New Grid of Porcelain," *Memories Acad. Sci., Paris*, 377–88 (1739).

Reck R., "Bioactive Glass-Ceramics in Ear Surgery: Animal Studies and Clinical Results," *Laryngoscope*, **94**, 1–9 (1984).

Reece M.J., Worrell C.A., Hill G.J., and Morrell R., "Microstructure and Dieclectric Properties of Ferroelectric Glass-Ceramics," *J. Am. Ceram. Soc.*, **79**, 17–26 (1966).

Reisfeld R., Kiselev A., Greenberg E., Buch A., and Ish-Shalom M., "Spectroscopy of Cr(III) in Transparent Glass-Ceramics Containing Spinel and Gahnite," *Chem. Phys. Lett.*, **104**, 153–56 (1984).

Richter E.J. and Hertel R.C., "Erste Klinische Erfahrungen mit der Dicor-Glaskeramik-Krone," *Die Quint.*, 1661–69 (1987).

Rieger W., "Aluminium-und Zirkonoxidkeramiken in der Medizin," *Ind. Diamanten Rundschau*, **2**, 2–6 (1993).

Rittler H.L., "Glass-Ceramic Coated Optical Waveguides," U.S. Pat. No. 4 209 229, 1980.

Ro R. and Osborn E.F., "The System Lithium Metasilicate–Spodumene–Silica," *J. Am. Ceram. Soc.,* **71**, 2086–95 (1949).

Robax® product information, "ROBAX® S for Use as Room Heater Window," SCHOTT Glass, 1998.

Rogier R. and Pernot F., "Phosphate Glass-Ceramic–Titanium Composite Materials," *J. Mater. Sci.*, **26**, 5664–70 (1991).

Rösler H.J., *Lehrbuch der Mineralogie*. Deutscher Verlag für Grundstoffindustrie, Leipzig, Germany, 1991.

Rothammel W., Burzlaff H., and Specht R., "Structure of Calcium Metaphoshate $Ca(PO_3)_2$," *Acta Crystallogr., Sect. C: Cryst. Struct. Commun.*, **C45**, 551–53 (1989).

Rüssel C., "Oriented Crystallization in Glasses," *J. Non-Cryst. Solids*, **219**, 212–18 (1997).

Russell C.K. and Bergeron C.G., "Structural Changes Preceding Growth of a Crystalline Phase in a Lead Silicate Glass," *J. Am. Ceram. Soc.*, **48**, 162–63 (1965).

Sack W., "Glas, Glaskeramik und Sinterglaskeramik," *Chemie-Ing-Tech.*, **37**, 1154–65 (1965).

Sack W. and Scheidler H., "Einfluss der Keimbildner TiO_2 und ZrO_2 auf die sich Ausscheidenden Kristallphasen bei der Bildung von Glaskeramik," *Glastech. Ber.*, **39**, 126–30 (1966).

Sack W. and Scheidler H., "Durchsichtige, in der Aufsicht Schwarze, in der Durchsicht Dunkelrote Glaskeramik des Systems SiO_2–Al_2O_3–Li_2O mit Hohem Wärmespannungsfaktor R Grösser als 1000, Insbesondere zur Herstellung von Beheizbaren Platten, Sowie Verfahren zur Herstellung der Glaskeramik," DE Pat. No. 24 29 563, 1974.

Sadanaga R., Tokonami M., and Takéuchi Y., "The Structure of Mullite, $2Al_2O_3 \cdot SiO_2$, and Relationship with the Structure of Sillimanite and Andalusit," *Acta Crystallogr.*, **15**, 65–68 (1962).

Saegusa K., "$PbTiO_3$–PbO–B_2O_3 Glass-Ceramics by a Sol-Gel Process," *J. Am. Ceram. Soc.*, **79**, 3282–88 (1996).

Saegusa K., "$PbTiO_3$–PbO–SiO_2 Glass-Ceramic Thin Film by a Sol-Gel Process," *J. Am. Ceram. Soc.*, **80**, 2510–16 (1997).

Sakamoto A. and Wada M., "Glass-Ceramic for Optical Connector Fabricated by Redrawing the Cerammed Preform," Congress on Glass XVIII, San Francisco, CA, 1998.

Salz U., "Adhesive Cementation of Full Ceramic Restorations," *Ivoclar-Vivadent Rep.*, **10**, 9 (1994).

Sänger A.T. and Kuhs W.F.Z., "Structural Disorder in Hydroxyapatite," *Z. Kristallogr.*, **199**, 123–48 (1992).

Sarno R.D., Tomozawa M., "Toughening Mechanisms for Zirconia–Lithium Aluminosilicate Glass-Ceramic," *J. Mater. Sci.*, **30**, 4380–88 (1995).

Schairer J.F. and Bowen N.L., "Melting Relations in the Systems Na_2O–Al_2O_3–SiO_2 and K_2O–Al_2O_3–SiO_2," *Am. J. Sci.*, **245**, 199 (1947).

Scherer G.W. and Uhlmann D.R., "Diffusion-Controlled Crystal Growth in K_2O–SiO_2 Compositions," *J. Non-Cryst. Solids*, **23**, 59-80 (1977).

Schiffer U., "Nucleation in Parent Glasses for Lithia Alumino-Silicate Glass-Ceramic"; pp. 25–38 in *Low Thermal Expansion Glass-Ceramics*. Edited by H. Bach. Springer, Berlin, Heidelberg, Germany, 1995.

Schmedt auf der Günne J., Meise-Gresch K., Eckert H., Höland W., and Rheinberger V., "Multinuclear Solid State NMR Investigations of Crystallization Processes in Glass-Ceramics of the SiO_2–Al_2O_3–K_2O–Na_2O–CaO–P_2O_5–F System," *Glastech. Ber. Glass Sci. Technol.*, **73** [C1] 98–103 (2000).

Schmelzer J., Möller J., Gutzow I., Pascova R., Müller R., and Pannhorst W., "Surface Energy and Structure Effects on Surface Crystallization," *J. Non-Cryst. Solids*, **183**, 215–33 (1995).

Schmid M., Fischer J., Salk M., and Strub J., "Microgefüge Leucit-Verstärkter Glaskeramiken," *Schweiz Monatsschr Zahnmed*, **102**, 1046–53 (1992).

Schmidt A. and Frischat G.H., "Atomic Force Microscopy of Early Stage Crystallization in $Li_2O \cdot SiO_2$ Glasses," *Phys. Chem. Glasses*, **38**, 161–66 (1997).

"FORTURAN," Schott/mikroglas, 1–4 (1999).

SCHOTT-Product information, "Leichtgewichts-Strukturen aus Zerodur-Glaskeramik," 1–4, Schott, 1988.

Schreyer W. and Schairer J.F., "Metastable Solid Solution with Quartz-Type Structure on the Join SiO_2–$MgAl_2O_4$," *Z. Kristallogr.*, **116**, 60–82 (1961).

Schreyer W. and Schairer J.F., "Metastable Osumilite-and Petalite-Type Phases in the System MgO–Al_2O_3–SiO_2, *Am. Min.*, **47**, 90–104 (1962).

Schubert T., Purath W., Liebscher P., and Schulze K.J., "Klinische Indikation für die Anwendung der Jenaer Bioaktiven Glaskeramik in Orthopädie und Traumatologie," *Beitr. Orthop. Traumatol*, **35**, 7–16 (1988).

Schulz C., Miehe G., Fuess H. Wange P., and Götz W., "X-ray Powder Diffraction of Crystalline Phases in Phosphate Bioglass Ceramics," *Z. Kristallogr.*, **209**, 249–55 (1994).

Schwarzenbach D., "Verfeinerung der Struktur der Tiefquarz-Modifikation von $AlPO_4$," *Z. Kristallogr.*, **123**, 161–85 (1966).

Schweiger M., Frank M., Rheinberger V., and Höland W., "New Sintered Glass-Ceramic Based on Apatite and Zirconia"; pp. 229–35 in *Proceedings International Symposium on Glass Problems*, Vol. 2 (Istanbul, Turkey) International Congress on Glass, 1996.

Schweiger M., Frank M., Cramer von Clausbruch S., Hoeland W., and Rheinberger V., "Microstructure and Properties of Pressed Glass-Ceramic Core to Zirconia Post," *Quint. Dent. Technol.*, **21**, 73–79 (1998).

Schweiger M., Höland W., Frank M., Drescher H., and Rheinberger V., "IPS Empress®2: A New Pressable High Strength Glass-Ceramic for Esthetic All Ceramic Restoration," *Quint. Dent. Technol.*, **22**, 143–52 (1999).

Schweiger M., Cramer von Clausbruch S., Höland W., and Rheinberger V., "Microstructure and Mechanical Properties of a Lithium Disilicate Glass-Ceramic in the SiO_2–Li_2O–K_2O–ZnO–P_2O_5 System," *Glastech. Ber. Glass Sci. Technol.*, **73** [C1] 43–50 (2000).

Seifert F. and Schreyer W., "Synthesis of a New Mica, $KMg_{2.5}(Si_4O_{10})(OH)_2$," *Am. Mineral.*, **50**, 1114–18 (1965).

Semar W. and Pannhorst W., "Dispersion-Strengthened Cordierite," *Silic. Ind.*, **56**, 71–75 (1991).

Semar W., Pannhorst W., Hare M.T. and Pulmour III H., "Sintering of a Crystalline Cordierite/ZrO_2 Composite," *Glastech. Ber.*, **62**, 74–78 (1989).

Sestak J., "Use of Phenomenological Kinetics and the Enthalpy versus Temperature Diagram (and Its Derivative—DTA) for a Better Understanding of Transition Processes in Glasses," *Thermochim. Acta.*, **280/281**, 175–90 (1996).

Seward III T.P., Uhlmann D.R., and Turnbull D., "Phase Separation in the System BaO–SiO_2," *J. Am. Ceram. Soc.*, **51**, 278–85 (1968).

Sglavo V.M. and Pacheri P., "Crack Decoration Technique for Fracture-Toughness Measurements in Alumina," *J. Eur. Ceram. Soc.*, **17**, 1697–706 (1997).

Shirk B.T. and Buessem W.R., "Magnetic Properties of Barium Ferrite Formed by Crystallization of a Glass," *J. Am. Ceram. Soc.*, **53**, 192–96 (1970).

Simmons W.B. and Peacor, "Refinement of the Crystal Structure of a Volcanic Nepheline," *Am. Min.*, **57**, 1711–19 (1972).

Smith G.P., "Some Recent Research and Development at Corning Glass Works, 100 Jahre Jena^{er}Glas, Tagungsband," *Jena*, 285–304 (1984).

Smith R.I., Howie R.A., West A.R., Aragon-Pina A., and Villuerte-Castrejon M.E., "The Structure of Metastable Lithium Disilicate, $Li_2Si_2O_5$," *Acta Crystallogr., Sect. C: Cryst. Struct. Commun.*, **C46**, 363–65 (1990).

Sokolova G.V., Kashaev A.A., Duts V.A., and Ilyukhin V.V., "The Crystal Structure of Fedorite," *Sov. Phys. Crystallogr.*, **28**, 95–96 (1983).

Sorensen J.A. and Mito W.T., "Rational and Clinical Technique for Esthetic Restoration of Endodentically Treated Teeth with the CosmoPost and IPS Empress Post System," *Quint. Dent. Technol.*, **21**, 81–90 (1998).

Sorensen J.A., Sultan E., and Condon J.R., "Three-body *In-Vitro* Wear of Enamel Antagonist Dental Ceramic," *J. Dent. Res.*, **78**, 219 (1999).

Sorensen J.A., "The IPS Empress®2 System: Defining the Possibilities," *Quint. Dent. Technol.*, **22**, 153–63 (1999).

Speit B., "Formätzteile aus Glas und Glaskeramik," *F & M*, **101**, 339–41 (1993).

Steinborn G., Berger G., and Büchting H., "*In-Vitro*-Untersuchungen zur Löslichkeit von Implantatmaterialien," *Sprechsaal*, **126**, 606–11 (1993).

Stewart D.R., "TiO_2 and ZrO_2 as Nucleants in a Lithia Aluminosilicate Glass-Ceramic"; pp. 83–92 in *Advances in Nucleation and Crystallization in Glasses*, Edited by L.L. Hench and S.W. Freiman. Special Publication 5, American Ceramic Society, Columbus, OH, 1971.

Stewart D.R., "Verfahren zur Herstellung eines Glaskeramikgegenstandes Hohen Bruchmoduls," DE Pat. No. 2 203 675, 1972.

Stookey S.D., "Thermal Expansion of Some Synthetic Lithia Minerals," *J. Am. Ceram. Soc.*, **34**, 235–39 (1951).

Stookey S.D., "Chemical Machining of Photosensitive Glass," *Ind. Eng. Chem.*, **45**, 115–18 (1953).

Stookey S.D., "Photosensitively Opacifiable Glass," U.S. Pat. No. 2 684 911, 1954.

Stookey S.D., "Catalyzed Crystallization of Glass in Theory and Practice," *Ind. Eng. Chem.*, **51**, 805–808 (1959).

Stookey S.D., "Ceramic Body and Method of Making It," U.S. Pat. No. 2 971 853, 1961.

Strnad Z., *Glass-Ceramic Materials*, Elsevier, Amsterdam, The Netherlands, 1986.

Sumikin Photon Ceramic Co. Machinable Glass-Ceramic products brochure, 1998.

Suzuki S., Tanaka M., and Kaneko T., "Glass-Ceramics from Sewage Sludge Ash," *J. Mater. Sci.*, **32**, 1775–79 (1997).

Szabo I., Pannhorst W., and Rappensberger M., "Investigation on the Effect of Surface Treatment and Annealing on the Surface Crystallization of the $MgO-Al_2O_3-SiO_2$ Glass," *Bol. Soc. Esp. Ceram. VID*, **31-C** [5] 119–24 (1992).

Szabo I., Barnab S., Völksch G., and Höland W., "Crystallization and Color of Apatite–Leucite Glass-Ceramic," *Glastech. Ber. Glass Sci. Technol.*, **73** [C1] 354–57 (2000).

Takeuchi Y., Mitachi S., and Nagase R., "High-Strength Glass-Ceramic Ferrule for SC-Type Single-Mode Optical Fiber Connector," *IEEE Photonics Techn. Lett.*, **9**, 1502–04 (1997).

Tammann, *Der Glaszustand*. Leonard Voss, Leipzig, Germany, 1933.

Tashiro M., "Crystallization of Glasses, Science and Technology," *J. Non-Cryst. Solids*, **73**, 575–84 (1985).

Tashiro T. and Wada M., "Glass-Ceramics Crystallized with Zirconia"; pp. 18–19 in *Advances in Glass Technology*, Plenum Press, New York, 1963.

Taubert J., Hergt R., Müller R., Ulbrich C., Schlüppel W., Schmidt H.G., and Görnert, "Phase Separation in Ba-ferrite Glass-Ceramics Investigated by Faraday Microscopy," *J. Magn. Magn. Mater.*, **168**, 187–95 (1996).

Thompson Jr J.B., "Metasomatism and the Phase Rule"; in *Researches in Geochemistry*, Edited by P.H. Abelson, Wiley, New York, 1959.

Tick P.A., Borrelli N.F., Cornelius L.K., and Newhouse MA, "Transparent Glass-Ceramics for 1300 nm Amplifier Applications," *J. Appl. Phys.*, **78**, 93–100 (1995).

Tick P.A., Borrelli N.F., and Reaney I.M., "The Relationship between Structure and Transparency in Glass-Ceramic Materials," *Opt. Mater.*, **15**, 81–91 (2000).

Tummala R.R., "Ceramic and Glass-Ceramic Packaging in the 1990s," *J. Am. Ceram. Soc.*, **74**, 895–908 (1991).

Ubassy G., "Metal-Free Restorations—Optimum Circulation of Light"; presented at Dental Competence 2000 (Berlin, Germany).

Ueda T., "The Crystal Structure of Monazite $CePO_4$," *Mem. Coll. Sci. Univ. Kyoto, Ser. B*, **20**, 227–46 (1953).

Uhlmann D.R., "Crystal Growth in Glass-Forming Systems—A Review"; pp. 91–115 in *Advances in Nucleation and Crystallization in Glasses*, Special Publication 5. Edited by L.L. Hench and F.W. Frieman, American Ceramic Society, Columbus, OH, 1971.

Uhlmann D.R., "Glass Formation," *J. Non-Cryst. Solids*, **25**, 43–85 (1977).

Uhlmann D.R., "On the Internal Nucleation of Melting," *J. Non-Cryst. Solids*, **41**, 347–57 (1980).

Uhlmann D.R., "Crystal Growth in Glass-Forming Systems: A Ten-Year Perspective"; pp. 80–124 in Advances in Ceramics, Vol. 4, *Nucleation and Crystallization in Glasses*. Edited by J.H. Simmons, D.R. Uhlmann, and G.H. Beall, American Ceramic Society, Columbus, OH, 1982.

Uhlmann D.R. and Kolbeck A.G., "Phase Separation and the Revolution in Concepts of Glass Structure," *Phys. Chem. Glasses*, **17**, 146–58 (1976).

Uhlmann D.R., Suratwala T., Davidson K., and Boulton J.M., "Sol-Gel-Derived Coatings on Glass," *J. Non-Cryst. Solids*, **218**, 113–22 (1997).

Uno T., Kasuga T. and Nakajima K., "High-Strength Mica-Containing Glass-Ceramics," *J. Am. Ceram. Soc.*, **74**, 3139–41 (1991).

Uno T., Kasuga T., Nakayama S., and Ikushima A., "Microstructure of Machinable Nanocomposite Glass-Ceramic," *J. Jpn. Powder Powder Metall.*, **39**, 1067–71 (1992).

Uno T., Kasuga T., Nakayama S., and Ikushima A.J., "Microstructure of Mica-Based Nanocomposite Glass-Ceramics," *J. Am. Ceram. Soc.*, **76**, 539–41 (1993).

Unterbrink G., "Clinical Aspects of Full Ceramic Systems," *Ivoclar-Vivadent Rept.*, **10**, 21–30 (1994).

Van Wazer J.R., *Phosphorous and Its Compounds*, Interscience, New York, 1959.

Völksch G., Szabo I., Höche T., and Höland W., "Microstructure of Apatite Glass-Ceramic," *Glastech. Ber. Glass Sci. Technol.*, **73** [C1] 358–61 (2000).

Vogel W., *Struktur und Kristallisation der Gläser*. Verlag fuer Grundstoffindustrie, Leipzig, Germany, 1963.

Vogel W., *Glaschemie*. Deutche Verlag für Grundstoffindustrie, Leipzig, Germany, 1978.

Vogel W., *Chemistry of Glass*. American Ceramic Society, Columbus, OH, 1985.

Vogel W., *Glasfehler*. Springer–Verlag, Berlin, Germany, 1992.

Vogel W. and Höland W., "Nucleation and Crystallization Kinetics of an $MgO–Al_2O_3–SiO_2$ Base Glass with Various Dopants"; pp. 125–45 in Advances in Ceramics, Vol. 4, *Nucleation and Crystallization in Glasses*. Edited by J.H. Simmons, D.R. Uhlmann, and G.H. Beall. American Ceramic Society, Columbus, OH, 1982.

Vogel W., and Höland W., "The Development of Bioglass Ceramics for Medical Application," *Angew Chem. Int. (Engl. Transl.)*, **26**, 527–44 (1987).

Vogel J., Carl G., and Völksch G., "Knochenersatz aus Glas und Keramik," *Zeiss Inf. Jenaer Rundschau*, **4**, 17 -20 (1995).

Vogel W., Vogel J., Höland W., and Wange P., "Zur Entwicklung Bioaktiver Kieselsäurefreier Phosphatglaskeramiken für die Medizin," *Wissenschaftliche Zeitschrift Friedrich-Schiller-Univ.t Jena Natur.*, R, **36**, 841–54 (1987).

Völksch G., Höche T., Szabo I., and Höland W., "Phase Content in a Glass-Ceramic from the System $SiO_2–Al_2O_3–K_2O–CaO–P_2O_5–F^2$, *Glastech. Ber. Glass Sci. Technol.*, **71C**, 500–503 (1998).

Volmer M., *Kinetik der Phasenbildung*. Th Steinkopf Verlag, 1939.

Wada M., personal information; 1998.

Wada M. and Nimomiya M., "Glass-Ceramic Architectural Cladding Materials," *Sci. Technol. Compos. Mater.*, **30**, 846–50 (1995).

Wakasa K., Yamaki M., and Matsui A., "An Experimental Study of Dental Ceramic Material: Differential Thermal Analysis," *J. Mater. Sci. Lett.*, **11**, 339–40

Wang Y. and Ohwaki J., "New Transparent Vitroceramics Co-doped with Er^{3+} and Yb^{3+} for Efficient Frequency Upconversion," *Appl. Phys. Lett.*, **63** [24] 3268–70 (1993).

Wange P., Vogel J., Horn L., Höland W., and Vogel W., "The Morphology of Phase Formations in Phosphate Glass Ceramics," *Silic. Ind.*, 231–36 (1990).

Weinberg M.C., "Transformation Kinetics of Particles with Surface and Bulk Nucleation," *J. Non-Cryst. Solids*, **142**, 126–232 (1992a).

Weinberg M.C., "Non-isothermal Surface Nucleation Transformation Kinetics," *J. Non-Cryst. Solids*, **151**, 81–87 (1992b).

Weinberg M.C. and Zanotta E.D., "Calculation of the Volume Fraction Crystallised in Non-isothermal Transformations," *Phys. Chem. Glasses*, **30**, 110–15 (1989).

Weinberg M.C. and Bernie III D.P., "Kinetics of Crystallization of Highly Anisotropic Particles," *Glastech. Ber. Glass Sci. Technol.*, **72 C**, 129–37 (2000).

Weinberg M.C., Bernie III D.P. and Shneidman, "Crystallization Kinetics and the J MAK Equation," *J. Non-Cryst. Solids*, **219**, 89–99 (1997).

Wilder J.A., Healey J.T., and Bunker B.C., "Phosphate Glass-Ceramics: Formation, Properties, and Application"; pp. 313–26 in Advances in Ceramics, Vol. 4, *Nucleation and Crystallization in Glasses*. Edited by J.H. Simmons, D.R. Uhlmann, and G.H. Beall. American Ceramic Society, Columbus, OH 1982.

Wilson J., Yli-Urpo A. and Risto-Pekka H., "Bioactive Glasses: Clinical Applications"; pp. 63–74 in *An Introduction to Bioceramics*. Edited by L.L. Hench and J. Wilson. World Scientific, Singapore, 1993.

Winand J.M., Rulmont A., and Tarte P., "Ionic Conductivity of the $Na_{1+x}M^{III}_xZr_{2-x}(PO_4)_3$ System (M = Al, Ga, Cr, Fe, Sc, In, Y, Yb)," *J. Mater. Sci.*, **25**, 4008–13 (1990).

Winkler H.G.F., "Synthese und Kristallstruktur des Eucryptite $LiAlSiO_4$, *Acta Crystallogr.*, **1**, 27–34 (1948).

Winter W., Berger A., Müller G., and Pannhorst W., "TEM Investigation of Cordierite Crystallization from a Glass Powder with Composition $Mg_2Al_4Si_{11}O_{30}$," *J. Eur. Ceram. Soc.*, **15**, 65–70 (1995).

Wohlwend A. and Schärer P., "Die Empress-Technik—Ein Neues Verfahren zur Herstellung von Vollkeramischen Kronen, Inlays und Facetten," *Quint. Zahntech*, **16**, 966–78 (1990).

Wolcott C.C., "Cansite-Apatite Glass-Ceramic," Eur. Pat. No. EP 0 641 556, 1994.

Wright A.F. and Lehmann M.S., "The Structure of Quartz at 25° and 590°C Determined by Neutron Diffraction," *J. Solid State Chem.*, **36**, 371–80 (1981).

Wu J.M., Cannon W.R., and Panzera C., "Castable Glass-Ceramic Composition Useful as Dental Restorative," U.S. Pat. No. 4 515 634, 1985.

Yamamuro T., "A/W Glass-Ceramic: Clinical Applications"; pp. 89–104 in *An Introduction to Bioceramics*, Edited by L.L. Hench and J. Wilson. World Scientific, Singapore, 1993.

Young R.A. and Elliot J.C., "Scale Bases for Several Properties of Apatites," *Arch. Oral Biol.*, **11**, 699–707 (1966).

Xu H., Heaney P.J., Yates D.M., Von Dreele R.B., and Bourke M.A., "Structural Mechanism Underlying Near-Zero Thermal Expansion in β-eucryptite: A Combined Synchrotron X-ray and Neutron Rietveld Analysis, *J. Mater. Res.*, **14**, [7] 3138–51 (1999).

Zanotto E.D., "Crystallization of Glass: a Ten Year Perspective" (1993 Vittorio Gottardi Prize Lecture), *Chim. Chronica, Newseries*, **23**, 3–17 (1994).

Zanotto E.D., "Metastable Phases in Lithium Disilicate Glasses," *J. Non-Cryst. Solids*, **219**, 42–48 (1997).

Zanotto E.D. and Weinberg M.C., "Trends in Homogeneous Crystal Nucleation in Oxide Glasses," *Phys. Chem. Glasses*, **30**, 186–92 (1988).

Zanotto E.D. and Galhardi A., "Experimental Test of the General Theory of Transformation Kinetics: Homogeneous Nucleation in a $Na_2O \cdot 2CaO \cdot 3SiO_2$ Glass," *J. Non-Cryst. Solids*, **104**, 73–80 (1988).

Zanotto E.D. and James P.F., "Experimental Test of the General Theory of Transformation Kinetics: Homogeneous Nucleation in a $BaO \cdot 2SiO_2$ Glass," *J. Non-Cryst. Solids*, **104**, 70–72 (1988).

ZERODUR information, ZERODUR-Präzision aus Glaskeramik; pp. 1–31, SCHOTT, 1991.

Additional References

Ermich M., Kunzmann K., and Assmann S., "Röntgenographische Untersuchungen im System leucit-haltiger Dental Keramiken," Tagung Dt. Gesell. f. Kristallog., Bayreuth, Poster, (2001).

Gaber M., Harder U., Hähnert M., and Geissler H., "Water Release Behavior of Soda–Lime–Silica Glass Melts," *Glastech Bir Glass Sci Technol.*, **68**, 339–345 (1995).

Mörmann W.H. and Bindl A., "The Cerec 3—A Quantum Leap for Computer-Aided

Restorations: Initial Clinical Results," *Quint. Int.*, **31**, 699–712 (2000).

Pannhorst W., "Surface nucleation," *Int comm on Glass* (2000).

Photoveel catalog, Sumikin Photon Ceramics Co. Ltd., Japan (1998).

Richerson D.W. and Hummel F.A., "Synthesis and Thermal Expansion of Polycrystalline Cesium Minerals," *J Am Ceram Soc*, **55** [5] 269 (1972).

Taylor D. and Henderson C.M.B., "Thermal Expansion of the Leucite Group of Minerals," *Am. Mineral*, **53** [9-10] 1476 (1968).

Index

361